NONLINEAR
AND COLLECTIVE
PHENOMENA
IN BEAM PHYSICS
1998 WORKSHOP

Nonlinear and Collective Phenomena in Beam Physics
Arcidosso, Italy September 1-5, 1998

NONLINEAR AND COLLECTIVE PHENOMENA IN BEAM PHYSICS 1998 WORKSHOP

International Committee on
Future Accelerators

Arcidosso, Italy   September 1998

EDITORS
Swapan Chattopadhyay
Lawrence Berkeley National Laboratory

Max Cornacchia
Stanford Linear Accelerator Center

Claudio Pellegrini
University of California, Los Angeles

**AIP CONFERENCE
PROCEEDINGS 468**

American Institute of Physics   **Woodbury, New York**

Editors:

Swapan Chattopadhyay
Lawrence Berkeley National Laboratory
Center for Beam Physics, MS 71-259
1 Cyclotron Road
Berkeley, CA 94720
U.S.A.

E-mail: chapon@lbl.gov

Max Cornacchia
Stanford Synchrotron Radiation Laboratory
Stanford Linear Accelerator Center
MS 69, P.O. Box 4349
Stanford, CA 94309
U.S.A.

E-mail: cornacchia@ssrl.slac.stanford.edu

Claudio Pellegrini
Department of Physics and Astronomy
University of California, Los Angeles
405 Hilgard Avenue, Box 951547
Los Angeles, CA 90095-1547
U.S.A.

E-mail: pellegrini@physics.ucla.edu

L.C. Catalog Card No. 99-61288
ISBN 1-56396-862-2
ISSN 0094-243X
DOE CONF- 980984

Printed in the United States of America

CONTENTS

WORKING GROUP REPORTS

WORKING GROUP ON SINGLE-PARTICLE NONLINEAR DYNAMICS

WORKING GROUP ON CREATION AND MANIPULATION OF HIGH PHASE DENSITY BEAMS

WORKING GROUP ON PHYSICS OF, AND PHYSICS WITH, HIGH ENERGY DENSITY BEAMS

PREFACE

The 16th Advanced ICFA Beam Dynamics Workshop on Nonlinear and Collective Phenomena in Beam Physics took place in Arcidosso (Italy) from the 1st to the 5th of September, 1998. The Workshop was sponsored by the International Committee on Future Accelerators, the US Department of Energy, the University of California at Los Angeles, the Stanford Linear Accelerator Center, Frascati National Laboratory-INFN (Italy), Lawrence Berkeley National Laboratory, and the KEK Laboratory in Japan. The Workshop's chairmen were M. Cornacchia (SLAC) and C. Pellegrini (UCLA). The meeting attracted 65 experts on nonlinear dynamics in particle accelerators as well as on the creation, manipulation and utilization of high brightness beams. Arcidosso is a medieval town in southern Tuscany, close to the city of Sienna. The meeting took place in the historically evocative scenario of an 11th century castle atop a hill dominating the nearby valley. The castle was restored in 1989, and preserves all the atmosphere and raggedness of medieval times.

There were three invited lectures on Tuesday, September 1st, the day that opened the subjects, and three summary talks in the afternoon of Saturday 5th. All the other presentations were either informal or in the form of posters.

The Group on Single Particle Nonlinear Dynamics was coordinated by David Robin (LBNL), with an introductory talk by Ezio Todesco (CERN). The Group discussed issues relevant to hadron colliders, storage ring light sources, and high current linacs. The Group on Creation and Manipulation of High Phase Density Beams was coordinated and its summary reported by Joerg Rossbach (DESY) with the introductory talk by Bruce Carlsten (LANL). Topics of discussions were those relevant to the production, transport and monitoring of high brightness beams, including coherent and radiation effects. The Group on Physics of, and Physics with, High Energy Density Beams was led by Ingolf Lindau (University of Lund and Stanford University). The wide field of discussion was opened by a talk by Andrew Sessler (LBNL) on the physics with photon and particle beams and on the physics of particle beams. This Group attracted contributions from different concepts and utilizations of intense beams.

On behalf of the Organizing Committee, we would like to thank the sponsoring institutions for supporting the Workshop. We would also like to thank the Mayor, Ing. Attilio Marino, and the Administration of Arcidosso for hosting the meeting. We are grateful to the Regional and Provincial Administrations, to the "Comunità Montana del Monte Amiata" and to the representatives of local authorities for providing valuable support. Special thanks are due to the Workshop Coordinator, Ms. Melinda Laraneta and to the Coordinator of the Local Organization, Sig. Gianfranco Nanni. The people of Arcidosso, as always, have contributed to creating a pleasant atmosphere with their welcome, warmth, and friendship. We thank the Program Committee and the Working Group Coordinators and Speakers who contributed to an exciting scientific program of presentations and discussions and worked hard and with enthusiasm to implement it.

M. Cornacchia and C. Pellegrini

ORGANIZING COMMITTEE

S. Chattopadhyay (LBNL)
M. Cornacchia (SLAC)
K. Hirata (KEK)
C. Pellegrini (UCLA)
A. Renieri (ENEA)
G. Vignola (INFN-LNF)

PROGRAM COMMITTEE

J. Bisognano (Jefferson Lab)
B. Carlsten (LANL)
J.-P. Delahaye (CERN)
J.-P. Koutchouk (CERN)
L. Palumbo (U. Rome/LNF-INFN)
J. Rossbach (DESY)
F. Ruggiero (CERN)
R. Ruth (SLAC)
W. Scandale (CERN)
R. Sheffield (LANL)
J. Sheppard (SLAC)
L. Teng (ANL)

WELCOME FROM THE MAYOR OF ARCIDOSSO

Opening the 16th ICFA Beam Dynamics Workshop, I wish to thank, first of all, the President of the Italian Republic, Oscar Luigi Scalfaro, who is kindly sponsoring this workshop, acknowledging the importance of the subjects discussed, the high value of the participants, and the worldwide scientific institutions that are present here.

On behalf of Arcidosso, I salute all the participants and the authorities present here today, and am proud to have all of you in our town for the third time.

I thank, in particular, the International Committee for Future Accelerators, a worldwide organization coordinating the research on particle beams, which by sponsoring this Workshop amplifies its importance in the scientific world.

I also thank very much all the people and institutions who supported in many ways this important meeting, in particular the Regione Toscana, the Provincial Administration of Grosetto, the Amiata APT, the Mountain Community and the Cassa di Risparmio de Firenze.

I also welcome and thank the Prefetto di Grosseto, who is honoring us with his presence, and therefore strengthens the attention toward this meeting.

The Arcidosso Organizing Committee, coordinated for us by Gianfranco Nanni, by the councilman Bramerini, and by the Head Librarian, Carlo Goretti, worked with passion and competence for over a year to make this event a success. Yet we are all aware of the difficulty for a small community like ours to successfully repeat what has been done in the two previous meetings.

The community of Arcidosso and I are very proud to have you here and to be the focus of attention in the worldwide scientific community. The quiet atmosphere of a small town like Arcidosso, and the beauty of the environment of the green Mount Amiata ideally surround our historical castle and the theater where we are now. These two historical buildings with their old architecture and the addition of modern comfort provide a good environment for cultural events like this one. I am sure that these elements will allow a small town like Arcidosso to use its own best resources to be a part of more cultural events. We already proved this to be true with the Biomathematics and Biophysics meetings, hosted here in 1987.

The public lectures on scientific subjects of general interest that were organized during the 1986 Workshop on Non Linear Dynamics were a pleasant and successful novelty. This year we will repeat the experiment because we believe that this is the right way to create cultural growth and a link between your work and our community.

I therefore wish that the intense scientific work you are to begin may go together with a pleasant stay in our town, and I hope that you will again be our guests in the future. The City Council will do its best for the success of this workshop, in the hope that there will be more in the future. We are proud to have scientists coming from all over the world to Arcidosso, working for our common future.

I wish you again a successful workshop and hope to welcome you back in the near future. Arrivederci and buon lavoro.

The Mayor
Ing. Attilio Marino

Participants List

Alexei V. Agafonov	Lebedev Physical Institute
Dan Anderson	Chalmers University of Technology
Armando Bazzani	University of Bologna
Ilan Ben-Zvi	Brookhaven National Laboratory
Vinod Bharadway	SLAC
Bruce Carlsten	LANL
Patrick L. Colestock	Fermilab
Jeff Corbett	SLAC
Max Cornacchia	SLAC
Martin Dohlus	DESY
Georges Dome	CERN
Renato Fedele	Universita di Napoli Federico II
Antonina Fedorova	Institute of Problems in Mechanical Engineering
Massimo Ferrario	INFN-LNF
Klaus Foettmann	DESY
Paolo Freguglia	Universita di Siena
Rainer W.	Hasse GSI
Patrick Krejcik	SLAC
Glen R. Lambertson	Lawrence Berkeley National Laboratory
Andrei N. Lebedev	P.N. Lebedev Physical Institute
Valeri A. Lebedev	Jefferson Laboratory
Torsten Limberg	DESY
Ingolf Lindau	Lund University
Alfredo Luccio	Brookhaven National Laboratory
Mauro Migliorati	Universita' da Roma
Akihiko Mizuno	Japan Synchrotron Radiation Research Institute
Sergio Monteiro	Moorpark College
King Y. Ng	Fermilab
Heinz-Dieter Nuhn	SLAC
Kazuhito Ohmi	KEK
Hywel Owen	CLRC Daresbury Laboratory
Luigi Palumbo	Universita' di Roma
Yannis Papaphilippou	CERN
Claudio Pellegrini	UCLA
Stefania Petracca	Univ. di Salerno
David Robin	Lawrence Berkeley National Laboratory
Joerg Rossbach	DESY
Alessandro G. Ruggiero	Brookhaven National Laboratory
Francesco Ruggiero	CERN
James Safranek	SLAC
Walter Scandale	CERN

Peter Schlein	UCLA
Evgenii Schneidmiller	DESY
Luca Serafini	INFN Sez. Milan
Nicholas S. Sereno	Argonne National Laboratory
Andrew Sessler	Lawrence Berkeley National Laboratory
Rich Sheffield	LANL
Gennady Stupakov	SLAC
Andrei Terebilo	SLAC
Kathleen Thompson	SLAC
Ezio Todesco	CERN
Marnix J. van der Weil	Eindhoven University of Technology
Alexander Varfolomeev	Russian Research Center
Robert L. Warnock	Stanford University
Michael Zeitlin	Institute of Problems in Mechanical Engineering
G. Turchetti	Università di Bologna

WORKING GROUP REPORTS

Report of the Working Group on Single-Particle Nonlinear Dynamics

A. Bazzani, L. Bongini, J. Corbett, G. Dome, A. Fedorova,
P. Freguglia, K. Ng, K. Ohmi, H. Owen, Y. Papaphilippou, D. Robin,
J. Safranek, W. Scandale, A. Terebilo, G. Turchetti, E. Todesco,
R. Warnock, and M. Zeitlin
(Reported by D. Robin[1])

I WHAT ARE THE ISSUES?

When beginning a discussion on the current understanding of single particle dynamics in particle accelerators it is important to first consider the issues that limit the accelerators' performance. In Table 1 an outline is made of the issues for three types of accelerators: hadron colliders, storage ring based light sources, and high current linacs. For each type of accelerator there is a list of the issues or concerns, the specific mechanisms causing the concerns, and the principal sources of the nonlinearities. Even though the issues are basically the same (primarily preventing particle beam loss), the mechanisms that cause the problems can be quite different, as are the sources of the nonlinearities.

In both hadron machines and light sources the main issues are injection rate and beam lifetime. In high current linacs the main issue is activation of the linac. The mechanisms that affect particle loss in hadron machines are both fast and slow diffusion, small angle intrabeam scattering and small dynamic apertures. Because of the lack of radiation damping, power supply ripple and high order resonance excitation are important. In light sources the mechanisms are fast diffusion, large angle intrabeam scattering, small transverse and longitudinal apertures. Due to radiation damping, one does not need to be as concerned with long term diffusion. On the other hand in low emittance light sources large angle intrabeam scattering limits the lifetime. Therefore it is necessary to have a large longitudinal aperture ($\approx 50\sigma_L$) for a long lifetime. In hadron machines large angle intrabeam scattering is not as important and one can get by with a much smaller longitudinal aperture ($\approx 6\sigma_L$). In high current linacs where there is a relatively short number of periods (< 1000) the main concern is fast diffusion.

---

[1] This work was supported by the Director, Office of Energy Research, Office of Basic Energy Sciences, Materials Sciences Division, of the U.S. Department of Energy, under Contract No. DE-AC03-76SF00098.

CP468, *Nonlinear and Collective Phenomena in Beam Physics–1998 Workshop,*
edited by S. Chattopadhyay, M. Cornacchia, and C. Pellegrini

	Hadron Colliders	Storage Ring Sources	High Current Linacs
Issues or Concerns	Injection Rate, Beam Lifetime,	Injection Rate, Beam Lifetime	Activation of Linac from Beam Halos
Mechanisms	Fast ($< 10^4$ turns) and Slow (10^7 turns) Diffusion, Intrabeam Scattering, Small Transverse Apertures, Collective Effects	Fast ($< 10^4$ turns) Diffusion, Intrabeam Scattering Small Transverse and Longitudinal Apertures Collective Effects	Very Fast ($< 10^3$ turns) Diffusion
Causes of Nonlinearities	Magnetic Field Imperfections, Beam-Beam Force	Sextupole Magnets for Chromaticity Correction	Space Charge Force

The sources of the nonlinearities in all three cases are quite different. In hadron machines it is mostly the magnetic field imperfections and the beam-beam force. It is the sextupole magnets in light sources and space charge in the linacs. As a result the tools and techniques used to study these problems can be different.

In the following sections a summary is given of the group discussions with regards to studies of hadron machines, light sources, high current linacs, and the interplay between single particle dynamics and collective effects.

II HADRON COLLIDERS

During the past 15 years a large number of sophisticated numerical tools have been developed to evaluate the stability of motion in a particle accelerator. Many of these tools were discussed in the proceedings of the last Arcidosso workshop [1]. People still rely heavily on element-by-element symplectic integrators that have been around for a many years. These integrators are used for brute force tracking. Through tracking one attempts to determine the dynamic aperture of the machine—the region separating stable and unstable particle motion. Having a sufficiently large dynamic aperture is one of the criteria that is commonly used to specify whether the design of a machine is acceptable.

There are several problems associated with tracking and determining the size of the dynamic aperture:

1. Element-by-element particle tracking is slow.
2. There are numerical errors that build up over many turns.
3. The concept of a dynamic aperture separating stable and unstable particles is an unrealistically simple one.
4. Little insight is gained as to the causes of particle loss through a calculation of the dynamic aperture.

A   Faster Tracking

Even with modern computing power, the time it takes to track element-by-element for 10^7 turns in a machine like LHC is extremely large [3]. Due to the time required as well as concerns about the build up of numerical errors, it would be nice to have a reliable faster alternative. Considerable work has gone into studying tracking with full-turn maps and this work has shown promise. Robert Warnock presented two papers. The first was an approach to map tracking using a spline expansion [4]. He takes a non simplectic Taylor series map and then constructs a symplectic map using a mixed function canonical generator. Because of the mixed variables, this technique requires using a Newton search. He felt that this is not a drawback. Taking the LHC as an example this technique resulted in an increase in speed of 1 to 2 orders of magnitude over element-by-element tracking. In addition to this work he also presented recent work where he attempts to impose a formal limit on the error bounds for the numerical solution of differential equations [5].

B   Early Predictors of Long Term Stability

Because of the time required for tracking it would be nice to be able to predict the long term stability of motion based upon short term tracking. In the previous two Arcidosso workshops several approaches were discussed: Lyapunov exponents, tune drifts (Frequency Map Analysis), long term lifetime bounds and $1/\ln(N)$ extrapolation. At this workshop Walter Scandale presented more recent results on the empirical $1/\ln(N)$ extrapolation method [6]. This method allows one to predict long term stability based on a limited set of tracking data. Results for the LHC based upon 10^5 turns extrapolated to 10^6 turns and the agreement was within 5%.

C   Understanding Tracking Data

Tracking by itself offers very little insight into the beam dynamics. It was once said that "the dynamic aperture is like a grave stone. It tells when a particle was born and when it dies but does not say anything about the particle's life [2]." Many of the recent advances in theoretical tools have been in the area of understanding tracking data. These tools have been effective at both understanding the mechanisms and giving guidance towards improving the dynamics. Presently there is a large arsenal of tools that have been used in conjunction with tracking data to help improve the design of accelerators. Even with these tools improving the design of the machine is somewhat of a trial and error endeavor.

Yannis Papaphiliphou presented one such tool that was effective at determining the cause of dynamic aperture degradation between two versions of the LHC lattice [7]. A numerical tool was developed based upon standard Lie algebraic approach and normal form analysis called the Graphical Representation of Resonances (GRR). This tool presented a graphical representation of the resonance strengths and detuning up to a desired order. With the help of GRR it was possible to show that when going from one version of the lattice to the other the strengths of a particular resonance had increased and was the cause of the reduction of the aperture. From this knowledge it was then possible to identify which magnet family was responsible for exciting the resonance. Reducing the higher order multipole component of that family then restored the dynamic aperture.

Other techniques were presented that help visualize the dynamics of the beam. One is Frequency Map Analysis a technique introduced by Jacques Laskar in a prevous workshop [9]. Ezio Todesco presented some frequency maps for the LHC in his opening talk [3]. It is possible to numerically calculated the invariant actions using FMA in a manner similar to that used to calculate the frequencies. A demonstration of this using the Henon map was shown by Lorenzo Bongini [8].

The use of quality factors for improving the dynamics has also proven to be quite useful. In other words some studies have shown that the limit to the dynamic aperture may be strongly related to a single quantity such as nonlinear tuneshift or phase space distortion. Trying to optimize such a simple quality factor was found to be effective in improving the dynamics [3].

III LIGHT SOURCES

As was mentioned previously the problems facing light sources are somewhat different than those of hadron colliders.

1. They tend to have simple structures and a high degree of periodicity. This helps to suppress nonlinear resonance excitation.
2. Sextupoles are the dominant nonlinear elements in the machine.
3. The radiation damping means that long term tracking is not necessary.
4. The vertical physical aperture is very small (few millimeters) which is necessary for insertion devices but affects particle loss.

Many new third generation light sources have turned on in recent years. There have not been significant problems injecting beam. Concerns about the effects of insertion devices strongly impacting operation have not been realized. The only issue that has arisen and will have a large impact on the next group of light sources is the momentum acceptance.

The beam lifetime is proportional to the square of the momentum acceptance. In several third generation machines (ALS, ESRF, APS) the momentum acceptance has been measured to be around 2% [10] [11]. This is smaller than was predicted during the machine design studies. The next group of light sources (Soleil, Diamond, SLS) hopes to obtain momentum acceptances as large as 5%. If the momentum acceptance in these machines only reaches 2% than this would result in a lifetime reduction of a factor of 6! So the questions are: Do we understand the limitations in the momentum acceptance in the present light sources and can we design future light sources to obtain larger momentum acceptances?

Measuring the momentum acceptance in such a machine is rather easy. This is done by measuring the beam lifetime as a function of the RF cavity voltage. At low RF voltage the momentum acceptance is determined by the RF bucket. In this region the lifetime goes up as one increases the RF bucket. At larger RF voltages the momentum acceptance is determined by the dynamic aperture. In this region the lifetime will drop as a function of the RF voltage. Observing the transition between the two regions allows one to measure the momentum acceptance.

During the machine design studies, the predicted size of the momentum acceptance in the ALS was 3%. Measurements of the momentum acceptance were made at the ALS and were compared to tracking simulations. The measured momentum acceptances were found to range from 1.4% to 2.7% depending on machine conditions (tune, chromaticity, coupling, insertion devices). When the momentum acceptance was modeled including the proper vertical aperture and the estimated coupling errors the agreement between tracking and measurement was within 20% under all circumstances [10].

In future machines such as the SPEAR-III upgrade and DIAMOND, there are more than 2 families of sextupoles. The intention is that by carefully adjusting these sextupole families it will be possible to obtain a larger momentum acceptance. Quality factors such minimizing nonlinear tuneshift with energy seem to be effective ways of enhancing the momentum acceptance [12] [13] on paper. Still it remains to be seen whether one can actually achieve such large momentum acceptances in practice.

IV BEAM HALO CALCULATIONS IN HIGH CURRENT LINACS

One of the more interesting computational challenges in accelerator physics today is the accurate calculation of beam halos in high current linacs. One application is for the transmutation of nuclear waste. In such a linac it is important that the beam loss power be less than 1 W/m in order not to have significant activation of the linac chamber. This requires that beam loss should remain below the 10^{-7} level. To calculate such a low loss it is necessary to track a great number of particles. This is a problem that is pushing the limits of the most powerful supercomputers.

A Frequency Map Analysis of the Dynamics in High Intensity Beams

Again the issue is gaining insight into the dynamics without tracking large numbers of particles. One interesting possibility was presented by Armando Bazzani [14]. The model he uses is a particle in core model (PIC) for a high intensity beam in a linac with about a 1000 periods. The very small beam losses which are tolerated require a very detailed analysis of the structure of the phase space. The presence of nonlinear resonances and large chaotic regions could create orbits with a large excursion in the amplitudes of the particle contributing to the beam halo. The dynamics in the model is described by a time dependent 4D symplectic map.

The frequency map analysis introduced by Jacques Laskar [9] to study the stability problem of Celestial Mechanics is a suitable tool to represent the phase space of the PIC model. It is found to be a useful technique for studying the possible mechanism of halo formation in a FODO cell in the presence of a high intensity beam. Tracking a relatively small number of particles the Bazzani group finds that they can define a criterium to determine regions of phase space where there is fast instability growth. Brute force tracking using large numbers of particles reveals that these suspicious regions are indeed responsible for populating the beam halos. Comparisons are made in the case of matched and mismatched beams with good results. This technique may prove to be a valuable way to minimize computational time by first pinpointing suspicious regions which can then be studied in detail to quantify the halo population.

V INTERPLAY BETWEEN NONLINEAR SINGLE DYNAMICS AND COLLECTIVE EFFECTS.

The title of the working group was "Single-particle Dynamics in Particle Accelerators." For the most part single particle dynamics , beam-beam effects and collective effects have remained separate fields. This has been done for simplicity, but in reality one can not separate the dynamics of the single particle and the collective dynamics. When one makes a measurement of the beam, all effects come into play. Results from experiments were presented at the workshop that shows the richness of combined dynamics.

A Turn-by-turn Dynamic Experiments.

The first group of turn-by-turn measurements were presented by Andre Terebillo at SPEAR [15]. The experiments involved a beam kicker and a turn-by-turn monitoring system. In his experiments he found that there was a strong influence of the head tail instability on the dynamics. By changing the bunch current and the chromaticity he was able to balance the radiation damping and the Landau damping with head tail growth. This enabled him to create a macroparticle that would effectively oscillate with a constant amplitude indefinitely. Also he could control the growth or damping rate of the amplitude by varying the chromaticity. He was able explain those measurements using a two particle macroparticle model.

Having such a macroparticle gives him a probe to further investigate of the dynamics. For instance one of the effects measured with this macroparticle was the presence of nonlinear resonances. This is done by adjusting the working point of the machine just below the location of the resonance. When the bunch is kicked to large amplitude, the instantaneous tune is shifted upward due to positive tuneshift with amplitude. In fact it is shifted to the other side of the resonance. As the amplitude decreased the tune slowly decreases and crosses the resonance. By measuring the tune as a function of turn number (using turn-by-turn data), it is possible to identify and measure the width of a resonance. In addition to resonance excitation he used the macroparticle to measure power supply jitter.

Other turn-by-turn measurements were done at SPEAR by putting local orbit distortions and then kicking the bunch. An interesting effect was observed when a large orbit bump (8.5mm) was placed in the closed orbit in the vicinity of the RF cavity. After the beam was kicked the coherent damping of the transverse oscillations was extremely fast—much faster than a damping time. Presumably this is due to a transverse cavity mode. From these results one sees that this is a possible method for characterizing local impedances in the machine.

Another group of turn-by-turn experiments was performed at the Photon Factory and presented by Kazuhito Ohmi [16]. His observations showed that at large currents one can not treat Landau damping and the head-tail effect independently. At low currents where the head-tail effect is small, Landau damping is affected by tuneshift with amplitude. However at large currents the damping is a strong function of the sign of the tuneshift with amplitude. For positive tuneshift with amplitude Landau damping is suppressed. For negative tuneshift with amplitude Landau damping is amplified. He was able to obtain good agreement with a two particle model that had a linear map and octupole and transverse wakefields.

VI   OTHER TOPICS

There were several other interesting and important topics presented to the working group that were not covered in this summary.

1. Henon map approach to transverse off-energy dynamics. (A simple model to study off-energy dynamics) [17].
2. Equivalence of nonlinear dynamics and space charge effects [18].
3. Wavelet approach to the Hamiltonian. [19].
4. A synthetic geometrical approach to betatron motion [20].

VII   SUMMARY

We have obtained a large arsenal of tools to study nonlinear dynamics in a particle accelerator. However it is still far from an automatic task to design rings with large dynamic apertures—there is no simple recipe to follow. As a result during the design stage, large efforts are spent trying to optimize the design of a storage ring. Unfortunatelly comparatively little effort is spent studying the machine after it has been built to see how well we have succeeded. Nevertheless experiments show that nonlinear dynamics limits the performance of accelerators.

A few new experiments show that in reality one can not separate the nonlinear dynamics from the collective effects and they need to be considered as a whole.

VIII ACKNOWLEDGEMENTS

We wish to thank the organizers, Max Cornacchia and Claudio Pellegrini for hosting such an important and useful workshop in such a wonderful location.

REFERENCES

1. A. Chao, AIP Conference Proceedings 395, pp. 3 - 10
2. Quote by Martin Lee.
3. Plenary talk by Ezio Todesco
4. Robert Warnock, presentation to the working group.
5. Robert Warnock, presentation to the working group.
6. Walter Scandale, presentation to the working group.
7. Yannis Papaphiliphou, presentation to the working group.
8. Lorenzo Bongini, presentation to the working group.
9. J. Laskar, AIP Conference Proceedings 344, pp. 130 - 159
10. David Robin, presentation to the working group.
11. Nick Sereno, comment made at the workshop.
12. Jeff Corbett, presentation to the working group.
13. Hewl Owen, presentation to the working group.
14. Armando Bazzani, presentation to the working group.
15. Andre Terebillo, presentation to the working group.
16. Kazuhito Ohmi, presentation to the working group.
17. Giorgio Turchetti, presentation to the working group.
18. Bill Ng, presentation to the working group.
19. Michael Zeitlin, presentation to the working group.
20. Paolo Freguglia, presentation to the working group.

Report of the Working Group on Physics of, and Physics with, High Energy Density Beams

Group Leader: Ingolf Lindau

This working group covered a broad range of topics. The discussion was opened with a talk by Andrew Sessler (LBL) on the physics with photon and particle beams and on the physics of particle beams. Among topics covered were coherent x-ray beams, gamma-gamma colliders, beams for spallation source, transmutation of nuclear waste, colliders, inertial fusion and power beaming. This introductory talk was followed with contributions of different concepts and utilization of intense beams. Free electron laser, FEL, physics was covered by three contributions (C. Pellegrini, H.-D. Nuhn and E. Schneidmiller). C. Pellegrini, UCLA, described an interesting application of FEL, namely the creation of real e^+-e^- pairs from virtual vacuum fluctuations when the electric field approaches the Schwinger field, 10^{18} V/m. A strength close to this value can in principle be obtained from an x-ray FEL when the coherent photon beam is focussed to the diffraction limit. H.-D. Nuhn described in detail the technical characteristics of the proposed linac coherent light source, LCLS, which would provide coherent x-rays down to 0.15 nm with unprecedented peak and average photon brightness'. E. Schneidmiller reported on numerical simulations of SASE (self amplified spontaneous emission) in excellent agreement with recent experimental results from the UCLA/LANL/RRCKI/SLAC collaboration. There are also presentations and discussions on heavy-ion fusion (R.W. Haase), spallation neutron sources (A. Luccio) and plasma acceleration (A. Lebedev). On the physics of particle beams, there were stimulating presentations and exchange of ideas on the quantum-like description of charged-particle beam dynamics (R. Fedele), on the thermal wave model description of high energy charged particle beam dynamics (D. Anderson), on the nonlinear pattern formation and turbulence in coasting beams (P. Colestock) and on a dynamical model for the sawtooth instability in storage rings (L. Palumbo). The group also engaged in a discussion led by I. Lindau on the properties and experimental requirements of third generation synchrotron radiation sources and their extrapolation to the next, 4th generation light sources, possible based on free electron lasers.

CP468, *Nonlinear and Collective Phenomena in Beam Physics–1998 Workshop*, edited by S. Chattopadhyay, M. Cornacchia, and C. Pellegrini

WORKING GROUP ON
SINGLE-PARTICLE
NONLINEAR DYNAMICS

Frequency Map Analysis for

High Intensity Beams

A.Bazzani*, M.Comunian[†], A.Pisent[†]

* *Dept. of Physics Univ. of Bologna and INFN sezione di Bologna, Italy*
[†]INFN Laboratory Nazionali di Legnaro, Italy

Abstract. The Frequency Map analysis is applied to the Particle-Core model for an intense beam in a FODO cell. We show that the Frequency Map allows to get an useful picture of the phase space even in the unmatched cases and we point out the importance of the resonances between the betatronic frequencies and the envelope frequencies in the stability problem of a test particle near the beam core. We introduce a dynamics aperture for the halo formation and we compare the results of Frequency Map analysis with the results of a direct tracking on a simple model of a FODO cell.

INTRODUCTION

The comprehension of the single particle stability in the space charge dominated beams is one crucial issue for the project of future high intensity accelerators. The full dynamics of the beam is a Poisson-Vlasov problem whose numerical solution for a realistic case is beyond the possibilities of the existing computers. In the case of a linear lattice a self-consistent solution for the beam density is known (Kapchinsky-Vladimirsky distribution) and a perturbation approach is possible.

The KV distribution is considered in the study of halo formation by using the particle in core (P-C) model; the beam-halo is due to the presence of a few particles at very large emittance. For high intensity beams a percentage loss less than 10^{-6} is necessary to allow the hand maintenance of the machine. A big effort

CP468, *Nonlinear and Collective Phenomena in Beam Physics–1998 Workshop*,
edited by S. Chattopadhyay, M. Cornacchia, and C. Pellegrini

has been devoted to the understanding of the mechanism of halo formation. In the P-C model one assumes that the beam core follows a KV distribution and considers the dynamics of a test particle under the influence of the space charge forces and the magnetic lattice. This is a short cut to the complete problem of halo formation since one assumes that the particles in the halo do not affect the core distribution, but it gives indications on the stability of the distribution when few particles are spilled out from the core.

If we consider the betatronic dynamics of the test particle in a FODO cell, it is possible to define a 4D-symplectic map (the Poincarè map), which gives the coordinates in the phase space at defined transverse sections of the magnetic lattice. The stability of the test particle reduces to the stability of the corresponding orbit in the phase space. This problem has been considerably developed for the circular accelerators where the stability for a large number of turns is necessary and several methods have been studied and applied. However one could think that most of the results obtained for circular accelerators cannot be applied to the stability problem in a Linac since a particle performs a single passage in the magnetic lattice and only a limited number of iterations of the Poincarè map are physically relevant.

In this paper we show that the Frequency Map analysis introduced by J.Laskar [1,2] in the problems of Celestial Mechanics and successively applied to Accelerator Physics, gives very useful informations on the transverse phase space of the P-C model in a Linac defined by 1,000 FODO cells. In particular the location and the strength of low order resonances and the detection of large chaotic regions or regular regions allow to define necessary conditions that a test particle has to satisfy in order to contribute to the halo. As a consequence the use of a tracking program can be optimized and a very detailed analysis of the phase-space can be performed in order to consider very small losses. Moreover the FM analysis can be applied also in the case of a mismatched beam, when the beam-envelope functions do not follow a periodic solution and the Poincarè map is a time dependent symplectic map.

In the second section we briefly introduce the equations of P-C model; in the third section we discuss the properties of the FM analysis and in the last section we present the numerical results for our model.

THE PARTICLE-CORE MODEL

We assume that the beam follows the KV distribution which is a self-consistent solution to the Poisson-Vlasov problem. The equations of the transverse motion for a test particle are:

$$x_j'' + K_j(s)x_j - \frac{\xi}{(\hat{a}_1 + \hat{a}_2)\hat{a}_j}x_j = 0 \qquad j = 1,2 \tag{1}$$

$$a_j'' + K_j(s)a_j - \frac{\xi}{a_1 + a_2} - \frac{\epsilon_j^2}{a_j^3} = 0 \qquad j = 1,2 \tag{2}$$

where (x_1, x_2) are the transverse coordinates of the particle, (a_1, a_2) are the envelopes of the beam and the suffix $'$ denotes the derivative with respect to the longitudinal coordinate s. Moreover $\xi = [e/(\pi\epsilon_0)][I/(mc^3\beta^3\gamma^3)] = I/(I_c\beta^3\gamma^3)$ is the space charge parameter, with I beam current (peak current for a bunched beam), $I_c = 7.8MA$ proton characteristic current and β, γ relativistic factors; $K_j(s)$ is the quadrupole magnetic field and (ϵ_1, ϵ_2) are the beam emittances. We define the quantities $\hat{a}_j = \sqrt{a_j^2 + \chi}$ where χ is 0 when $x_1^2/a_1^2 + x_2^2/a_2^2 < 1$ (i.e. the test particle is inside the beam) and solution of $x_1^2/(a_1^2 + \chi) + x_2^2/(a_2^2 + \chi) = 1$ otherwise. As a consequence the dynamics is linear inside the beam core and non-linear outside; the space-charge force decreases as $1/\sqrt{x_1^2 + x_2^2}$ at large amplitudes.

In the case of a periodic focusing channel $(K_j(s + L) = K_j(s))$ when the beam is perfectly matched the envelopes a_j are periodic and it is possible to study the single particle equations (2) by using a Poincarè section. But in a more realistic case the beam envelopes follow solutions which are perturbed of an amount $\vec{\delta}$ respect to the matched solution $\vec{a}$ and we cannot directly use the Poincarè section to plot the phase space of the test particle.

In the smooth approximation The linearization of the envelope equations around the matched solution gives the envelope mode frequencies:

$$\alpha_\pm = \sqrt{\frac{H_1 + H_2}{2} + h \pm \sqrt{\left(\frac{H_1 - H_2}{2}\right)^2 + h^2}} \tag{3}$$

with $H_j = \nu_{0j}^2 + 3\nu_j^2$, and $h = \frac{\xi}{4\pi^2}\frac{L^2}{(a_1+a_2)^2}$. The eigenmodes can be written according to $\vec{\delta}_- = (-\sin\phi, \cos\phi)$ and $\vec{\delta}_+ = (\cos\phi, \sin\phi)$ with:

$$\phi = \frac{1}{2}\arctan\frac{2h}{H_1 - H_2} \tag{4}$$

mode mixing angle. In particular, if the focusing strength is equal in the two directions, the mixing angle is $\pi/4$ (taking the limit of eq.(4) for positive $H_1 - H_2$) and the two modes, called respectively odd and even envelope modes [3], have frequencies $\alpha_- = \sqrt{\nu_0^2 + 3\nu^2}$ and $\alpha_+ = \sqrt{2(\nu_0^2 + \nu^2)}$. If instead the difference in focusing strength is large the mixing angle tends to zero. In the magnetic lattices

17

of practical interest the two modes calculated in smooth approximation can be recognized.

Our reference model is a FODO cell defined by a focusing and a defocusing quadrupole of equal length (0.2 m) which are separated by two equal drift spaces (0.3 m long). The main parameters are listed in Table I. The small asymmetry in the two planes allows to avoid a strong coupling resonance. The deviation from the linear description of the envelope motion will be evaluated in the next sections. The tracking code, which has been used for numerical simulations, is a kick code, which integrates both the envelope equations (2) and the single particle equations (1). Each magnetic element is divided into 10 segments and the nonlinear force due to space charge is computed by means of a kick map which uses the envelope amplitude at the center of each segment. The linear motion is computed exactly.

TABLE 1 FODO parameters.

$\epsilon_1 = 10^{-6} \, m$	$K_f = 12.3 \, m^{-2}$	$\nu_1 = .148$	$\nu_{01} = .176$	$\alpha_- = .277$
$\epsilon_2 = 10^{-6} \, m$	$K_D = 11.7 \, m^{-2}$	$\nu_2 = .125$	$\nu_{02} = .166$	$\alpha_+ = .324$
$\xi = 10^{-6}$	$\phi = .34 \, \pi/4$			

THE FM ANALYSIS

The FM associates the betatronic frequencies $\vec{\nu}$ to each invariant torus in the transverse phase space. The frequencies are computed by using an analytical interpolation of the FFT of an orbit on the invariant torus. The introduction of an Hanning filter allows to achieve a numerical precision in the frequencies which scales $\propto 1/N^4$[4] where N is the iterations number. The betatronic frequencies are usually identified at the maximal values of the FFT for the projection of the orbits on the coordinate planes (x_i, p_i) $i = 1, 2$. Then the FM associates a point in the frequency space to a $2D$-torus in the phase space; the informations concerning the deformation of the torus (smear) are neglected.

In order to get an useful picture of the phase space one has to choose a transverse section of the phase space: i.e. a section that intersects transversally each invariant tori in the interesting region in a single point. Then we consider an uniform grid of initial conditions on this section and we compute the FM for each orbit for a fixed number of iterations of the Poincaré map. The image of the grid in the frequency space contains a lot of informations on the stability properties of the orbits. In the regular regions according to the KAM theory where the invariant tori are dense, the FM is a smooth map and we get a deformed grid of points in the frequency space. In these regions the diffusion through the phase space (Arnold diffusion) [5,6] is extremely slow and can be neglected in the halo formation problem. When we consider orbits in a resonant region, the betatronic frequencies are locked on a resonance line $\vec{k} \cdot \vec{\nu} = n$ where $\vec{k}$ and n are integers. Moreover at the border of the resonant regions the tune-shift is divergent. As a consequence the FM maps a resonant region in a empty channel with a straight line in the center; the amplitude of the channel is an indirect measure of the strength of the resonance. Finally the result of the FM for the grid points which correspond to chaotic orbits, is a fuzzy cloud of points in the frequency space since the chaoticity implies a sensitive dependence of the FM on the initial condition.

A typical result of the FM is plotted in fig. 1 (right) where we consider the phase space of the Poincarè map of our reference FODO when an intense matched beam is present. The initial grid of point is taken on the $(x - y)$ plane and it is shown in fig. 1 (left) where we have plot with different markers the points which correspond to different resonant regions. The low order resonances are well separated in the phase space so that we do not have any large chaotic regions: in such a case a large diffusion in the phase space is impossible after a small number of iterations of the Poincarè map .

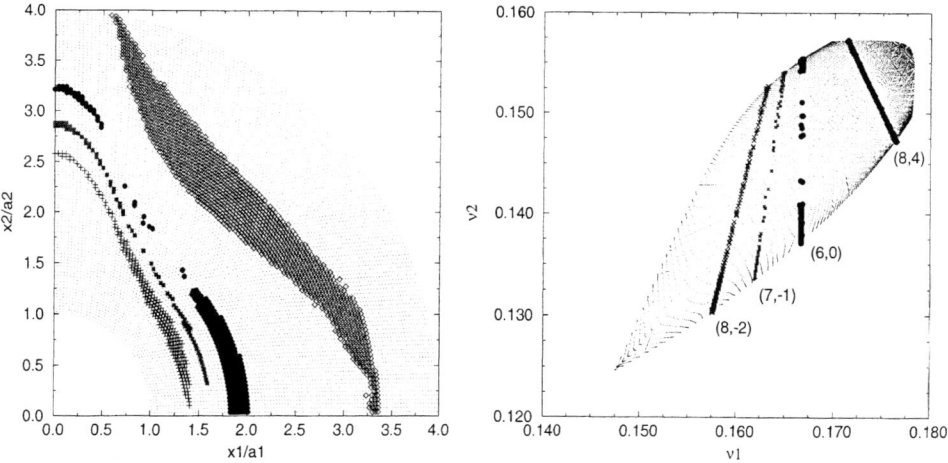

FIGURE 1.FM analysis for the P-C model in the matched case. The grid of initial conditions is plotted in the left part; the resonant points are plotted with different markers; the axis unities are normalized to the beam matched envelopes. The FM is shown in the right part: some resonance lines are indicated with the same markers as the left part; the integer numbers define the linear combination of the betatronic frequencies.

In the case of a mismatched beam, the FFT of an orbit for a test particle contains the integer combinations of the betatronic frequencies $\vec{\nu}$ and the envelope frequencies $\vec{\alpha}$, but still the betatronic frequencies correspond to the maximal values of the FFT of the projections of the orbits on the coordinate planes. Then even if the Poincarè map is no more an autonomous symplectic map, we can compute the FM in the same manner as in the matched case. The presence of the envelope frequencies creates a much richer resonance web since the resonance conditions are given by the linear integer combination between both the envelope and the betatronic frequencies

$$\vec{k} \cdot \vec{\nu} + \vec{h} \cdot \vec{\alpha} = n \qquad \vec{k}, \vec{h}, n \qquad \text{integers.} \tag{5}$$

Then, due to the mechanism of resonances overlapping, large chaotic regions appear in the phase space. The location and the strength of resonances play a crucial role in a possible explanation of the beam halo formation in the P-C model. The results of the FM analysis are shown in fig. 2 (right) where the appearance of the new resonances (5) is clear. The initial conditions which correspond to resonant orbits are plotted in fig. 2 (left) with different markers and the overlapping of resonances creates a large chaotic region.

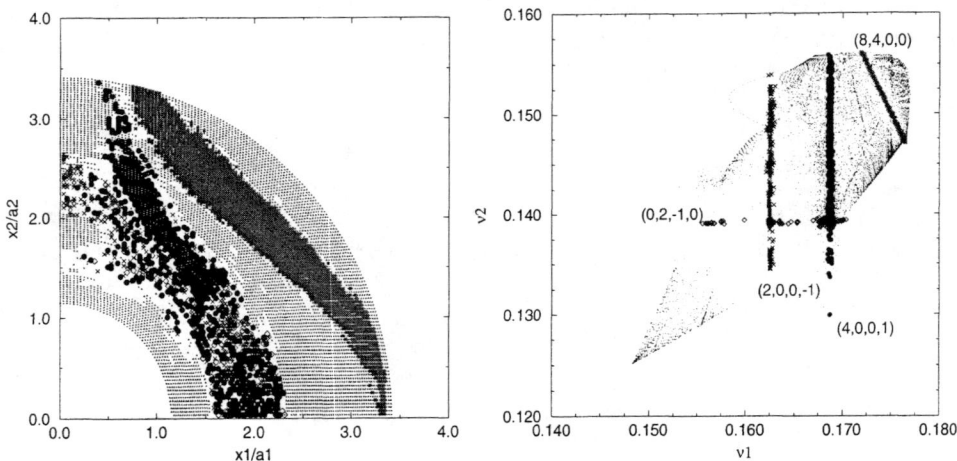

FIGURE 2. FM of the P-C model in the mismatched case: a 20% of mismatch is inserted in the envelope solutions. In the left part we plot the regular initial conditions according to our criterium (dots) and the low order resonance initial conditions with different markers; the axis unities are normalized to the beam matched envelopes. The FM is shown in the right part where the low order resonances between the betatronic frequencies and the envelope frequencies are indicated by the integer coefficients of the resonance relation; the appearance of a chaotic region due to resonances overlapping is clear.

This could be a possible mechanism to explain the halo formation: indeed a test particle that is spilled out from the core at the lower border of the chaotic region can perform a large excursion in the phase space even if we consider a limited number of iterations. By using the results of the FM it is possible to introduce a criterium to distinguish the regular regions from the chaotic unstable regions. The idea of the criterium is the following: the signal due to an orbit on an invariant torus is stationary so that the results of the FM obtained by using two different pieces of an orbit should be the same within the numerical errors; on the contrary if we consider an unstable chaotic orbit the FM results should change due to the nonstationary character of a chaotic signal. Therefore for each point of the grid we have evaluated the difference

$$\Delta\nu = N\|\vec{\nu}_N - \vec{\nu}_{2N}\| \qquad (6)$$

where N is the number of iterations and $\vec{\nu}_N$ are the frequencies computed by using the first N iterations whereas $\vec{\nu}_{2N}$ are the frequencies computed by using the iterations between N and $2N$. Moreover we have considered the gradient of the quantity (6) to take into account the sensitive dependence on the initial condition typical of the chaotic regions. Then we have introduced a threshold to distinguish regular and chaotic orbits, whose value is checked by the numerical simulations[7,8]. The stable points selected by this procedure are shown in the fig. 2 (right), which suggests the possibility of defining a *dynamic aperture*. We observe a annulus of regular orbits around the beam core which means that the diffusion is impossible in this region; the border of the annulus defines also the border of the large chaotic region due to the resonance overlapping. We define the dynamics aperture of our model as the lower radial distance from the origin (reference orbit) at which it is possible to find an orbit with a large excursion in a small number of iterations $N \simeq 1000$. The calculation of the dynamic aperture with a tracking program can be very cumbersome since we have to check an great number of initial conditions in order to avoid the possibility of very small losses. As a consequence the FM map analysis turns out to be useful to detect the border between the regular and the chaotic regions and to select the *dangerous* initial conditions for a tracking program. We shall discuss this point in the next section.

NUMERICAL RESULTS

In the choice of the transverse section for the mismatched case one has to fix the values and the derivatives of the beam envelopes a_j together with a section in the phase space (x_j, x'_j). Then one could think that there exists a dependence of the FM analysis from the previous choices. This is not the case if two different transverse sections intersect the same dynamical structures in the phase space: indeed the betatronic frequencies are nonlinear invariants that depend on the structures of the phase space like KAM tori and resonances. In the numerical simulations we have considered the section defined by $(a'_j = 0, x'_j = 0)$ $j = 1, 2$, where the envelope functions take the maximum value so that the test particles are at the largest distance from the origin. The mismatched solution is computed by using the initial envelopes

$$
\begin{aligned}
a_1^{(mis)} &= a_1(1 + \delta \cos \theta) \\
a_2^{(mis)} &= a_2(1 + \delta \sin \theta)
\end{aligned}
\tag{7}
$$

where a_j are the matched solution and δ is the total mismatch. The angle θ distributes the mismatch on the transverse plane. In order to check the dependence on θ of the FM we have computed the envelope frequencies and the linear betatronic frequencies for a uniform grid of points obtained by varying ϵ in the interval $[0,.3]$ and θ in the sector $[\phi, \phi + \pi/2]$ (see eq. (4) for the definition of ϕ). The results for our model are shown in fig. 3: we observe that the nonlinear character of the envelope equations (2) changes the envelope frequencies and the betatronic frequencies so that the resonance web of the P-C model has a weak dependence on the choice of both δ and θ. An exhaustive study of the mismatched case should consider different choices for δ and θ, however we restrict our analysis to the dependence on δ and we fix $\theta = \pi/4$, to excite both the even and the odd modes.

We remark also that fig. 3 (left) corresponds to the FM analysis applied to the envelope equations (1) and shows that for a value $\leq .3$ of the mismatched parameter δ the envelope solutions are regular.

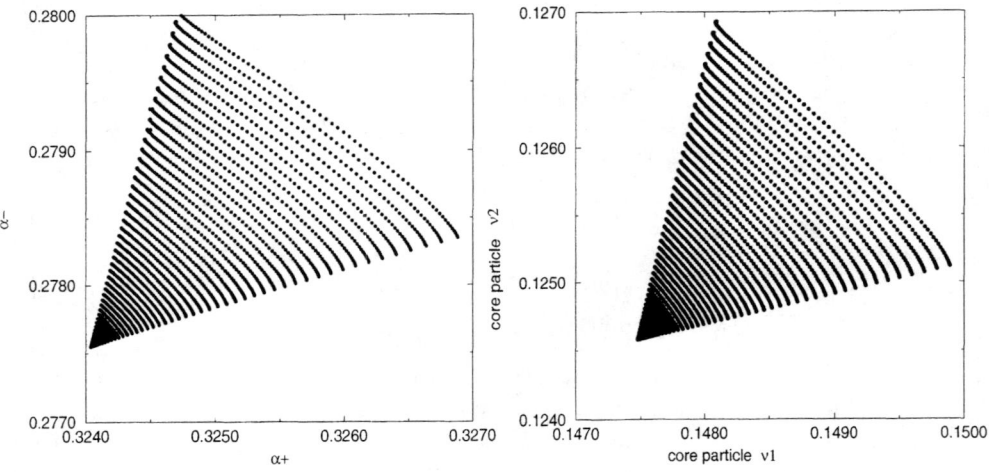

FIGURE 3. FM of the envelope modes (left part) and of the linear tunes (right part) when we consider an uniform grid of initial points in the plane of the horizontal and vertical mismatches.

22

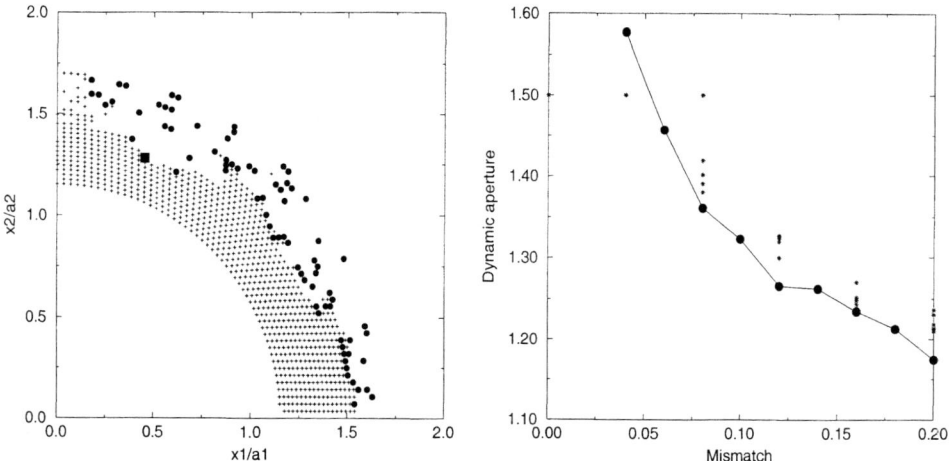

FIGURE 4. Left part: plot of the regular orbits (crosses) in the mismatched case $\delta=.2$; the black circles denotes the points with a nearby unstable orbits; the square is the point corresponding to the dynamics aperture; the axis unities are normalized to the beam matched envelopes. Right part: dynamics aperture as a function of the mismatch parameter δ: the black circles are the values computed by using the FM analysis, the stars are the results of a tracking program with an initial populations of 50,000 particles divided into 5 clusters randomly distributed between 1 and 1.5 beam envelopes; the spread of the stars measures the role of statistic in the evaluation of the dynamics aperture.

For the case $\delta = .2$ we have selected the regular initial conditions according to the criterium (6) with $N = 1024$ and a threshold $\Delta\nu_{crit} = 1.5$. In order to check that the border of the regular region defines really a dynamics aperture, for each regular point, we have considered a neighborhood of radius .01, which corresponds to $1/2$ of the radial step of the grid of the initial conditions, and we have chosen randomly 10 particles for each neighborhood. The particles have been iterated for 2,000 times and we have plotted with a different marker the initial conditions which have at least one nearby particle that performs a excursion in the phase space larger than one rms of the beam core. The results are shown in fig. 4 (left) where we observe that all the unstable particle are placed at the border of the regular region. Then we have selected the unstable particle with the lowest distance from the origin and we have defined the dynamics aperture. By using the same procedure we have computed the dynamics aperture for different values of ϵ and we have plotted the results in the fig. 4 (right). To compare our results with a standard tracking we have computed the dynamic aperture by using a single particle kick-code for P-C model, which tracks a population of 50,000 test particles randomly distributed in an annulus of maximal radius 1.5 beam envelopes around the beam core. The particles have been divided in 5 clusters in order to see the dependence of the dynamics aperture on the statistics; each particle has been iterated 1,000 times. The fig. 4 (right) shows that the tracking results confirm the FM analysis and points out the role of the statistics on the computation of dynamics aperture for our model. We remark that the FM reduces considerably the CPU time with respect to the standard tracking approach since it is not necessary to track an very large number of particles.

CONCLUSIONS

We have shown the possibility to use the FM analysis in the problem of halo formation for high intensity beams in a linac both in the matched and in the mismatched case. Despite of the limited iterations number of the Poincarè map, the FM provides informations on the regular, resonant and chaotic regions in transverse the phase space and allows to define a dynamic aperture by using the border of the regular regions around the beam core. In the mismatched case the FM analysis for the P-C model points out the role of the resonances between the betatronic frequencies and the envelope frequencies for the stability of the orbits, through the mechanism of resonance overlapping.

REFERENCES

1. Laskar J., *Icarus*, **88** p. 266, (1990)
2. Laskar J., *AIP Conference Proceedings* , **344**, p. 130-159 (1995)
3. Struckmeier J., Reiser N., *Part. Acc.* **14**, p. 227-260 (1984)
4. Bazzani A., Giovannozzi M., Scandale W., Todesco E., *Particle Accelerators* **52**, p. 147, (1996).
5. Poeshel J., *Commun. Pure Appl. Math.*, **35** n.1, p.653 (1982).
6. Arnold V.I., *Russ. Math. Surv.*, **18**, n. 6, p. 85, (1963).
7. Bazzani A., Comunian M., Pisent A., "Frequency Map Analysis of an intense mismatched beam in a FODO channel", submitted to *Particle Accelerator* and *LNL-INFN(Rep) 127/98 Internal preprint*, (1998).
8. Pisent A., Bazzani A.,Comunian M., submitted to *Il Nuovo Cimento*,(1998).

Dynamic Aperture Studies for SPEAR 3 [1]

Y. Nosochkov and J. Corbett

SLAC, Stanford University, Stanford, CA 94309

The Stanford Synchrotron Radiation Laboratory is investigating an accelerator upgrade project that would replace the present 130 nm·rad FODO lattice with an 18 nm·rad double bend achromat (DBA) lattice: SPEAR 3. The low emittance design yields a high brightness beam, but the stronger focusing in the DBA lattice increases chromaticity and beam sensitivity to machine errors. To ensure efficient injection and long Touschek lifetime, an optimization of the design lattice and dynamic aperture has been performed. In this paper, we review the methods used to maximize the SPEAR 3 dynamic aperture including necessary optics modifications, choice of tune and phase advance, optimization of sextupole and coupling correction, and modeling effects of machine errors, wigglers and lattice periodicity.

INTRODUCTION

SPEAR 3 is the 3 GeV upgrade project under study at SSRL [1]. It aims at replacing the current 130 nm·rad FODO lattice with an 18 nm·rad low emittance double bend achromat (DBA) lattice. To minimize the cost of the project and to use the existing synchrotron light beam lines, the new design [2,3] closely follows the racetrack configuration of the SPEAR tunnel, with the magnet positions fit to the 18 magnet girders shown in Fig. 1. As in the current design, the SPEAR 3 lattice has two-fold symmetry and periodicity with two identical arcs and two long straight sections. Each arc in the new lattice has 7 identical symmetric cells, and each straight section consists of two mirror symmetric matching cells.

The lattice functions of one quarter of the ring are shown in Fig. 2. The two bends and a quadrupole in the middle of each DBA cell compensate the dispersion, while the two quadrupole doublets at each end control the tune and cell β functions. Since the new lattice cells have to fit to the existing 11.7 m cell length, it results in a compact DBA design, and hence increases the focusing. Similar to other light source lattices, it has been found advantageous to add vertical focusing to the bends [5–8] to relax the optics and reduce the strength of cell sextupoles by

---

[1] Work supported by the Department of Energy Contract DE-AC03-76SF00515 and the Office of Basic Energy Sciences, Division of Chemical Sciences.

CP468, *Nonlinear and Collective Phenomena in Beam Physics–1998 Workshop,*
edited by S. Chattopadhyay, M. Cornacchia, and C. Pellegrini
1999 The American Institute of Physics 1-56396-862-2
25

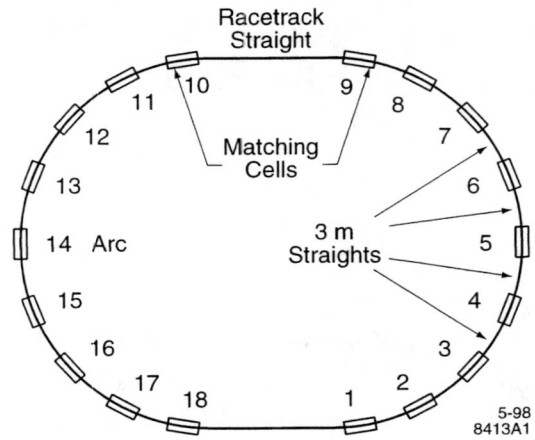

FIGURE 1. Schematic of SPEAR tunnel girders.

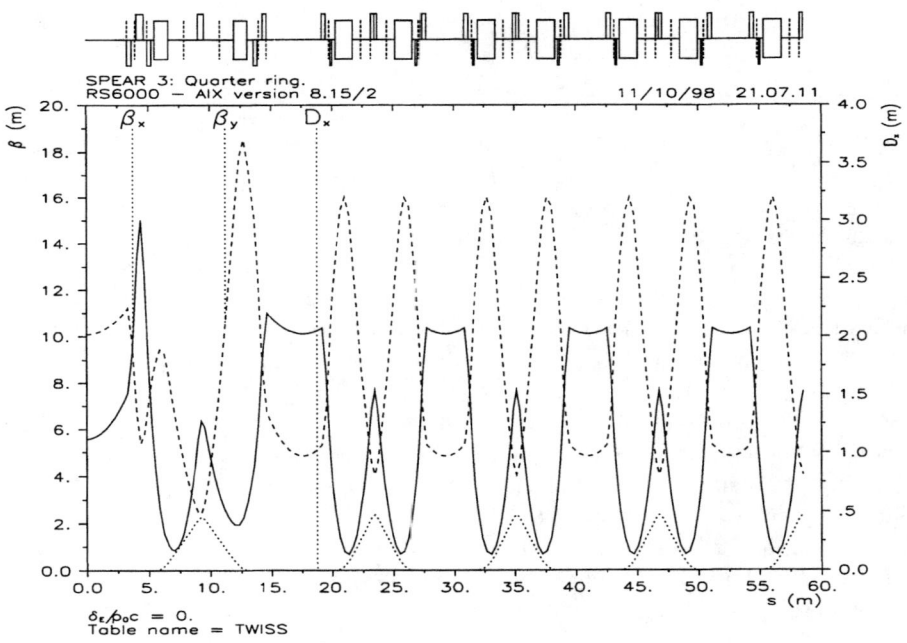

FIGURE 2. Optics of One Quadrant of SPEAR 3 (from the center of the long straight section to the middle of the arc).

26

increasing the difference in their β functions [9]. Each matching cell has an extra quadrupole for a better optics matching, two 3/4 bends, with magnet positions adjusted to maintain the ring circumference [4] (the current 358.53 MHz RF system will be used), and quadrupole strengths adjusted to optimize β functions and phase advance.

Though the DBA design has an advantage of a high brightness beam, its effect on the beam dynamics has to be verified. First, the lower emittance results in a higher particle density which increases probability of particle collisions inside the bunches and makes the Touschek effect the limiting factor for the beam lifetime. Secondly, the stronger focusing in the DBA lattice increases beam sensitivity to magnetic and chromatic errors and generates larger chromaticity. The latter requires strong sextupole correctors which increase the amplitude dependent and non-linear chromatic aberrations. These effects tend to reduce the dynamic aperture, and if the aperture is not sufficient for momentum errors up to $\delta = 3\%$ it will further reduce the Touschek beam lifetime. Adequate dynamic aperture is also important in order to minimize losses during horizontal oscillations of the injected beam.

Consequently, the low emittance design has to be optimized to achieve the maximum dynamic aperture. It is especially important to maximize the horizontal size of dynamic aperture to minimize the Touschek effect and allow large injection oscillations in SPEAR 3. The improvement of dynamic aperture starts from optimization of linear optics and correction systems. In the following sections we review the modifications made to the initial lattice design and present tracking studies including effects of magnetic and chromatic errors, perturbation due to wigglers, and the effect of lattice periodicity.

All of the tracking study has been done using the LEGO code [10] which employs element-by-element tracking based on symplectic integration techniques [11]. In a few cases we tested LEGO results against other available tracking codes and found good agreement. In our study we calculated dynamic aperture at injection point located at the symmetry point between arc cells where $\beta_x = 10.1$ m and $\beta_y = 4.8$ m. The other typical parameters were: number of tracking turns $N = 1024$, linear chromaticity corrected to zero, and synchrotron oscillations included.

ERROR FREE DYNAMIC APERTURE

Typically, the error free dynamic aperture serves as an upper limit for the aperture with machine errors or with insertion devices (ID). It is therefore important to maximize first the dynamic aperture for the ideal lattice without any magnetic or misalignment errors. Maximizing the error free dynamic aperture necessarily involves optimization of linear optics and the chromaticity correction system, minimization of chromatic and high order effects, and optimizing the betatron tune.

Cell Optics

The DBA cell optics was made to fit the existing 11.7 m cell length with the magnet positions constrained to provide ~ 3 m space for the insertion devices and with bend positions kept to fit to the current synchrotron light beam lines. This results in a compact DBA design which leads to stronger focusing and, hence, increased beam sensitivity to machine errors. Though the achromat lattice eliminates dispersion in the insertion devices and at injection, it limits the available positions for chromatic sextupoles to rather short dispersive regions between the bends and the middle quadrupole QFC. The close proximity of the SF and SD sextupoles reduces their effectiveness and requires larger strengths.

To relax the cell optics it has been found advantageous to add vertical focusing to the cell bends. This results in a better separation of horizontal and vertical focusing and reduces the quadrupole strength in the doublets. Most importantly, due to increased β_y at SD sextupoles it provides better optical separation between the SF and SD and reduces their strength. To further reduce the strengths of the sextupoles and QFC quadrupole, the bends were placed as far from the cell center as it is possible within existing geometric constraints.

Working Tune and Phase Advance

The choice for the phase advance in the arc cells was to be near $\mu_x \approx 0.75 \times 2\pi$ and $\mu_y \approx 0.25 \times 2\pi$. This provides favorable conditions for local cancellation of: 1) geometric aberrations from arc sextupoles located $-I$ apart, and 2) first order chromatic beta waves from sextupoles and quadrupoles located 90° apart, as well as systematic quadrupole errors. It is worth to note that the high horizontal phase advance is due to the achromat design which requires $\mu_x = \pi$ between the bends.

Initially, the matching cells were designed to provide I-transformation between the two arcs in both planes [12]. This would give the advantage of having effectively a 14-period lattice since the matching cells would be virtually invisible in the first order to the on-momentum particles. Generally, the high periodicity optics provides better cancellation of systematic errors.

With the above choices the total tune would be near $\nu_x \approx 14.5$ and $\nu_y \approx 5.5$. To move the working tune away from the half integer resonance, the phase advance in arc and/or matching cells has to be adjusted. To minimize the resistive wall impedance effects [13] it is favorable to move the tune into the lower quarter on the tune plane ($\nu < 1/2$). Tracking studies showed that relaxing the phase advance through the matching cells improves the SPEAR 3 off-momentum dynamic aperture since it reduces the chromaticity and the strengths of matching quads. On the other hand, relaxing the arc cells would increase the arc sextupole strengths because of unfavorable change of the cell β functions at the sextupoles.

The optimum phase conditions have been identified by performing a horizontal dynamic aperture scan across the matching cell phase advance μ_x and μ_y. The on-

28

TABLE 1. Horizontal Dynamic Aperture (in number of $\sigma_x = 0.45$ mm) versus ν_x, ν_y. The Tracking Included a Set of Random Machine Errors and 1% Momentum Error.

$\nu_x \rightarrow$ $\downarrow \nu_y$	.16	.17	.18	.19	.20	.21	.22	.23	.24	.25	.26	.27	.28	.29	.30
.30	49	49	49	49	51	46	47	47	46	44	41	38	36	49	48
.29	51	46	52	51	48	46	47	47	47	44	41	39	46	49	48
.28	49	49	50	51	49	48	51	49	46	44	42	40	36	48	48
.27	50	49	49	51	50	48	46	49	46	42	40	40	37	38	37
.26	49	46	47	51	49	46	44	47	43	42	42	41	41	39	41
.25	47	47	49	51	49	48	44	43	46	45	42	42	37	41	49
.24	47	47	46	50	47	49	45	44	48	46	43	41	45	43	50
.23	48	49	46	49	48	48	46	50	48	45	44	49	47	42	42
.22	48	48	45	44	49	48	50	48	48	45	46	42	42	43	43
.21	49	45	46	46	48	50	50	51	48	46	44	44	46	46	44
.20	48	48	44	46	41	47	44	50	48	45	46	42	47	48	45
.19	48	46	45	50	48	45	47	49	46	47	47	48	47	47	45
.18	47	45	47	51	49	47	47	42	46	47	43	49	46	43	40
.17	53	49	48	49	50	48	47	45	45	43	46	40	45	44	42
.16	51	49	48	50	49	50	47	47	44	42	41	36	42	41	40

momentum and off-momentum dynamic aperture were maximized at about $\mu_x = 0.78 \times 2\pi$ and $\mu_y = 0.42 \times 2\pi$ per matching cell.

To minimize the effect of strong low order betatron resonances the location of the working tune on the tune plane has been chosen slightly below .25, away from the 3rd and 4th order resonance lines. The final choice ($\nu_x = 14.19$, $\nu_y = 5.23$) was based on favorable horizontal injection conditions and the results of dynamic aperture tune scan. The two dimensional diagram of horizontal dynamic aperture (in number of σ_x) versus x and y tune is shown in Table 1, where $\sigma_x \approx 0.45$ mm is the horizontal rms beam size at injection point. During the aperture scan the tune was varied by changing arc phase advance, and the lattice was kept matched at all tunes. The above scan also included a set of random machine errors which will be described in the following sections. With the chosen tune, the phase advance per arc cell is $\mu_x = 0.7907 \times 2\pi$ and $\mu_y = 0.2536 \times 2\pi$. The location of the working tune on the tune plane along with betatron resonance lines up to 4th order is shown in Fig. 3.

Chromatic Correction

Efficient chromatic correction is essential for a large off-momentum dynamic aperture and for a long Touschek lifetime. Since the sextupoles also give rise to non-linear geometric aberrations it is important to minimize these effects by using compensation techniques and reducing the sextupole strengths.

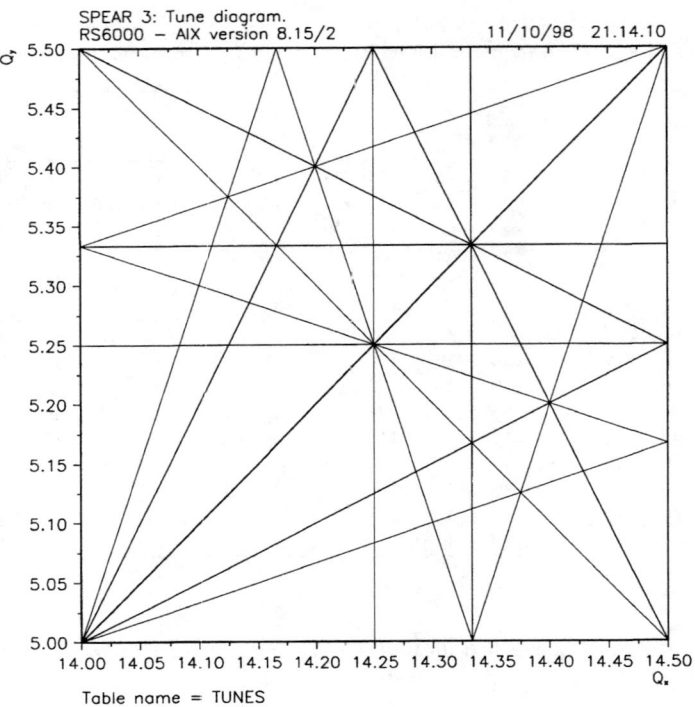

FIGURE 3. SPEAR 3 Working Tune on the Tune Plane.

As mentioned previously, the choice for phase advance in the arc cells provides conditions for local compensation of sextupole geometric aberrations and chromatic beta waves from arc quadrupoles and sextupoles. This scheme would work optimally if the number of arc cells was $4 \times integer$. With 7 arc cells in the design, however, this correction is not complete, and the geometric constraints do not allow for 8 identical arc cells.

The chromaticity correction using only two sextupole families in the arc cells does not provide adequate dynamic aperture for off-momentum particles. Since the two family sextupoles can only compensate the linear chromaticity, the off-momentum aperture is mostly affected by the non-linear chromatic effects. The reason for the large high order chromaticity is due to the matching cells which contribute about 20% to the total chromaticity and break periodicity of the 14 arc cells. Two additional families of sextupoles (SFI, SDI) placed in the matching cells, similar to the arc cells, help to reduce the non-linear terms and significantly improve the off-momentum dynamic aperture. Table 2 compares HARMON [14] calculations of the high order chromaticity for the lattice with and without SFI, SDI sextupoles.

The matching cell sextupoles also generate geometric aberrations and therefore

TABLE 2. 2nd and 3rd Order Chromaticity for SPEAR 3 with and w/o Matching Cell Sextupoles.

2nd and 3rd Order Chromaticity	$\dfrac{d\nu_x}{d\delta^2}$	$\dfrac{d\nu_y}{d\delta^2}$	$\dfrac{d\nu_x}{d\delta^3}$	$\dfrac{d\nu_y}{d\delta^3}$
w/o SFI/SDI	-117	-52	-674	-301
with SFI/SDI	-48	-12	-228	74

have to be kept relatively weak in order to preserve the on-momentum aperture. The optimum strengths of the matching cell sextupoles have been evaluated by performing a horizontal aperture scan versus SFI, SDI strengths while using the arc sextupoles to keep the ring chromaticity constant. Other chromatic correction schemes which use more sextupole families in the arcs did not result in a better aperture.

As it was mentioned earlier, the close proximity of the SF, SD sextupoles in the short dispersive region between bends and QFC quadrupole reduces the effectiveness of the sextupoles and increases their strength. To increase the optical separation of the sextupoles two other options were studied. In one option, the SD sextupole was moved away from the SF by combining with part of the adjacent bend. This increased the β_y function at the SD, but the dispersion inside the bend was rather low. The tracking study showed that the dynamic aperture reduces in this option. In the second study, the SF sextupole was combined with the QFC quadrupole. Due to higher dispersion and β_x at the QFC, this led to weaker SF sextupoles and potentially reduced high order sextupole effects. The dynamic aperture, however, did not improve using this option. Consequently, in the current design the sextupoles are kept separate from the bends and QFC quadrupole.

The resultant error free dynamic aperture for on-momentum and $\delta = 3\%$ particles is shown in Fig. 4. The axes refer to the initial particle amplitude at the injection point, and the two curves show the boundary for the stable motion for on and off-momentum particles. Clearly, the off-momentum aperture is very robust against momentum errors and provides favorable conditions for a long Touschek lifetime.

EFFECT OF MACHINE ERRORS

Magnetic field and alignment errors introduce optics perturbations and enhance effects of resonances that limit dynamic aperture. For conservative results, we included several different classes of magnet errors in the tracking studies including random main field errors, random and systematic multipole errors, and random alignment errors. Since the skew quadrupoles combined with sextupoles in SPEAR 3 generate a skew octupole field, and the large orbit variation in the 10.6° bend samples high order multipole fields, these effects were added to the error set. The effect of positive linear chromaticity, large β function distortions, orbit dis-

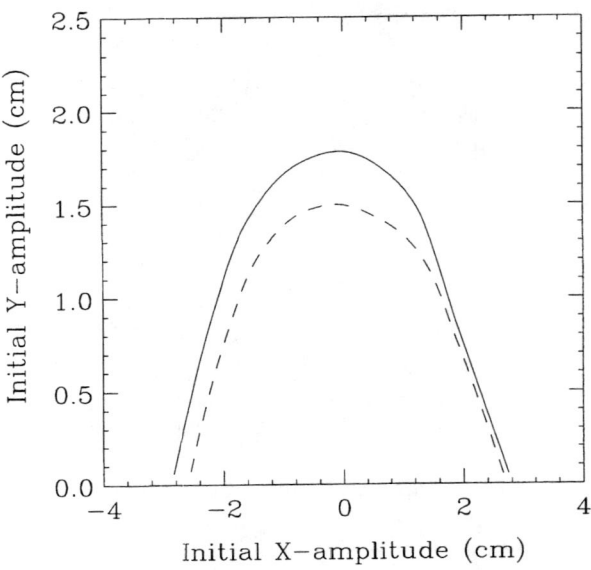

FIGURE 4. Error Free Dynamic Aperture for on-Momentum (solid) and $\delta = 3\%$ off-Momentum (dash) Particles.

tortions, large amplitude coupling and insertion devices were studied. Harmonic sextupole correction has been tried to improve the dynamic aperture, and the effect of lattice periodicity was analyzed.

Alignment and Field Errors

The magnet error and alignment specifications used for the SPEAR 3 tracking studies can be met with standard manufacturing techniques. For tracking simulations, the following values for rms errors were used with 2σ truncation to simulate realistic quality control.

Alignment

The alignment rms errors shown in Table 3 can be achieved with survey techniques used in practice and are large enough to yield conservative tracking results. The dipole alignment specification is the same as for quadrupoles since the SPEAR 3 dipoles include a strong quadrupole field.

32

TABLE 3. Alignment rms Errors Used in Tracking Studies.

Element	$\Delta x\ (\mu m)$	$\Delta y\ (\mu m)$	Roll (μrad)
Dipole	200	200	500
Quadrupole	200	200	500
Sextupole	200	200	500

TABLE 4. Systematic rms Multipole Field Errors.

Magnet	$r\,(mm)$	n	$\Delta B_n/B$
Dipole	30	2	1×10^{-4}
		3-14	5×10^{-4}
Quadrupole	32	6,10,14	5×10^{-4}
Sextupole	32	4	-8.8×10^{-4}
		5	-6.6×10^{-4}
		9	-1.6×10^{-3}
		15	-4.5×10^{-4}

Systematic Multipole Errors

Systematic multipole field errors are field components of higher order than the main field which apply to all magnets with a common core design. In LEGO, the multipole errors are defined in terms of a ratio of the multipole field ΔB_n (normal or skew) to the main magnet field B at radius r, where $n = 1, 2, 3, \ldots$ is the multipole order starting with a bend. The normal rms values $\Delta B_n/B$ used for the SPEAR 3 magnets are listed in Table 4. For the quadrupole magnets, only the allowed multipoles were used. For the gradient dipole magnets and sextupoles combined with skew quads, all systematic multipoles were used with the largest values shown in the Table 4.

In the SPEAR 3 design the skew quadrupoles used for coupling correction are combined with sextupoles. Such combined magnets also generate a skew octupole field proportional to the skew quad strength. At radius of 32 mm the magnitude of the octupole field will be 57% of the skew quad field. In the tracking, the skew octupole component was systematically added in proportion to the skew quadrupole field. It is worth to note that typical skew quad strengths required for coupling correction are less than 1% of the main ring quadrupole strengths.

TABLE 5. Random rms Multipole Field Errors.

Magnet	$r(mm)$	n	$\Delta B_n/B$
Dipole	30	2	1×10^{-4}
Quadrupole	32	3,6,10,14	5×10^{-4}
		4,5,7-9,11-13	1×10^{-4}
Sextupole	32	5	1.5×10^{-3}
		7	4.8×10^{-4}

Random Field Errors

Differences in magnetic core length will give rise to $\sim 10^{-3}$ random main field errors. Normal random multipole errors, introduced by magnet assembly imperfections, are listed in Table 5. To achieve conservative tracking results, large values were specified for the random n =3,6,10,14 multipoles on the quadrupole magnets.

Coupling Correction

The initial design of coupling correction employed four independent skew quadrupoles placed in non-dispersive regions in the matching cells. The four skew quads were used to uncouple the 4×4 one turn transfer matrix. However, the skew quad placement in the matching cells did not provide enough variation of the sum and difference phase advance $(\mu_x \pm \mu_y)$ for efficient orthogonal correction. Depending on the set of random machine errors, this occasionally led to strong skew quadrupoles and reduced dynamic aperture.

Further study showed that it was beneficial for dynamic aperture to use skew quadrupole components on the chromatic sextupoles located in the dispersive regions of the arc cells [3]. This configuration provided a more orthogonal set of the skew quad positions with reduced strengths. The negative effects, such as induced vertical dispersion, were small compared to improved aperture and robustness of the correction. In total, 24 skew quads were arranged in four families and placed at their optimum phase positions. One other negative effect is that a skew quadrupole in a sextupole magnet gives rise to a systematic skew octupole field. In our tracking study, the effect of this field did not reduce the dynamic aperture. As a future option, the large number of skew quads in the above scheme allows expansion of the number of independent families with correction of the vertical dispersion as well.

Dynamic Aperture with Errors

For a realistic simulation with errors, LEGO first generates and adds the chosen set of errors to the magnets, then iteratively applies correction schemes to minimize the optics perturbation, and finally tracks particles with a variety of initial amplitudes to define the dynamic aperture. The basic set of correction schemes in LEGO includes tune, orbit, linear chromaticity and coupling correction systems. In the SPEAR 3 simulation, the tune was corrected by using two families of doublet quads in the arc cells. The nominal orbit correction routine in LEGO is based on a three corrector bump scheme, but other techniques can be implemented as well. Typically in this study, the linear chromaticity was adjusted to zero with the two families of sextupoles in the arc cells, while the matching cell sextupole strength was kept constant. The coupling correction was done by using the four family skew quad correction scheme described in the previous section. An RF voltage of 3.2 MV was used to generate synchrotron oscillations for off-momentum particles.

The resultant dynamic aperture for 6 random seeds of machine errors for on-momentum and $\delta = 3\%$ off-momentum particles with the described correction schemes is shown in Fig. 5. The linear chromaticity was set to zero in this case. Compared to Fig. 4 (no errors), the dynamic aperture has reduced by about 20-30%. Since insertion devices were not included in this calculation, this reduction is solely due to the machine errors and quality of correction procedures described above. As in the case of the error free lattice, the off-momentum aperture for machine with errors is comfortably large. As the Fig. 5 shows, the horizontal dynamic aperture is in the range of 18 to 20 mm for all particles within $\delta = \pm 3\%$ momentum range. This provides favorable conditions for a long Touschek lifetime and sufficient room

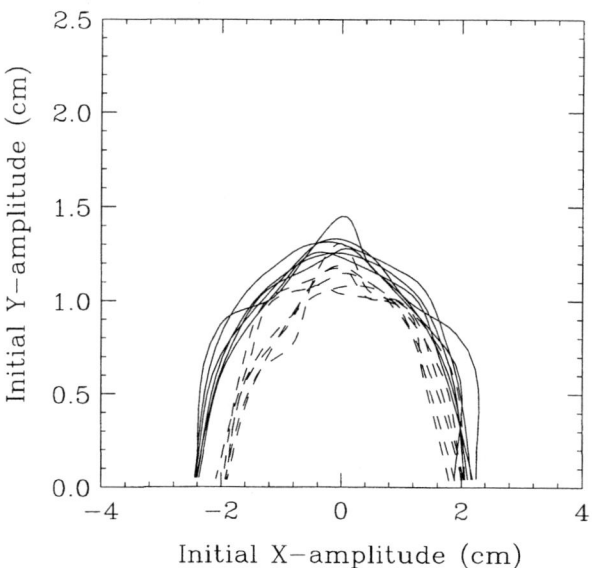

FIGURE 5. Dynamic Aperture for 6 Seeds of SPEAR 3 Machine Errors for $\delta = 0$ (solid) and 3% (dash) Momentum Oscillations.

for horizontal injection oscillations.

Positive Linear Chromaticity

Though most of this study was done with the linear chromaticity corrected to zero, in real machines the value of $\xi = \Delta\nu/\delta$ is typically set slightly positive, up to several units, to avoid effects such as head-tail instability. The main impact of the positive chromaticity on the dynamic aperture is from the increased tune spread in the beam. Due to synchrotron oscillations the particles with large momentum errors would sample a larger area on the tune plane and might cross more harmful betatron resonances. As a result, the momentum aperture can be significantly reduced if the linear chromaticity is too large. The effect on the on-momentum aperture is typically smaller due only to the increased strength of sextupole correctors.

Fig. 6,7 show dependence of linear chromaticity on the dynamic aperture for the particles with momentum oscillations of $\delta = 0, 1\%$ and 3%. The tracking included a full random set of machine errors, and the chromaticity was set equal in the x and y planes. Fig. 6,7 show that the particles with momentum errors of $\delta = 1\%$ and 3% lose stability at $\xi \approx 15$ and 6, respectively. Clearly, this is the effect of a half integer resonance. For the SPEAR 3 working tune ($\nu_x = 14.19$, $\nu_y = 5.23$) the off-momentum particles would likely be lost when momentum dependent tune shift approaches to $\Delta\nu \approx -0.2$. The dynamic aperture for the core beam (low momentum error) is not significantly reduced even for the large positive chromaticity. However, the beam lifetime can be reduced for $\xi > 5$ due to Touschek effect, since the scattered particles with large δ may not survive.

Feed-down Studies

Since the electron beam in a SPEAR 3 dipole magnet follows a 10.6° arc orbit with 16.6 mm sagitta, even the particles traveling along the ideal trajectory will sample the full set of high order multipole field. Taking into account the large horizontal excursion of the beam orbit, the dipoles are specified to have a full 92 mm good field region [3]. Since normally the Taylor expansion of the multipole field is specified around the central magnetic axis, each individual multipole of this field expanded about the curved beam orbit in a dipole will generate (feed-down) an additional full set of lower order multipoles proportional to the orbit displacement. Both the nominal and feed-down multipoles have to be combined to realistically estimate the effect of dipole errors. Though this feed-down effect might be negligible for large machines, it has to be verified for smaller machines with large sagitta in the dipoles.

In LEGO and most other codes, the multipole field in a dipole would be expanded with respect to the ideal orbit, not the magnetic axis. Therefore, when applying the multipole errors in a dipole, the feed-down terms were included in addition to

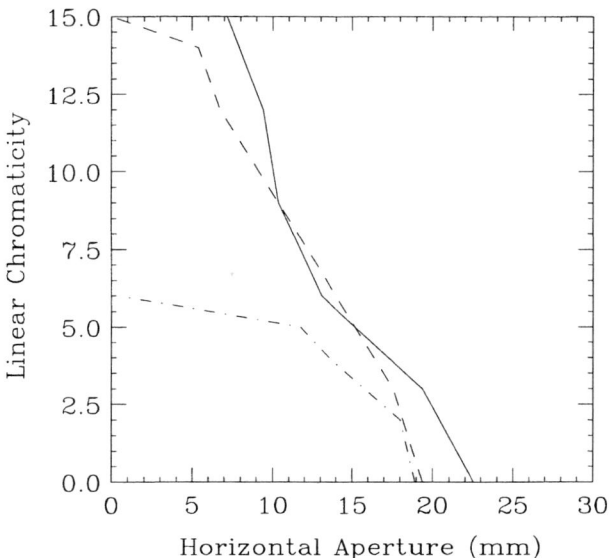

FIGURE 6. Horizontal Dynamic Aperture versus Linear Chromaticity for $\delta = 0$ (solid), 1% (dash) and 3% (dot-dash) Momentum Oscillations.

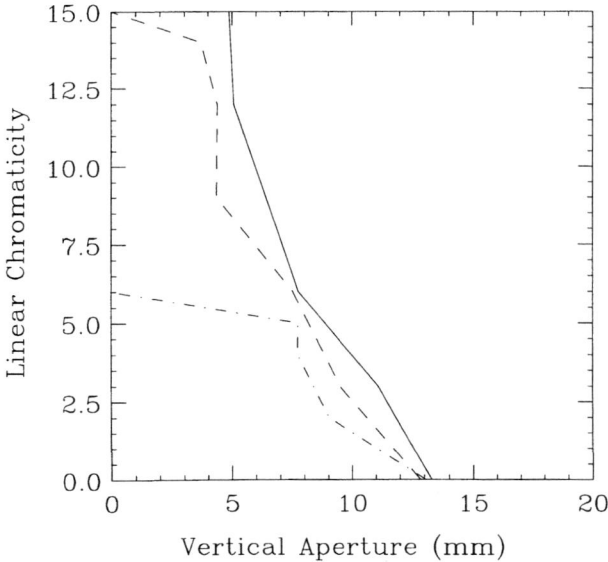

FIGURE 7. Vertical Dynamic Aperture versus Linear Chromaticity for $\delta = 0$ (solid), 1% (dash) and 3% (dot-dash) Momentum Oscillations.

the nominal set of multipoles specified around the magnetic center. Since the feed-down terms depend on the orbit displacement, we used the following technique to evaluate this effect. The 1.45 m dipole was 'sliced' in a reasonably short pieces and the average orbit displacement was calculated for each slice. Based on the orbit displacement in each slice and the nominal set of multipole fields, the systematic feed-down terms were calculated for each slice. The tracking simulation was then done including sliced dipoles with multipole feed-down terms. The results showed no degradation of the dynamic aperture due to this effect.

Since the sliced dipoles would significantly increase the computer time for element-by-element tracking, normally this model was not used in tracking studies. However, the magnitude of the nominal dipole multipole field used in tracking runs without explicit feed-down effects was set conservatively large to produce comparable field around the orbit.

Large Beta Distortion

In the tracking, typical β distortions caused by the SPEAR 3 specification errors after correction were on the order of $\Delta\beta/\beta = \pm(5 - 10)\%$. However, in a real machine it is not unusual to observe much larger modulations since some of the design specifications may not be achieved, especially during commissioning. The β distortions lead to a larger beam size and may increase the effects of high order field and resonances.

To verify the effect of large β modulation on the SPEAR 3 dynamic aperture, the quadrupole errors in two quad families in the matching cells were increased to the level of a few percent to produce $\Delta\beta_x/\beta_x \approx \pm30\%$ and $\Delta\beta_y/\beta_y \approx \pm20\%$. The calculated dynamic aperture for 5 seeds of random machine errors is shown in Fig. 8. The average reduction of the aperture due to this β distortion is about 15% compared to Fig. 5. Though this aperture is still adequate to operate the machine, it is clear that such large quadrupole errors have to be identified during initial operation and corrected.

Orbit Distortion

Large orbit distortions can cause a reduction of dynamic aperture since the electron beam would experience much larger effects from non-linear sextupole and multipole fields. Normally, in this study, the rms orbit was corrected very well, down to the level of $\sim 100\mu m$. To investigate large orbit effects, an additional set of magnet alignment errors was introduced in LEGO just prior to tracking. Using additional uncorrected random alignment errors with $\sigma_{\Delta x} = 80\mu m$, $\sigma_{\Delta y} = 40\mu m$, and $50\mu rad$ roll, rms orbit distortions of up to 3 mm in the horizontal and 1.5 mm in the vertical plane were generated. In each case, the on-energy dynamic aperture was reduced by a maximum of 2 mm in the horizontal and vertical planes for 6

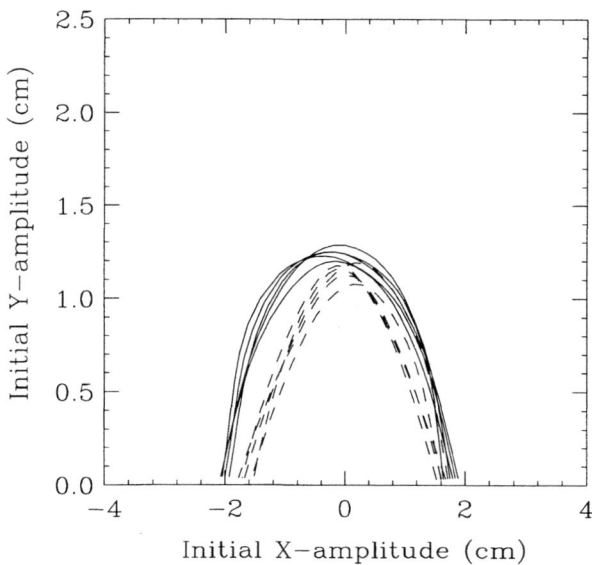

FIGURE 8. Dynamic Aperture with Large β Distortion ($\pm30\%$ (x), $\pm20\%$ (y)) for on-Momentum (solid) and $\delta = 3\%$ off-Momentum (dash) Particles for 5 Random Error Seeds.

different seeds. The off-energy aperture was reduced similarly compared to the off-energy dynamic aperture without orbit distortions.

For safe machine operation, the peak vertical orbit excursion must be held to < 1 mm and the horizontal excursion should be < 3 mm, so orbit induced reduction of dynamic aperture is negligible. Absolute BPM reading errors are expected to be on the order of a few hundred μm or less. The low sensitivity of the aperture to orbit distortions should simplify initial injection and machine commissioning at low currents.

In a related test, we studied large sextupole misalignment while the orbit was well corrected. The sextupole misalignments generate random quadrupole errors and result in optics distortion and coupling. To simulate this effect, rms sextupole misalignments of 1 mm were assigned in both planes, and the orbit was corrected down to a few hundred μm level. The 1 mm sextupole displacements generate an order of magnitude larger quadrupole errors compared to the specified field errors in the ring quadrupoles. Of the 6 seeds studied, 5 cases showed > 17 mm horizontal dynamic aperture for on-momentum particles. The vertical aperture was always larger than the ±6 mm ID chamber size. At $\delta = 3\%$, the horizontal aperture remained above 13 mm. The one 'bad' seed produced 13 mm horizontal dynamic aperture on-momentum and about the same result for 3% off-momentum particles. In practice, with sextupole alignments much better than 1 mm rms, no reduction of dynamic aperture is expected.

Large Amplitude Coupling

In order to monitor the full extent of the dynamic aperture, the SPEAR 3 tracking simulations did not include physical apertures. In practice, however, the vacuum chamber has horizontal and physical apertures that can limit beam lifetime. In the vertical plane, for example, SPEAR has two insertion devices with $y = \pm 6$ mm vacuum chambers that define the vertical acceptance. Although the height of the ID chambers yield acceptable gas scattering lifetime [3], they can limit the Touschek lifetime if strong coupling is present. In the presence of machine errors and strong sextupole fields, for instance, particles with large horizontal amplitudes can reach resonances which couple the horizontal motion into the vertical plane. This effect has been observed in operational machines [15].

To study the effect of large amplitude coupling, we launched particles with variable horizontal and synchrotron oscillation amplitudes and monitored the maximum vertical excursion. Fig. 9 shows the degree of x-y coupling as a function of initial horizontal amplitude for particles with $\delta = 0, 1, 2$, and 3% energy oscillations. Each particle was launched with a small initial vertical amplitude of $100 \mu m$ and the peak vertical amplitude was monitored for 1024 turns at the ID location. The plot shows the average value taken over 6 machines with different error seeds. Based on these results, one can conservatively anticipate an effective reduction of horizontal aperture from 20 mm to about 18 mm, and a corresponding reduction in Touschek lifetime from ≈ 135 hrs to ≈ 125 hrs [3]. The 10 mm injection oscillations in the horizontal plane should not be effected by this coupling.

Harmonic Sextupoles

The first non-linear field magnets typically introduced in the optics are sextupoles which are placed in dispersive regions to correct chromaticity. In addition to their chromatic effect, sextupoles generate geometric aberrations such as amplitude dependent tune shift and high order resonances. Even without machine errors these effects may significantly limit dynamic aperture. Therefore, it is important to verify this limitation and use available techniques to minimize a reduction of aperture.

Though the choice of cell phase advance in the SPEAR 3 arcs helps to reduce the sextupole geometric aberrations, they are not completely canceled. To further minimize these effects, additional 'harmonic' sextupoles can be used [16]. These sextupoles are usually placed in non-dispersive regions to avoid their chromatic effect, and their strengths are optimized to minimize the total amplitude dependent tune shift. Since this tune shift can be mathematically decomposed into a series of tune harmonic components generated by sextupoles [16], one technique to reduce the tune shift is to minimize the strongest harmonic components.

Obviously, the upper limit for dynamic aperture with sextupoles is the aperture where sextupole aberrations are not present. One way to evaluate this limit is to track on-momentum particles in the lattice with sextupoles turned off. The RF

40

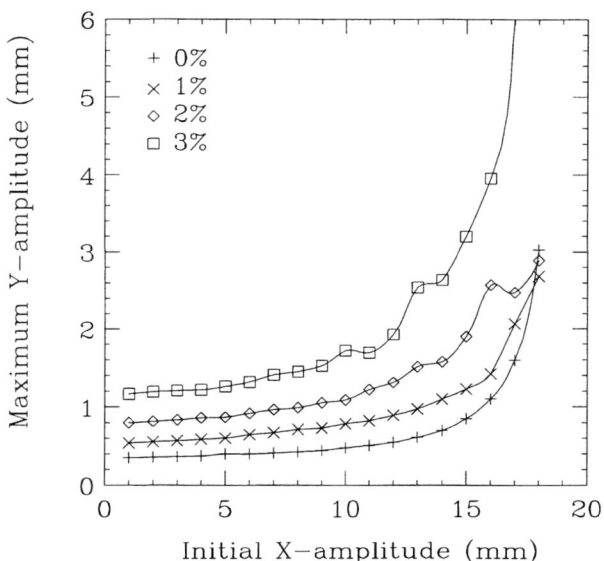

FIGURE 9. Peak Vertical Excursion as a Function of Initial Horizontal Amplitude for $\delta = 0, 1, 2$, and 3% Momentum Oscillation. Results Averaged over 6 Seeds of Machine Errors.

cavities have to be turned off as well to eliminate any chromatic effects. The result of this tracking for SPEAR 3 with 6 random seeds of machine errors is shown in Fig. 10. It follows that if the sextupoles are perfectly compensated the horizontal aperture could be as large as 28 mm or 40% larger compared to Fig. 5. Realistically, this limit may not be achieved since all sextupole aberrations have to be canceled all at once, and even with perfect global compensation the local effects would be present. Effectiveness of a harmonic sextupole system would also depend on the phase advance of a particular optics configuration.

Based on formulas in [16], we analyzed the magnitude of harmonic components in the amplitude dependent tune shift generated by the chromatic sextupoles. The analysis showed that one of the largest contributions comes from 14th and 18th horizontal tune harmonics. To minimize these contributions we tested a scheme which has two family harmonic sextupoles in each of the 14 arc cells. Because of very limited space in non-dispersive region in the cells, in this test we used thin lens harmonic sextupoles attached to cell quadrupoles in a doublet. The strength of these sextupoles was optimized by scanning and maximizing dynamic aperture. The reduction of tune shift with amplitude due to harmonic sextupoles was verified using HARMON and the results are presented in Table 6, where ϵ is the rms beam emittance. The dynamic aperture calculation with the harmonic sextupoles for error free lattice and machine with errors is shown in Fig. 11,12. Compared to Fig. 4, the improvement of error free dynamic aperture is about 10-15%. However,

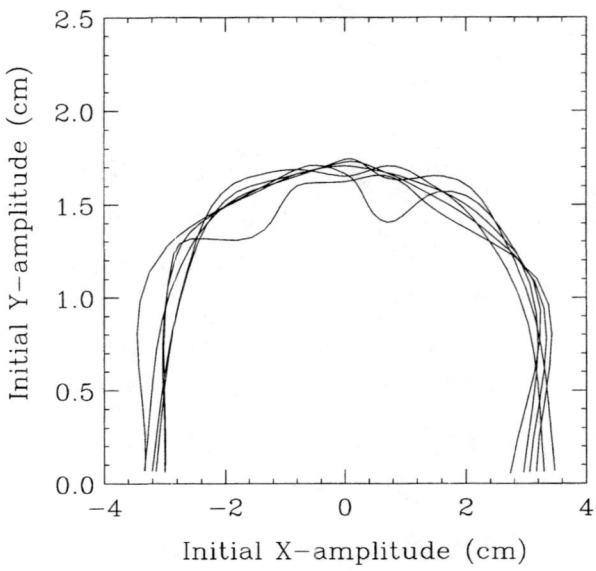

FIGURE 10. On-Momentum Dynamic Aperture without Sextupoles for 6 Random Error Seeds with RF Cavities Turned off.

with machine errors included the improvement of horizontal aperture reduces to a minimum. Taking into account the cost and design complications associated with additional sextupoles as well as marginal aperture improvement, the harmonic sextupole correction was not included in the current design.

Insertion Devices

At present, seven horizontally deflecting wigglers are planned in the 3 m drifts between arc cells. The typical parameters of these insertion devices are: the peak field B_y up to 2 T, and the total length up to 2.3 m. Though the wiggler magnetic

TABLE 6. Amplitude Dependent Tune Shift for SPEAR 3 with and w/o Two Family Harmonic Sextupoles in Arc Cells.

Amplitude dependent tune shift	$\dfrac{d\nu_x}{d\epsilon_x}$	$\dfrac{d\nu_y}{d\epsilon_y}$	$\dfrac{d\nu_y}{d\epsilon_x}$
w/o harmonic sextupoles	878	1392	643
with harmonic sextupoles	446	866	479

42

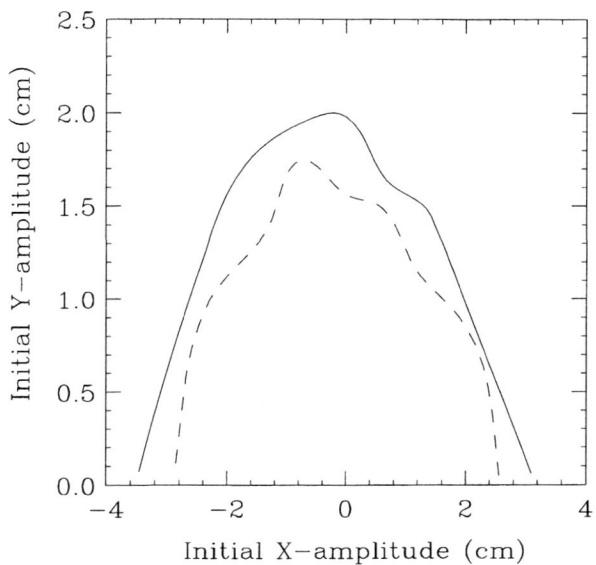

FIGURE 11. Error Free Dynamic Aperture with Two Family Harmonic Sextupoles in Arc Cells for 0 (solid) and 3% (dash) Momentum Oscillations.

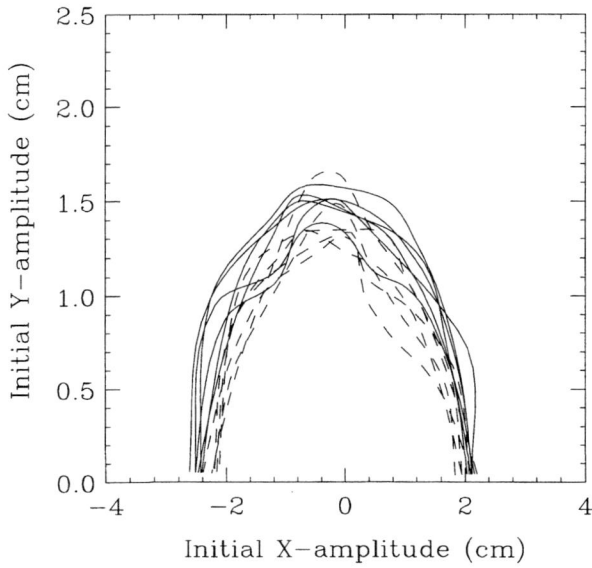

FIGURE 12. Dynamic Aperture with Two Family Harmonic Sextupoles in Arc Cells with SPEAR 3 Machine Errors for 0 (solid) and 3% (dash) Momentum Oscillations.

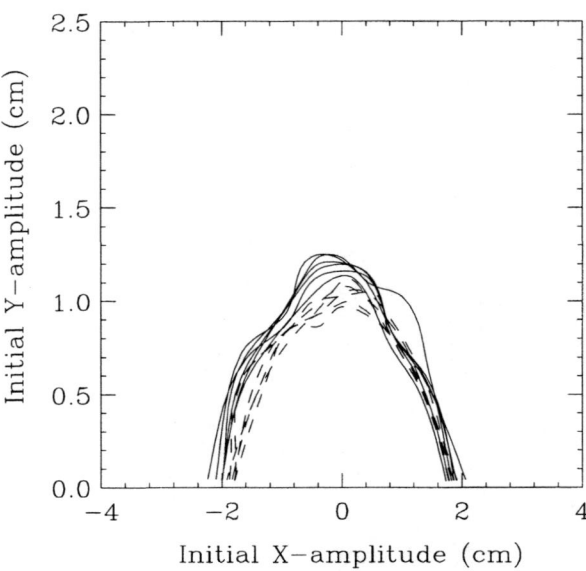

FIGURE 13. Dynamic Aperture with 7 Wigglers, Corrected Wiggler Focusing, Systematic Wiggler Multipole Errors and 6 Seeds of Random Machine Errors for 0 (solid) and 3% (dash) Momentum Oscillations. The Intrinsic High Order Wiggler Fields not Included.

field is highly non-linear, some of its high order effects are locally canceled due to wiggler periodicity. The remaining lowest order perturbations to the beam optics are vertical focusing and amplitude-dependent vertical tune shift due to octupole-like horizontal field. In summary, the wiggler effect on dynamic aperture can arise from perturbation of β functions and tune, high order field effects, reduced periodicity and symmetry of the lattice, as well as wiggler field errors and misalignment.

In SPEAR 3 the wiggler perturbation of β functions and phase advance will be locally corrected using doublet quadrupoles in the cells adjacent to either side of the wiggler [17]. Tracking studies with seven wigglers showed that without high order wiggler effects, the corrected wiggler focusing alone does not reduce dynamic aperture. Without this correction the vertical aperture reduces by about 20%, though it is still well outside the $y = \pm 6$ mm wiggler physical aperture.

The study of wiggler multipole errors was based on recently measured systematic field errors in one of the strongest wigglers (Beamline 11). The measured field data was fit to a set of normal and skew field multipoles up to 12th order and used as wiggler errors in the tracking. Fig. 13 shows that when these systematic multipole errors are included, the dynamic aperture with seven wigglers reduces by about 10%.

The effects of the non-linear wiggler field can further reduce the aperture. Since these intrinsic high order fields are in the horizontal plane, they mostly affect the vertical aperture. Simulation of the first two non-linear terms (octupole and dodecapole-like field) showed rather modest reduction of vertical aperture from 11 mm to 9 mm with the above systematic multipole errors included [17]. This aperture is still well outside the wiggler physical aperture.

Effect of Lattice Periodicity

The advantage of a high periodicity lattice is that the number of resonances excited by systematic multipole field errors is reduced in proportion to the number of periods. In SPEAR 3 the matching cells break the 14 periodic arc cells and reduce the machine periodicity to 2. To verify the effect of periodicity on dynamic aperture, we tracked particles in a lattice which had only 14 identical arc cells and compared the results to the aperture of the full lattice. The results for on-momentum particles are shown in Fig. 14. The two solid curves correspond to aperture without machine errors, and the dash curves define the aperture for 6 seeds of machine errors. In both cases the pure 14 periodic cells provide a larger dynamic aperture compared to the nominal lattice with matching cells. Since without errors the only non-linear fields are from sextupoles, one can conclude that cancellation of systematic sextupole aberrations is much better in a more periodic lattice. Another observation is that either breaking periodicity or including random machine errors will reduce the SPEAR 3 aperture to about similar size.

Due to geometric constraints in the existing SPEAR tunnel it was unavoidable to break the periodicity of arc cells. To keep the effective periodicity of 14, one of the earliest proposals suggested a I-transformation for the matching cell lattice between the arcs [12]. However, the matching cells also contribute a significant amount of chromaticity, and it was found that relaxing the matching cell optics improved the off-momentum dynamic aperture without compromising the on-momentum aperture.

CONCLUSIONS

A long Touschek lifetime and adequate injection conditions were the primary motivations to maximize the SPEAR 3 dynamic aperture. The described optimization included linear optics, working tune, chromaticity and coupling correction, and compensation of wiggler focusing. Other potential improvements were analyzed as well. The effects of large momentum oscillations, realistic machine errors, insertion devices and larger optics perturbations on dynamic aperture were verified. The results consistently showed that the dynamic aperture for up to $\delta = 3\%$ momentum errors is not significantly affected by these effects. The presented analysis shows that SPEAR 3 dynamic aperture will provide adequate injection conditions and result in > 100 hrs of Touschek beam lifetime.

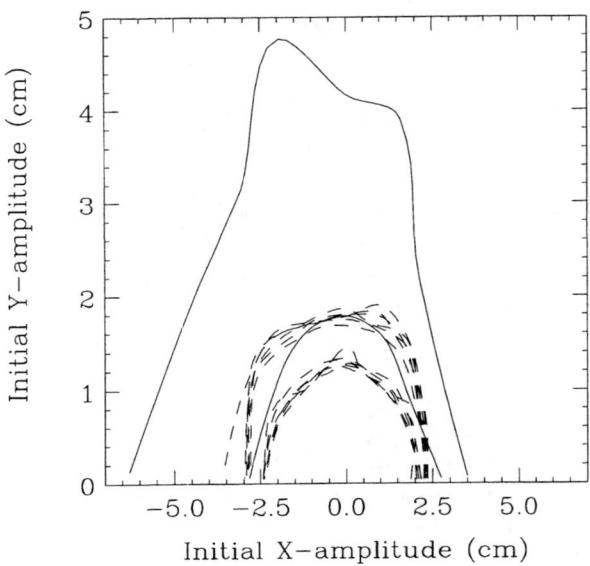

FIGURE 14. On-Momentum Dynamic Aperture: 1) 14 Cells w/o Errors (bigger solid curve), 2) Full Lattice w/o Errors (smaller solid), 3) 14 Cells with Errors (bigger dash), 4) Full Lattice with Errors (smaller dash).

ACKNOWLEDGEMENTS

The authors would like to thank many people who contributed to the SPEAR 3 lattice design. In particular, the authors would like to thank M. Cornacchia, Y. Cai, A. Garren, R. Hettel, J. Safranek and H. Wiedemann for many fruitful discussions.

REFERENCES

1. Hettel, R., and Brennan, S., *Synchrotron Radiation News* **11**, No. 1 (1998).
2. Corbett, J., et al, "Design of the SPEAR 3 Magnet Lattice", presented at the 6th European Part. Acc. Conf., Stockholm, Sweden, 1998.
3. SPEAR 3 Conceptual Design Report, in press.
4. Corbett, J., "Matching Cell Dipole Locations in the SPEAR 3 Lattice", SPEAR 3 Engineering Note 11, September 1998.
5. 1-2 GeV Synchrotron Radiation Source/Conceptual Design Report, LBL, PUB-5172, July 1986.
6. Hsue, C. S., et al., "Lattice Design of the SRRC 1.3 GeV Storage Ring", in *Proceedings of the 1991 IEEE Part. Acc. Conf.*, San Francisco, 1991, pp. 2670-2672.
7. Bassetti, M., Bocchetta, C. J., Wrulich, A., "Magnet Lattice for the Sincrotrone Trieste", ST/M-87/10 (1987).

8. The Proposal for Construction of a National Synchrotron Light Source for Canada (CLS), Centre du Rayonnement Synchrotron Canadien, September 1996.

9. The use of gradient dipoles for SPEAR 3 was proposed by Safranek, J.

10. Cai, Y., et al., "LEGO: A Modular Accelerator Design Code," SLAC-PUB-7642 (1997).

11. Forest, E.,"Canonical Integrators as Tracking Codes", SSC-138 (1987).

12. I-transformers were suggested by Garren, A. See "SPEAR 3 Upgrade Project/Director's Review", Nov. 3-5, 1997.

13. Laclare, J. L., "Introduction to Coherent Instabilities - Coasting Beam Case", in *CERN Accelerator School*, CERN 85-19, 1985, pp. 377-414.

14. Grote, H., and Iselin, F. C., "The MAD Program", CERN/SL/90-13 (AP) Rev.4 (1994).

15. Robin, D., these *Proceedings*.

16. Crosbie, E. A., "Improvement of the Dynamic Aperture in Chasman Green Lattice Design Light Source Storage Rings", in *Proceedings of the 1987 IEEE Part. Acc. Conf.*, Washington, DC, 1987, pp. 443-445.

17. Corbett, J., and Nosochkov, Y., "Effect of Insertion Devices in SPEAR-3", SPEAR 3 Engineering Note-010, August 1998.

Variational Approach in Wavelet Framework to Polynomial Approximations of Nonlinear Accelerator Problems

A. Fedorova*, M. Zeitlin* and Z. Parsa† **

*Institute of Problems of Mechanical Engineering, Russian Academy of Sciences, 199178, Russia, St. Petersburg, V.O., Bolshoj pr., 61.   E-mail: zeitlin@math.ipme.ru
†Physics Department, Bldg. 901A, Brookhaven National Laboratory, Upton, NY 11973-5000, USA.   E-mail: parsa@bnl.gov

Abstract. In this paper we present applications of methods from wavelet analysis to polynomial approximations for a number of accelerator physics problems. According to a variational approach in the general case we have the solution as a multiresolution (multiscales) expansion on the base of compactly supported wavelet basis. We give an extension of our results to the cases of periodic orbital particle motion and arbitrary variable coefficients. Then we consider more flexible variational method which is based on a biorthogonal wavelet approach. Also we consider a different variational approach, which is applied to each scale.

I   INTRODUCTION

This is the first part of our two-part presentation in which we consider applications of methods from wavelet analysis to nonlinear accelerator physics problems. This is a continuation of results from [1]-[6], which is based on our approach to investigation of nonlinear problems – general, with additional structures (Hamiltonian, symplectic or quasicomplex), chaotic, quasiclassical, quantum, which are considered in the framework of local (nonlinear) Fourier analysis, or wavelet analysis. Wavelet analysis is a relatively novel set of mathematical methods, which gives us the possibility of working with well-localized bases in functional spaces and with the general type of operators (differential, integral, pseudodifferential) in such bases.

We consider the application of multiresolution representation to a general nonlinear dynamical system with the polynomial type of nonlinearities. In

**) This work was performed under the auspices of the U.S. Departmentof Energy under Contract No. DE-AC02-98CH10886.

CP468, *Nonlinear and Collective Phenomena in Beam Physics–1998 Workshop,*
edited by S. Chattopadhyay, M. Cornacchia, and C. Pellegrini
1999 The American Institute of Physics 1-56396-862-2

part II we consider this very useful approximation in the cases of orbital motion in a storage ring, a particle in the multipolar field, effects of insertion devices on beam dynamics, and spin orbital motion. Starting in part III A from variational formulation of initial dynamical problem we construct via multiresolution analysis (part III B) explicit representation for all dynamical variables in the base of compactly supported (Daubechies) wavelets. Our solutions (part III C) are parametrized by solutions of a number of reduced algebraical problems, one of which is nonlinear with the same degree of nonlinearity, and the rest are the linear problems which correspond to a particular method of calculation of scalar products of functions from wavelet bases and their derivatives. Then we consider the further extension of our previous results. In part V we consider modification of our construction to the periodic case; in part VI we consider generalization of our approach to variational formulation in the biorthogonal bases of compactly supported wavelets, and in part VII to the case of variable coefficients. In part IV we consider the different variational approach which is based on ideas of para-products (A) and approximation for a multiresolution approach, which gives us the possibility for computations in each scale separately (B).

II PROBLEMS AND APPROXIMATIONS

We consider below a number of examples of nonlinear accelerator physics problems which are from the formal mathematical point of view not more than nonlinear differential equations with polynomial nonlinearities and variable coefficients.

A   Orbital Motion in Storage Rings

We consider as the main example the particle motion in storage rings in a standard approach, which is based on consideration of [7]. Starting from Hamiltonian, which described classical dynamics in storage rings,

$$\mathcal{H}(\vec{r}, \vec{P}, t) = c\{\pi^2 + m_0^2 c^2\}^{1/2} + e\phi \quad, \tag{1}$$

and using Serret–Frenet parametrization, we have the following Hamiltonian for orbital motion in machine coordinates:

$$\mathcal{H}(x, p_x, z, p_z, \sigma, p_\sigma; s) = p_\sigma - [1 + f(p_\sigma)] \cdot [1 + K_x \cdot x + K_z \cdot z] \times \tag{2}$$
$$\left\{1 - \frac{[p_x + H \cdot z]^2 + [p_z - H \cdot x]^2}{[1 + f(p_\sigma)]^2}\right\}^{1/2}$$
$$+ \frac{1}{2} \cdot [1 + K_x \cdot x + K_z \cdot z]^2 - \frac{1}{2} \cdot g \cdot (z^2 - x^2) - N \cdot xz$$

$$+\frac{\lambda}{6} \cdot (x^3 - 3xz^2) + \frac{\mu}{24} \cdot (z^4 - 6x^2z^2 + x^4)$$

$$+\frac{1}{\beta_0^2} \cdot \frac{L}{2\pi \cdot h} \cdot \frac{eV(s)}{E_0} \cdot \cos\left[h \cdot \frac{2\pi}{L} \cdot \sigma + \varphi\right] \quad .$$

Then, after standard manipulations with truncation of power series expansion of square root, we arrive at the following approximated Hamiltonian for particle motion:

$$\mathcal{H} = \frac{1}{2} \cdot \frac{[p_x + H \cdot z]^2 + [p_z - H \cdot x]^2}{[1 + f(p_\sigma)]} + p_\sigma - [1 + K_x \cdot x + K_z \cdot z] \quad (3)$$

$$\cdot f(p_\sigma) + \frac{1}{2} \cdot [K_x^2 + g] \cdot x^2 + \frac{1}{2} \cdot [K_z^2 - g] \cdot z^2 - N \cdot xz +$$

$$\frac{\lambda}{6} \cdot (x^3 - 3xz^2) + \frac{\mu}{24} \cdot (z^4 - 6x^2z^2 + x^4)$$

$$+\frac{1}{\beta_0^2} \cdot \frac{L}{2\pi \cdot h} \cdot \frac{eV(s)}{E_0} \cdot \cos\left[h \cdot \frac{2\pi}{L} \cdot \sigma + \varphi\right] \quad ,$$

and the corresponding equations of motion:

$$\frac{d}{ds}x = \frac{\partial \mathcal{H}}{\partial p_x} = \frac{p_x + H \cdot z}{[1 + f(p_\sigma)]};$$

$$\frac{d}{ds}p_x = -\frac{\partial \mathcal{H}}{\partial x} = \frac{[p_z - H \cdot x]}{[1 + f(p_\sigma)]} \cdot H - [K_x^2 + g] \cdot x + N \cdot z +$$

$$K_x \cdot f(p_\sigma) - \frac{\lambda}{2} \cdot (x^2 - z^2) - \frac{\mu}{6}(x^3 - 3xz^2); \quad (4)$$

$$\frac{d}{ds}z = \frac{\partial \mathcal{H}}{\partial p_z} = \frac{p_z - H \cdot x}{[1 + f(p_\sigma)]};$$

$$\frac{d}{ds}p_z = -\frac{\partial \mathcal{H}}{\partial z} = -\frac{[p_x + H \cdot z]}{[1 + f(p_\sigma)]} \cdot H - [K_z^2 - g] \cdot z + N \cdot x +$$

$$K_z \cdot f(p_\sigma) - \lambda \cdot xz - \frac{\mu}{6}(z^3 - 3x^2z);$$

$$\frac{d}{ds}\sigma = \frac{\partial \mathcal{H}}{\partial p_\sigma} = 1 - [1 + K_x \cdot x + K_z \cdot z] \cdot f'(p_\sigma) -$$

$$\frac{1}{2} \cdot \frac{[p_x + H \cdot z]^2 + [p_z - H \cdot x]^2}{[1 + f(p_\sigma)]^2} \cdot f'(p_\sigma)$$

$$\frac{d}{ds}p_\sigma = -\frac{\partial \mathcal{H}}{\partial \sigma} = \frac{1}{\beta_0^2} \cdot \frac{eV(s)}{E_0} \cdot \sin\left[h \cdot \frac{2\pi}{L} \cdot \sigma + \varphi\right] \quad .$$

Then we use series expansion of function $f(p_\sigma)$ from [1]:

$$f(p_\sigma) = f(0) + f'(0)p_\sigma + f''(0)\frac{1}{2}p_\sigma^2 + \ldots = p_\sigma - \frac{1}{\gamma_0^2} \cdot \frac{1}{2}p_\sigma^2 + \ldots \quad (5)$$

and the corresponding expansion of RHS of equations (4). In the following we take into account only arbitrary polynomial (in terms of dynamical variables) expressions and neglect all nonpolynomial types of expressions; i.e. we consider such approximations of RHS which are not more than polynomial functions in dynamical variables and arbitrary functions of independent variable s ("time" in our case, if we consider our system of equations as a dynamical problem).

B Particle in the Multipolar Field

The magnetic vector potential of a magnet with $2n$ poles in Cartesian coordinates is

$$A = \sum_n K_n f_n(x, y), \tag{6}$$

where f_n is a homogeneous function of x and y of order n.

The real and imaginary parts of the binomial expansion of

$$f_n(x, y) = (x + iy)^n \tag{7}$$

correspond to regular and skew multipoles. The cases $n = 2$ to $n = 5$ correspond to low-order multipoles: quadrupole, sextupole, octupole, decapole.

Then we have, in this particular case, the following equations of motion for a single particle in a circular magnetic lattice in the transverse plane (x, y) ([8] for designation):

$$\frac{d^2x}{ds^2} + \left(\frac{1}{\rho(s)^2} - k_1(s)\right) x = \mathcal{R}e \left[\sum_{n \geq 2} \frac{k_n(s) + ij_n(s)}{n!} \cdot (x + iy)^n\right] \tag{8}$$

$$\frac{d^2y}{ds^2} + k_1(s)y = -\mathcal{J}m \left[\sum_{n \geq} \frac{k_n(s) + ij_n(s)}{n!} \cdot (x + iy)^n\right]$$

and the corresponding Hamiltonian:

$$H(x, p_x, y, p_y, s) = \frac{p_x^2 + p_y^2}{2} + \left(\frac{1}{\rho(s)^2} - k_1(s)\right) \cdot \frac{x^2}{2} + k_1(s)\frac{y^2}{2} \tag{9}$$

$$- \mathcal{R}e \left[\sum_{n \geq 2} \frac{k_n(s) + ij_n(s)}{(n + 1)!} \cdot (x + iy)^{(n+1)}\right]$$

Then we may take into account arbitrary but finite number in expansion of RHS of Hamiltonian (9) and from our point of view the corresponding Hamiltonian equations of motion are also not more than nonlinear ordinary differential equations with polynomial nonlinearities and variable coefficients.

C   Effects of Insertion Devices on Beam Dynamics

Assuming a sinusoidal field variation, we may consider, according to [9], the analytical treatment of the effects of insertion devices on beam dynamics. One of the major detrimental aspects of the installation of insertion devices is the resulting reduction of dynamic aperture. Introduction of non-linearities leads to enhancement of the amplitude-dependent tune shifts and distortion of phase space. The nonlinear fields will produce significant effects at large betatron amplitudes.

The components of the insertion device magnetic field used for the derivation of equations of motion are as follows:

$$
\begin{aligned}
B_x &= \frac{k_x}{k_y} \cdot B_0 \sinh(k_x x) \sinh(k_y y) \cos(kz) \\
B_y &= B_0 \cosh(k_x x) \cosh(k_y y) \cos(kz) \\
B_z &= -\frac{k}{k_y} B_0 \cosh(k_x x) \sinh(k_y y) \sin(kz),
\end{aligned}
\tag{10}
$$

with $k_x^2 + k_y^2 = k^2 = (2\pi/\lambda)^2$, where λ is the period length of the insertion device, B_0 its magnetic field, and ρ the radius of the curvature in the field B_0. After a canonical transformation to change to betatron variables, the Hamiltonian is averaged over the period of the insertion device, and hyperbolic functions are expanded to the fourth order in x and y (or an arbitrary order). Then we have the following Hamiltonian:

$$
\begin{aligned}
H ={} & \frac{1}{2}[p_x^2 + p_y^2] + \frac{1}{4k^2\rho^2}[k_x^2 x^2 + k_y^2 y^2] \\
& + \frac{1}{12k^2\rho^2}[k_x^4 x^4 + k_y^4 y^4 + 3k_x^2 k^2 x^2 y^2] \\
& - \frac{\sin(ks)}{2k\rho}[p_x(k_x^2 x^2 + k_y^2 y^2) - 2k_x p_y xy] \quad .
\end{aligned}
\tag{11}
$$

We also have in this case nonlinear (polynomial with degree 3) dynamical system with variable (periodic) coefficients. As a related case we may consider wiggler and undulator magnets. We have in the horizontal $x - s$ plane the following equations:

$$
\ddot{x} = -\dot{s}\frac{e}{m\gamma}B_z(s) \quad,
\tag{12}
$$

$$
\ddot{s} = \dot{x}\frac{e}{m\gamma}B_z(s),
$$

where the magnetic field has periodic dependence on s and hyperbolic on z.

D   Spin-Orbital Motion

Let us consider the system of equations for orbital motion

$$\frac{dq}{dt} = \frac{\partial H_{orb}}{\partial p}, \qquad \frac{dp}{dt} = -\frac{\partial H_{orb}}{\partial q} \qquad (13)$$

and the Thomas-BMT equation for classical spin vector (see [10] for designation)

$$\frac{ds}{dt} = w \times s \ , \qquad (14)$$

Here,

$$H_{orb} = c\sqrt{\pi^2 + m_0 c^2} + e\Phi, \qquad (15)$$

$$w = -\frac{e}{m_0 \gamma c}\left((1 + \gamma G)\vec{B} - \frac{G(\vec{\pi} \cdot \vec{B})\vec{\pi}}{m_0^2 c^2(1 + \gamma)} - \frac{1}{m_0 c}\left(G + \frac{1}{1+\gamma}\right)[\pi \times E]\right),$$

where $q = (q_1, q_2, q_3), p = (p_1, p_2, p_3)$ the canonical position and momentum, $s = (s_1, s_2, s_3)$ the classical spin vector of length $\hbar/2$, and $\pi = (\pi_1, \pi_2, \pi_3)$ is the kinetic momentum vector. We may introduce in 9-dimensional phase space $z = (q, p, s)$ the Poisson brackets

$$\{f(z), g(z)\} = f_q g_p - f_p g_q + [f_s \times g_s] \cdot s \ , \qquad (16)$$

and the corresponding Hamiltonian equations:

$$\frac{dz}{dt} = \{z, H\}, \qquad (17)$$

with Hamiltonian

$$H = H_{orb}(q, p, t) + w(q, p, t) \cdot s. \qquad (18)$$

More explicitly we have

$$\frac{dq}{dt} = \frac{\partial H_{orb}}{\partial p} + \frac{\partial(w \cdot s)}{\partial p}$$

$$\frac{dp}{dt} = -\frac{\partial H_{orb}}{\partial q} - \frac{\partial(w \cdot s)}{\partial q} \qquad (19)$$

$$\frac{ds}{dt} = [w \times s]$$

We will consider this dynamical system also in our second paper in this volume via an invariant approach, based on consideration of Lie-Poisson structures on semidirect products of groups.

But from the point of view used in this paper we may consider approximations similar to preceding examples and then also arrive to at some type of polynomial dynamics.

III POLYNOMIAL DYNAMICS

The first main part of our consideration is some variational approach to this problem, which reduces the initial problem to the problem of solving functional equations at the first stage and some algebraical problems at the second stage. We have the solution in a compactly supported wavelet basis. Multiresolution expansion is the second main part of our construction. The solution is parameterized by solutions of two reduced algebraical problems, one being nonlinear and the second being some linear problem, which is obtained from one of the next wavelet constructions: Fast Wavelet Transform (FWT), Stationary Subdivision Schemes (SSS), the method of Connection Coefficients (CC).

A Variational Method

Our problems may be formulated as the systems of ordinary differential equations

$$\mathrm{dx_i/dt} = f_i(x_j, t), \quad (i, j = 1, ..., n) \quad , \tag{20}$$

with fixed initial conditions $x_i(0)$, where f_i are not more than polynomial functions of dynamical variables x_j and have arbitrary dependence of time. Because of time dilation we can consider only the next time interval: $0 \le t \le 1$. Let us consider a set of functions,

$$\Phi_i(t) = x_i \mathrm{dy_i/dt} + f_i y_i \tag{21}$$

and a set of functionals

$$F_i(x) = \int_0^1 \Phi_i(t) dt - x_i y_i \mid_0^1, \tag{22}$$

where $y_i(t)(y_i(0) = 0)$ are dual variables. It is obvious that the initial system and the system

$$F_i(x) = 0 \tag{23}$$

are equivalent. In the last part we consider a more general approach, which is based on the possibility of taking into account underlying symplectic structure and using a more useful and flexible analytical approach, related to bilinear structure of initial function.

Now we consider formal expansions for x_i, y_i:

$$x_i(t) = x_i(0) + \sum_k \lambda_i^k \varphi_k(t) \quad y_j(t) = \sum_r \eta_j^r \varphi_r(t), \tag{24}$$

where, because of initial conditions, we need only $\varphi_k(0) = 0$. Then we have the following reduced algebraical system of equations on the set of unknown coefficients λ_i^k of expansions (24):

$$\sum_k \mu_{kr} \lambda_i^k - \gamma_i^r(\lambda_j) = 0 \qquad (25)$$

Its coefficients are

$$\mu_{kr} = \int_0^1 \varphi_k'(t)\varphi_r(t)dt, \quad \gamma_i^r = \int_0^1 f_i(x_j, t)\varphi_r(t)dt. \qquad (26)$$

Now, when we solve system (25) and determine unknown coefficients from formal expansion (24) we therefore obtain the solution of our initial problem. It should be noted that, if we consider only the truncated expansion (24) with N terms, then we have from (25) the system of $N \times n$ algebraical equations; and the degree of this algebraical system coincides with the degree of initial differential system. So, we have the solution of the initial nonlinear (polynomial) problem in the form

$$x_i(t) = x_i(0) + \sum_{k=1}^N \lambda_i^k X_k(t), \qquad (27)$$

where coefficients λ_i^k are roots of the corresponding reduced algebraical problem (25). Consequently, we have a parametrization of the solution of the initial problem by solution of the reduced algebraical problem (25). The first main problem is a problem of computations of coefficients of the reduced algebraical system. As we will see, these problems may be explicitly solved in the wavelet approach.

Next we consider the construction the of explicit time solution for our problem. The obtained solutions are given in the form (27), where $X_k(t)$ are the basis functions and λ_k^i are roots of the reduced system of equations. In our first wavelet case, $X_k(t)$ are obtained via multiresolution expansions and represented by compactly supported wavelets, and λ_k^i are the roots of the corresponding general polynomial system (25) with coefficients, which are given by FWT, SSS or CC constructions. According to the variational method giving the reduction from the differential to the algebraical system of equations, we need to compute the objects γ_a^j and μ_{ji}, which are constructed from objects:

$$\sigma_i \equiv \int_0^1 X_i(\tau)d\tau,$$

$$\nu_{ij} \equiv \int_0^1 X_i(\tau)X_j(\tau)d\tau,$$

$$\mu_{ji} \equiv \int X_i'(\tau)X_j(\tau)d\tau, \qquad (28)$$

$$\beta_{klj} \equiv \int_0^1 X_k(\tau)X_l(\tau)X_j(\tau)d\tau$$

55

for the simplest case of Riccati systems, where the degree of nonlinearity equals two. For the general case of arbitrary n we have analogous to (28) iterated integrals with the degree of monomials in integrand, which is one bigger than the degree of the initial system.

B   Wavelet Framework

Our constructions are based on a multi-resolution approach. Because affine group of translations and dilations are part of the approach, this method resembles the action of a microscope. We have a contribution to the final result from each scale of resolution from the whole infinite scale of spaces. More exactly, the closed subspace $V_j (j \in \mathbf{Z})$ corresponds to level j of resolution, or to scale j. We consider a r-regular multiresolution analysis of $L^2(\mathbf{R}^n)$ (of course, we may consider any different functional space), which is a sequence of increasing closed subspaces V_j:

$$...V_{-2} \subset V_{-1} \subset V_0 \subset V_1 \subset V_2 \subset ... \ , \tag{29}$$

satisfying the following properties:

$$\bigcap_{j \in \mathbf{Z}} V_j = 0, \quad \overline{\bigcup_{j \in \mathbf{Z}} V_j} = L^2(\mathbf{R}^n),$$
$$f(x) \in V_j <=> f(2x) \in V_{j+1},$$
$$f(x) \in V_0 <=> f(x-k) \in V_0, \quad , \forall k \in \mathbf{Z}^n. \tag{30}$$

There exists a function $\varphi \in V_0$ such that $\{\varphi_{0,k}(x) = \varphi(x-k), k \in \mathbf{Z}^n\}$ forms a Riesz basis for V_0.

The function φ is regular and localized: φ is C^{r-1}; $\varphi^{(r-1)}$ is almost everywhere differentiable and for almost every $x \in \mathbf{R}^n$, for every integer $\alpha \leq r$; and for all integers p there exists constant C_p such that

$$| \partial^\alpha \varphi(x) | \leq C_p (1 + |x|)^{-p} \quad . \tag{31}$$

Let $\varphi(x)$ be a scaling function, $\psi(x)$ a wavelet function and $\varphi_i(x) = \varphi(x-i)$. Scaling relations that define φ, ψ are

$$\varphi(x) = \sum_{k=0}^{N-1} a_k \varphi(2x - k) = \sum_{k=0}^{N-1} a_k \varphi_k(2x), \tag{32}$$

$$\psi(x) = \sum_{k=-1}^{N-2} (-1)^k a_{k+1} \varphi(2x + k). \tag{33}$$

Let indices ℓ, j represent translation and scaling, respectively and

$$\varphi_{jl}(x) = 2^{j/2} \varphi(2^j x - \ell) \quad ; \tag{34}$$

56

then the set $\{\varphi_{j,k}\}, k \in \mathbf{Z}^n$ forms a Riesz basis for V_j. The wavelet function ψ is used to encode the details between two successive levels of approximation. Let W_j be the orthonormal complement of V_j with respect to V_{j+1}:

$$V_{j+1} = V_j \bigoplus W_j. \tag{35}$$

Then just as V_j is spanned by dilation and translations of the scaling function, so are W_j spanned by translations and dilation of the mother wavelet $\psi_{jk}(x)$, where

$$\psi_{jk}(x) = 2^{j/2}\psi(2^j x - k). \tag{36}$$

All expansions which we used are based on the following properties:

$$\{\psi_{jk}\}, \quad j, k \in \mathbf{Z} \quad \text{is a Hilbertian basis of } L^2(\mathbf{R})$$
$$\{\varphi_{jk}\}_{j \geq 0, k \in \mathbf{Z}} \quad \text{is an orthonormal basis for} L^2(\mathbf{R}),$$
$$L^2(\mathbf{R}) = V_0 \overline{\bigoplus_{j=0}^{\infty} W_j}, \tag{37}$$

or

$$\{\varphi_{0,k}, \psi_{j,k}\}_{j \geq 0, k \in \mathbf{Z}} \quad \text{is an orthonormal basis for} L^2(\mathbf{R}).$$

C   Wavelet Computations

Now we give construction for computations of objects (28) in the wavelet case. We use a compactly supported wavelet basis: an orthonormal basis for functions in $L^2(\mathbf{R})$.

Let $f : \mathbf{R} \longrightarrow \mathbf{C}$ and the wavelet expansion be

$$f(x) = \sum_{\ell \in \mathbf{Z}} c_\ell \varphi_\ell(x) + \sum_{j=0}^{\infty} \sum_{k \in \mathbf{Z}} c_{jk} \psi_{jk}(x) \quad . \tag{38}$$

If in formulae (38) $c_{jk} = 0$ for $j \geq J$, then $f(x)$ has an alternative expansion in terms of dilated scaling functions only $f(x) = \sum_{\ell \in \mathbf{Z}} c_{J\ell} \varphi_{J\ell}(x)$. This is a finite wavelet expansion, and it can be written solely in terms of translated scaling functions. Also we have the shortest possible support: scaling function DN (where N is even integer) will have support $[0, N-1]$ and $N/2$ vanishing moments. There exists $\lambda > 0$ such that DN has λN continuous derivatives; for small $N, \lambda \geq 0.55$. To solve our second associated linear problem we need to evaluate derivatives of $f(x)$ in terms of $\varphi(x)$. Let $\varphi_\ell^n = \mathrm{d}^n \varphi_\ell(\mathrm{x})/\mathrm{dx}^n$. We consider computation of the wavelet - Galerkin integrals. If $f^d(x)$ is a d-derivative of function $f(x)$, then we have $f^d(x) = \sum_\ell c_l \varphi_\ell^d(x)$, and values $\varphi_\ell^d(x)$ can be expanded in terms of $\varphi(x)$,

57

$$\phi_\ell^d(x) = \sum_m \lambda_m \varphi_m(x), \tag{39}$$

$$\lambda_m = \int\limits_{-\infty}^{\infty} \varphi_\ell^d(x)\varphi_m(x)\mathrm{d}x,$$

where λ_m are wavelet-Galerkin integrals. The coefficients λ_m are 2-term connection coefficients. In general we need to find $(d_i \geq 0)$,

$$\Lambda_{\ell_1\ell_2...\ell_n}^{d_1 d_2...d_n} = \int\limits_{-\infty}^{\infty} \prod \varphi_{\ell_i}^{d_i}(x)dx \quad . \tag{40}$$

For Riccati case we need to evaluate two and three connection coefficients

$$\Lambda_\ell^{d_1 d_2} = \int_{-\infty}^{\infty} \varphi^{d_1}(x)\varphi_\ell^{d_2}(x)dx, \quad \Lambda^{d_1 d_2 d_3} = \int\limits_{-\infty}^{\infty} \varphi^{d_1}(x)\varphi_\ell^{d_2}(x)\varphi_m^{d_3}(x)dx \quad . \tag{41}$$

According to the CC method [11] we use the next construction. When N in the scaling equation is a finite even positive integer, the function $\varphi(x)$ has compact support contained in $[0, N-1]$. For a fixed triple (d_1, d_2, d_3) only some $\Lambda_{\ell m}^{d_1 d_2 d_3}$ are nonzero: $2 - N \leq \ell \leq N - 2$, $2 - N \leq m \leq N - 2$, $|\ell - m| \leq N - 2$. There are $M = 3N^2 - 9N + 7$ such pairs (ℓ, m). If $\Lambda^{d_1 d_2 d_3}$ is an M-vector, whose components are numbers $\Lambda_{\ell m}^{d_1 d_2 d_3}$, then we have the first reduced algebraical system : Λ satisfy the system of equations $(d = d_1 + d_2 + d_3)$,

$$A\Lambda^{d_1 d_2 d_3} = 2^{1-d}\Lambda^{d_1 d_2 d_3}, \quad A_{\ell,m;q,r} = \sum_p a_p a_{q-2\ell+p} a_{r-2m+p} \quad . \tag{42}$$

By moment equations we have created a system of $M + d + 1$ equations in M unknowns. It has rank M and we can obtain unique solution by combination of LU decomposition and QR algorithm. The second reduced algebraical system gives us the 2-term connection coefficients:

$$A\Lambda^{d_1 d_2} = 2^{1-d}\Lambda^{d_1 d_2}, \quad d = d_1 + d_2, \quad A_{\ell,q} = \sum_p a_p a_{q-2\ell+p} \tag{43}$$

For a nonquadratic case we have additional analogously linear problems for objects (40). Solving these linear problems, we obtain the coefficients of a nonlinear algebraical system (25), and after that we obtain the coefficients of wavelet expansion (27). As a result we obtained the explicit time solution of our problem in the base of compactly supported wavelets. We use for modelling D6, D8, and D10 functions and programs RADAU and DOPRI for testing.

In the following we consider the extension of this approach to the case of periodic boundary conditions, the case of presence of arbitrary variable coefficients and a more flexible biorthogonal wavelet approach.

IV EVALUATION OF NONLINEARITIES SCALE BY SCALE

A   Para-product and Decoupling between Scales

Before we consider two different schemes of modification of our variational approach we consider different scales separately. For this reason we need to compute errors of approximations. The main problems come of course from nonlinear terms. We follow the approach from [12].

Let P_j be the projection operators on the subspaces $V_j, j \in \mathbf{Z}$:

$$P_j \; : \; L^2(\mathbf{R}) \to V_j \tag{44}$$
$$(P_j f)(x) = \sum_k < f, \varphi_{j,k} > \varphi_{j,k}(x) \quad ,$$

and Q_j are projection operators on the subspaces W_j:

$$Q_j = P_{j-1} - P_j \quad . \tag{45}$$

So, for $u \in L^2(\mathbf{R})$ we have $u_j = P_j u$ and $u_j \in V_j$, where $\{V_j\}, j \in \mathbf{Z}$ is a multiresolution analysis of $L^2(\mathbf{R})$. It is obvious that we can represent u_0^2 in the following form:

$$u_0^2 = 2 \sum_{j=1}^n (P_j u)(Q_j u) + \sum_{j=1}^n (Q_j u)(Q_j u) + u_n^2 \quad . \tag{46}$$

In this formula there is no interaction between different scales. We may consider each term of (46) as bilinear mappings:

$$M_{VW}^j : V_j \times W_j \to L^2(\mathbf{R}) = V_j \oplus_{j' \geq j} W_{j'} \tag{47}$$

$$M_{WW}^j : W_j \times W_j \to L^2(\mathbf{R}) = V_j \oplus_{j' \geq j} W_{j'} \quad . \tag{48}$$

For numerical purposes we need formula (46) with a finite number of scales, but when we consider limits $j \to \infty$ we have

$$u^2 = \sum_{j \in \mathbf{Z}} (2P_j u + Q_j u)(Q_j u), \tag{49}$$

which is para-product of Bony, Coifman and Meyer.

Now we need to expand (46) into the wavelet bases. To expand each term in (46) into wavelet basis, we need to consider the integrals of the products of the basis functions, e.g.,

$$M_{WWW}^{j,j'}(k, k', \ell) = \int_{-\infty}^{\infty} \psi_k^j(x) \psi_{k'}^j(x) \psi_\ell^{j'}(x) \mathrm{d}x, \tag{50}$$

where $j' > j$ and

$$\psi_k^j(x) = 2^{-j/2}\psi(2^{-j}x - k) \tag{51}$$

are the basis functions. If we consider compactly supported wavelets then

$$M_{WWW}^{j,j'}(k, k', \ell) \equiv 0 \quad \text{for} \quad |k - k'| > k_0, \tag{52}$$

where k_0 depends on the overlap of the supports of the basis functions and

$$|M_{WWW}^r(k - k', 2^r k - \ell)| \le C \cdot 2^{-r\lambda M} \quad . \tag{53}$$

Let us define j_0 as the distance between scales so that for a given ε all the coefficients in (53) with labels $r = j - j'$, $r > j_0$ have absolute values less than ε. For the purposes of computing with accuracy ε, we replace the mappings in (47), (48) by

$$M_{VW}^j : V_j \times W_j \to V_j \oplus_{j \le j' \le j_0} W_{j'} \tag{54}$$

$$M_{WW}^j : W_j \times W_j \to V_j \oplus_{J \le j' \le j_0} W_{j'} \quad . \tag{55}$$

Since

$$V_j \oplus_{j \le j' \le j_0} W_{j'} = V_{j_0 - 1} \tag{56}$$

and

$$V_j \subset V_{j_0 - 1}, \qquad W_j \subset V_{j_0 - 1} \quad , \tag{57}$$

we may consider bilinear mappings (54), (55) on $V_{j_0 - 1} \times V_{j_0 - 1}$. For the evaluation of (54), (55) as mappings $V_{j_0 - 1} \times V_{j_0 - 1} \to V_{j_0 - 1}$, we need significantly fewer coefficients than for mappings (54), (55). It is enough to consider only coefficients

$$M(k, k', \ell) = 2^{-j/2}\int_\infty^\infty \varphi(x - k)\varphi(x - k')\varphi(x - \ell)\mathrm{d}x, \tag{58}$$

where $\varphi(x)$ is the scale function. Also we have

$$M(k, k', \ell) = 2^{-j/2}M_0(k - \ell, k' - \ell), \tag{59}$$

where

$$M_0(p, q) = \int \varphi(x - p)\varphi(x - q)\varphi(x)\mathrm{d}x \quad . \tag{60}$$

Now, as in section (3C), we may derive and solve a system of linear equations to find $M_0(p, q)$.

B Non-regular Approximation

We use the wavelet function $\psi(x)$, which has k vanishing moments $\int x^k \psi(x) dx = 0$, or equivalently $x^k = \sum c_\ell \varphi_\ell(x)$ for each k, $0 \le k \le K$.

Let P_j again be the orthogonal projector on space V_j. By tree algorithm we have for any $u \in L^2(\mathbf{R})$ and $\ell \in \mathbf{Z}$, that the wavelet coefficients of $P_\ell(u)$, i.e. the set $\{< u, \psi_{j,k} >, j \le \ell - 1, k \in \mathbf{Z}\}$, can be computed using hierarchical algorithms from the set of scaling coefficients in V_ℓ, i.e. the set $\{< u, \varphi_{\ell,k} >, k \in \mathbf{Z}\}$ [13]. Because for scaling function φ we have in general only $\int \varphi(x) dx = 1$, therefore we have for any function $u \in L^2(\mathbf{R})$:

$$\lim_{j \to \infty, k2^{-j} \to x} | 2^{j/2} < u, \varphi_{j,k} > -u(x) | = 0 \quad . \tag{61}$$

If the integer $n(\varphi)$ is the largest one so that

$$\int x^\alpha \varphi(x) dx = 0 \qquad \text{for} \qquad 1 \le \alpha \le \text{n} \quad , \tag{62}$$

then if $u \in C^{(n+1)}$ with $u^{(n+1)}$ is bounded we have for $j \to \infty$ uniformly in k:

$$| 2^{j/2} < u, \varphi_{j,k} > -u(k2^{-j}) | = O(2^{-j(n+1)}). \tag{63}$$

Such scaling functions with zero moments are very useful for us from the point of view of time-frequency localization, because we have for the Fourier component $\hat{\Phi}(\omega)$ of them, that exists some $C(\varphi) \in \mathbf{R}$, so that for $\omega \to 0$ $\hat{\Phi}(\omega) = 1 + C(\varphi) | \omega |^{2r+2}$ (remember that we consider r-regular multiresolution analysis). Using this type of scaling functions lead to superconvergence properties for general Galerkin approximation [13]. Now we need some estimates in each scale for non-linear terms of type $u \mapsto f(u) = f \circ u$, where f is C^∞ (in previous and future parts we consider only truncated Taylor series action). Let us consider the non-regular space of approximation $\tilde{V}$ of the form

$$\tilde{V} = V_q \oplus \sum_{q \le j \le p-1} \widetilde{W_j}, \tag{64}$$

with $\widetilde{W_j} \subset W_j$. We need an efficient and precise estimate of $f \circ u$ on $\tilde{V}$. Let us set for $q \in \mathbf{Z}$ and $u \in L^2(\mathbf{R})$

$$\prod f_q(u) = 2^{-q/2} \sum_{k \in \mathbf{Z}} f(2^{q/2} < u, \varphi_{q,k} >) \cdot \varphi_{q,k} \quad . \tag{65}$$

We have the following (important for us) estimation (uniformly in q) for $u, f(u) \in H^{(n+1)}$ [13]:

$$\| P_q \left(f(u) \right) - \prod f_q(u) \|_{L^2} = O \left(2^{-(n+1)q} \right) \quad . \tag{66}$$

For non-regular spaces (64) we set

$$\prod f_{\widetilde{V}}(u) = \prod f_q(u) + \sum_{\ell=q,p-1} P_{\widetilde{W_j}} \cdot \prod f_{\ell+1}(u) \qquad (67)$$

Then we have the following estimate:

$$\|P_{\widetilde{V}}(f(u)) - \prod f_{\widetilde{V}}(u)\|_{L^2} = O(2^{-(n+1)q}) \quad , \qquad (68)$$

uniformly in q and $\widetilde{V}$ (64).

This estimate depends on q, not p, i.e. on the scale of the coarse grid, not on the finest grid used in definition of $\widetilde{V}$. We have for total error

$$\|f(u) - \prod f_{\widetilde{V}}(u)\| = \|f(u) - P_{\widetilde{V}}(f(u))\|_{L^2} + \|P_{\widetilde{V}}(f(u) - \prod f_{\widetilde{V}}(u))\|_{L^2} \quad , $$
$$(69)$$

and since the projection error in $\widetilde{V}$: $\|f(u) - P_{\widetilde{V}}(f(u))\|_{L^2}$ is much smaller than the projection error in V_q, we have the improvement (68) of (66). In our concrete calculations and estimates it is very useful to consider our approximations in the particular case of c-structured space:

$$\widetilde{V} = V_q + \sum_{j=q}^{p-1} span\{\psi_{j,k}, k \in [2^{(j-1)} - c, 2^{(j-1)} + c] \mod 2^j\} \quad . \qquad (70)$$

V   VARIATIONAL WAVELET APPROACH FOR PERIODIC TRAJECTORIES

We start with an extension of our approach to the case of periodic trajectories. The equations of motion corresponding to Hamiltonians (from part II) may also be formulated as a particular case of the general system of ordinary differential equations $dx_i/dt = f_i(x_j, t)$, $(i, j = 1, ..., n)$, $0 \leq t \leq 1$, where f_i are not more than polynomial functions of dynamical variables x_j and have arbitrary dependence of time but with periodic boundary conditions. According to our variational approach we have the solution in the following form:

$$x_i(t) = x_i(0) + \sum_k \lambda_i^k \varphi_k(t), \qquad x_i(0) = x_i(1), \qquad (71)$$

where λ_i^k are again the roots of reduced algebraical systems of equations with the same degree of nonlinearity, and $\varphi_k(t)$ corresponds to useful types of wavelet bases (frames). It should be noted that coefficients of reduced algebraical system are the solutions of additional linear problem and also depend on a particular type of wavelet construction and type of bases.

This linear problem is our second reduced algebraical problem. We need to find in general situation objects

$$\Lambda_{\ell_1 \ell_2 \ldots \ell_n}^{d_1 d_2 \ldots d_n} = \int_{-\infty}^{\infty} \prod \varphi_{\ell_i}^{d_i}(x) \mathrm{dx}, \tag{72}$$

but now in the case of periodic boundary conditions. Now we consider the procedure of their calculations in the case of periodic boundary conditions in the base of periodic wavelet functions on the interval [0,1] and corresponding expansion (71) inside our variational approach. Periodization procedure gives us

$$\hat{\varphi}_{j,k}(x) \equiv \sum_{\ell \in Z} \varphi_{j,k}(x - \ell) \tag{73}$$

$$\hat{\psi}_{j,k}(x) = \sum_{\ell \in Z} \psi_{j,k}(x - \ell) \quad .$$

So, $\hat{\varphi}, \hat{\psi}$ are periodic functions on the interval [0,1]. Because $\varphi_{j,k} = \varphi_{j,k'}$ if $k = k' \mathrm{mod}(2^j)$, we may consider only $0 \le k \le 2^j$, and, as a consequence, our multiresolution has the form $\bigcup_{j \ge 0} \hat{V}_j = L^2[0,1]$, with $\hat{V}_j = \mathrm{span}\{\hat{\varphi}_{j,k}\}_{k=0}^{2j-1}$ [14].
Integration by parts and periodicity gives useful relations between objects (72), in particular the quadratic case ($d = d_1 + d_2$):

$$\Lambda_{k_1,k_2}^{d_1,d_2} = (-1)^{d_1} \Lambda_{k_1,k_2}^{0,d_2+d_1}, \quad \Lambda_{k_1,k_2}^{0,d} = \Lambda_{0,k_2-k_1}^{0,d} \equiv \Lambda_{k_2-k_1}^{d} \quad . \tag{74}$$

So, any 2-tuple can be represented by Λ_k^d. Then our second additional linear problem is reduced to the eigenvalue problem for $\{\Lambda_k^d\}_{0 \le k \le 2^j}$ by creating a system of 2^j homogeneous relations in Λ_k^d and inhomogeneous equations. So, if we have a dilation equation in the form $\varphi(x) = \sqrt{2} \sum_{k \in Z} h_k \varphi(2x - k)$, then we have the following homogeneous relations:

$$\Lambda_k^d = 2^d \sum_{m=0}^{N-1} \sum_{\ell=0}^{N-1} h_m h_\ell \Lambda_{\ell+2k-m}^d, \tag{75}$$

or in such form $A\lambda^d = 2^d \lambda^d$, where $\lambda^d = \{\Lambda_k^d\}_{0 \le k \le 2^j}$. Inhomogeneous equations are:

$$\sum_\ell M_\ell^d \Lambda_\ell^d = d! 2^{-j/2}, \tag{76}$$

where objects $M_\ell^d(|\ell| \le N - 2)$ can be computed by a recursive procedure

$$M_\ell^d = 2^{-j(2d+1)/2} \tilde{M}_\ell^d, \quad \tilde{M}_\ell^k = <x^k, \varphi_{0,\ell}> = \sum_{j=0}^{k} \binom{k}{j} n^{k-j} M_0^j, \quad \tilde{M}_0^\ell = 1. \tag{77}$$

So, we reduced our last problem to a standard linear algebraical problem. Then we used the same methods as in part III C. As a result we obtained for closed trajectories of orbital dynamics described by Hamiltonians from part II the explicit time solution (71) in the base of periodized wavelets (73).

VI VARIATIONAL APPROACH IN BIORTHOGONAL WAVELET BASES

Now we consider further generalization of our variational wavelet approach. In [1]-[3] we consider different types of variational principles which give us weak solutions to our nonlinear problems.

Before this we consider the generalization of our wavelet variational approach to the symplectic invariant calculation of closed loops in Hamiltonian systems [3]. We also have the parametrization of our solution by some reduced algebraical problem; but in contrast to the general case where the solution is parametrized by construction based on scalar refinement equation, in the symplectic case we have parametrization of the solution by matrix problems – Quadratic Mirror Filters equations [3]. But because integrand of variational functionals is represented by a bilinear form (scalar product), it seems more reasonable to consider wavelet constructions [15] which take into account all advantages of this structure.

The action functional for loops in the phase space is [16],

$$F(\gamma) = \int_\gamma pdq - \int_0^1 H(t, \gamma(t))dt \tag{78}$$

The critical points of F are those loops γ, which solve the Hamiltonian equations associated with the Hamiltonian H and hence are periodic orbits. By the way, all critical points of F are the saddle points of the infinite Morse index, but surprisingly this approach is very effective. This will be demonstrated using several variational techniques starting from minimax due to Rabinowitz and ending with Floer homology. So, (M, ω) is equal to symplectic manifolds, $H : M \to R$, H is Hamiltonian, X_H is the unique Hamiltonian vector field defined by

$$\omega(X_H(x), v) = -dH(x)(v), \quad v \in T_x M, \quad x \in M, \tag{79}$$

where ω is the symplectic structure. A T-periodic solution $x(t)$ of the Hamiltonian equations,

$$\dot{x} = X_H(x) \quad \text{on } M \quad , \tag{80}$$

is a solution, satisfying the boundary conditions $x(T) = x(0), T > 0$. Let us consider the loop space $\Omega = C^\infty(S^1, R^{2n})$, where $S^1 = R/\mathbf{Z}$, of smooth loops in R^{2n}. Let us define a function $\Phi : \Omega \to R$ by setting

$$\Phi(x) = \int_0^1 \frac{1}{2} < -J\dot{x}, x > dt - \int_0^1 H(x(t))dt, \quad x \in \Omega \quad . \tag{81}$$

The critical points of Φ are the periodic solutions of $\dot{x} = X_H(x)$. Computing the derivative at $x \in \Omega$ in the direction of $y \in \Omega$, we find

$$\Phi'(x)(y) = \frac{d}{d\epsilon}\Phi(x+\epsilon y)|_{\epsilon=0} = \int_0^1 < -J\dot{x} - \bigtriangledown H(x), y > dt \qquad (82)$$

Consequently, $\Phi'(x)(y) = 0$ for all $y \in \Omega$ if the loop x satisfies the equation

$$-J\dot{x}(t) - \bigtriangledown H(x(t)) = 0 \quad ; \qquad (83)$$

i.e., $x(t)$ is a solution of the Hamiltonian equations, which also satisfies $x(0) = x(1)$, i.e., the periodic of period 1. Periodic loops may be represented by their Fourier series:

$$x(t) = \sum_{k\in\mathbf{Z}} e^{k2\pi Jt}x_k, \quad x_k \in R^{2k}, \qquad (84)$$

where J is the quasicomplex structure. We give relations between the quasicomplex structure and wavelets in our second paper in this volume (see also [3]). But now we need to take into account underlying bilinear structure via wavelets.

We started with two hierarchical sequences of approximations spaces [15]:

$$\dots V_{-2} \subset V_{-1} \subset V_0 \subset V_1 \subset V_2 \dots, \qquad \dots \tilde{V}_{-2} \subset \tilde{V}_{-1} \subset \tilde{V}_0 \subset \tilde{V}_1 \subset \tilde{V}_2 \dots, \quad (85)$$

and as usual, W_0 is a complement to V_0 in V_1, but now not necessarily an orthogonal complement. New orthogonality conditions now have the following form:

$$\widetilde{W}_0 \perp V_0, \qquad W_0 \perp \tilde{V}_0, \qquad V_j \perp \widetilde{W}_j, \qquad \tilde{V}_j \perp W_j \quad , \qquad (86)$$

translates of ψ span W_0, translates of $\tilde{\psi}$ span $\widetilde{W}_0$. Biorthogonality conditions are

$$< \psi_{jk}, \tilde{\psi}_{j'k'} >= \int_{-\infty}^{\infty} \psi_{jk}(x)\tilde{\psi}_{j'k'}(x)\mathrm{d}x = \delta_{\mathrm{kk'}}\delta_{\mathrm{jj'}}, \qquad (87)$$

where $\psi_{jk}(x) = 2^{j/2}\psi(2^j x - k)$. Functions $\varphi(x), \tilde{\varphi}(x-k)$ form a dual pair:

$$< \varphi(x-k), \tilde{\varphi}(x-\ell) >= \delta_{kl}, \quad < \varphi(x-k), \tilde{\psi}(x-\ell) >= 0 \quad \text{for} \quad \forall k, \forall \ell. \qquad (88)$$

Functions $\varphi, \tilde{\varphi}$ generate a multiresolution analysis. $\varphi(x-k)$, $\psi(x-k)$ are synthesis functions, and $\tilde{\varphi}(x-\ell)$, $\tilde{\psi}(x-\ell)$ are analysis functions. Synthesis functions are biorthogonal to analysis functions. Scaling spaces are orthogonal to dual wavelet spaces. Two multiresolutions are intertwining $V_j + W_j = V_{j+1}$, $\tilde{V}_j + \widetilde{W}_j = \tilde{V}_{j+1}$. These are direct sums but not orthogonal sums.

So, our representation for a solution now has the form

$$f(t) = \sum_{j,k} \tilde{b}_{jk}\psi_{jk}(t), \qquad (89)$$

65

where synthesis wavelets are used to synthesize the function. But $\tilde{b}_{jk}$ comes from inner products with analysis wavelets. Biorthogonality yields

$$\tilde{b}_{\ell m} = \int f(t)\tilde{\psi}_{\ell m}(t)\mathrm{dt}. \tag{90}$$

So, now we can introduce this more complicated construction into our variational approach. We have a modification only on the level of computing coefficients of a reduced nonlinear algebraical system. This new construction is more flexible. The biorthogonal point of view is more stable under the action of a large class of operators, while the orthogonal (one scale for multiresolution) is fragile all computations are much simpler and we accelerate the rate of convergence. In all types of Hamiltonian calculation, which are based on some bilinear structures (symplectic or Poissonian structures, bilinear form of integrand in variational integral), this framework leads to greater success.

VII VARIABLE COEFFICIENTS

In the case when we have a situation where our problem is described by a system of nonlinear (polynomial)differential equations, we need to consider extension of our previous approach, which can take into account any type of variable coefficients (periodic, regular or singular). We can produce such an approach if we add in our construction an additional refinement equation, which would encode all information about variable coefficients [17]. According to our variational approach we need to compute integrals of the form

$$\int_D b_{ij}(t)(\varphi_1)^{d_1}(2^m t - k_1)(\varphi_2)^{d_2}(2^m t - k_2)\mathrm{dx}, \tag{91}$$

where now $b_{ij}(t)$ are arbitrary functions of time, where trial functions φ_1, φ_2 satisfy a refinement equation:

$$\varphi_i(t) = \sum_{k \in \mathbf{Z}} a_{ik}\varphi_i(2t - k) \tag{92}$$

If we consider all computations in the class of compactly supported wavelets, then only a finite number of coefficients do not vanish. To approximate the non-constant coefficients, we need to choose a different refinable function φ_3, along with some local approximation scheme,

$$(B_\ell f)(x) := \sum_{\alpha \in \mathbf{Z}} F_{\ell,k}(f)\varphi_3(2^\ell t - k), \tag{93}$$

where $F_{\ell,k}$ are suitable functionals supported in a small neighborhood of $2^{-\ell}k$, and then replace b_{ij} in (91) by $B_\ell b_{ij}(t)$. In this particular case, one can take a characteristic function and can thus approximate non-smooth coefficients locally. To guarantee sufficient accuracy of the resulting approximation to (91) it is important to have the flexibility of choosing φ_3 different from φ_1, φ_2. In the case when D is some domain, we can write

$$b_{ij}(t) \mid_D = \sum_{0 \le k \le 2^\ell} b_{ij}(t) \chi_D(2^\ell t - k), \tag{94}$$

where χ_D is a characteristic function of D. So, if we take $\varphi_4 = \chi_D$, which is again a refinable function, then the problem of the computation of (91) is reduced to the problem of calculation of the integral

$$H(k_1, k_2, k_3, k_4) = H(k) = \tag{95}$$
$$\int_{\mathbf{R}^s} \varphi_4(2^j t - k_1) \varphi_3(2^\ell t - k_2) \varphi_1^{d_1}(2^r t - k_3) \varphi_2^{d_2}(2^s t - k_4) \mathrm{dx} \quad .$$

The key point is that these integrals also satisfy some sort of refinement equation:

$$2^{-|\mu|} H(k) = \sum_{\ell \in \mathbf{Z}} b_{2k-\ell} H(\ell), \qquad \mu = d_1 + d_2. \tag{96}$$

This equation can be interpreted as the problem of computing an eigenvector. Thus, we reduced the problem of the extension of our method to the case of variable coefficients to the same standard algebraical problem as in the preceding sections. So, the general scheme is the same one, and we have only one more additional linear algebraic problem by which we, in the same way, can parameterize the solutions of the corresponding problem.

An extended version and related results may be found in [1]-[6].

ACKNOWLEDGMENTS

We would like to thank Professors M. Cornacchia, C. Pellegrini, L. Palumbo, Mrs. M. Laraneta, J. Kono, and G. Nanula for the nice hospitality, help and support, and all the participants of the Arcidosso meeting for interesting discussions.

REFERENCES

1. Fedorova, A.N., Zeitlin, M.G. 'Wavelets in Optimization and Approximations', *Math. and Comp. in Simulation*, **46**, 527-534 (1998).
2. Fedorova, A.N., Zeitlin, M.G., 'Wavelet Approach to Mechanical Problems', Proc. Cornell Meeting, Chaos'97, Kluwer, 1998.
3. Fedorova, A.N., Zeitlin, M.G., 'Wavelet Approach to Mechanical Problems. Symplectic Group, Symplectic Topology and Symplectic Scales', Proc. Cornell Meeting, Chaos'97, Kluwer, 1998.

4. Fedorova, A.N., Zeitlin, M.G., 'Nonlinear Dynamics of Accelerator via Wavelet Approach', AIP Conf. Proc., vol. 405, ed. Z. Parsa, 87-102, 1997, Los Alamos preprint, physics/9710035.

5. Fedorova, A.N., Zeitlin, M.G, Parsa, Z., 'Wavelet Approach to Accelerator Problems', parts 1-3, Proc. PAC97 **2**, 1502-1504, 1505-1507, 1508-1510, IEEE, 1998.

6. Fedorova, A.N., Zeitlin, M.G, Parsa, Z., 'Nonlinear Effects in Accelerator Physics: from Scale to Scale via Wavelets', 'Wavelet Approach to Hamiltonian, Chaotic and Quantum Calculations in Accelerator Physics', Proc. EPAC'98, 930-932, 933-935, Institute of Physics, 1998.

7. Dragt, A.J., Lectures on Nonlinear Dynamics, CTP, 1996; Heinemnn,K., Ripken, G., Schmidt, F., DESY 95-189.

8. Bazzarini, A., e.a., CERN 94-02.

9. Ropert, A., CERN 98-04.

10. Balandin, V., NSF-ITP-96-155i.

11. Latto, A., Resnikoff, H.L., and Tenenbaum, E., Aware Technical Report AD910708, 1991.

12. Beylkin, G., Colorado preprint, 1992.

13. Liandrat, J., Tchamitchian, Ph., *Advances in Comput. Math.* (1996).

14. Schlossnagle, G., Restrepo, J.M., Leaf, G.K., Technical Report ANL-93/34.

15. Cohen, A., Daubechies, I., Feauveau, J.C., *Comm. Pure. Appl. Math.*, **XLV**, 485-560 (1992).

16. Hofer, E., Zehnder, E., *Symplectic Topology*: Birkhauser, 1994.

17. Dahmen, W., Micchelli, C., *SIAM J. Number Anal.*, **30**, no. 2, 507-537 (1993).

Symmetry, Hamiltonian Problems and Wavelets in Accelerator Physics

A. Fedorova*, M. Zeitlin* and Z. Parsa† **

*Institute of Problems of Mechanical Engineering, Russian Academy of Sciences, 199178, Russia, St. Petersburg, V.O., Bolshoj pr., 61.   E-mail: zeitlin@math.ipme.ru
†Physics Department, Bldg. 901A, Brookhaven National Laboratory, Upton, NY 11973-5000, USA.   E-mail: parsa@bnl.gov

Abstract. In this paper we consider applications of methods from wavelet analysis to nonlinear dynamical problems related to accelerator physics. In our approach we take into account underlying algebraical, geometrical and topological structures of corresponding problems.

I   INTRODUCTION

This paper is the sequel to our first paper in this volume [1], in which we considered the applications of a number of analytical methods from nonlinear (local) Fourier analysis, or wavelet analysis, to nonlinear accelerator physics problems. This paper is the continuation of results from [2]–[7], which is based on our approach to investigation of nonlinear problems, both general and with additional structures (Hamiltonian, symplectic or quasicomplex), chaotic, quasiclassical, and quantum. Wavelet analysis is a relatively novel set of mathematical methods, which gives us the possibility of working with well-localized bases in functional spaces and with the general type of operators (differential, integral, pseudodifferential) in such bases. In contrast with paper [1], in this paper we try to take into account before using power analytical approaches underlying algebraical, geometrical, and topological structures related to the kinematical, dynamical and hidden symmetry of physical problems. In this paper we give a review of a number of the corresponding problems and describe the key points of some possible methods by which we can find the full solutions of the initial physical problem. We describe a few concrete problems in [1, part II]. The most interesting case is the dynamics of spin-orbital motion [1, II D]. Related problems may be found in [8].

**) This work was performed under the auspices of the U.S. Departmentof Energy under Contract No. DE-AC02-98CH10886.

CP468, *Nonlinear and Collective Phenomena in Beam Physics–1998 Workshop*, edited by S. Chattopadhyay, M. Cornacchia, and C. Pellegrini

The content of this paper is nothing more than an attempt to extract the most complicated formal, mathematical or principal parts of the world of nonlinear accelerator physics, which is today beyond the mainstream, in our opinion.

In part II we consider dynamical consequences of covariance properties regarding relativity (kinematical) groups and continuous wavelet transform as a method for the solution of dynamical problems. In part II A we introduce the semidirect product structure, which allows us to consider, from a general point of view all relativity groups such as Euclidean, Galilei, and Poincare. Then in part II B we consider the Lie-Poisson equations and obtain the manifestation of semiproduct structure of the (kinematic) symmetry group on a dynamical level. So, correct description of dynamics is a consequence of correct understanding of real symmetry of the concrete problem. In part II C we consider the technique for simplification of dynamics related to semiproduct structure by using reduction to corresponding orbit structure. As result we have simplified Lie-Poisson equations. In part II D we consider the Lagrangian theory related to semiproduct structure and an explicit form of variation principle and corresponding (semidirect) Euler-Poicare equations. In part II E we introduce a continuous wavelet transform and corresponding analytical technique, which allow covariant wavelet analysis. In part II F we consider, in the particular case, the affine Galilei group with the semiproduct structure, also the corresponding orbit technique for constructing different types of invariant wavelet bases. In part III we consider, instead of kinematical symmetry, dynamical symmetry. In part III A, according to the orbit method and by using construction from the geometric quantization theory, we construct the symplectic and Poisson structures associated with generalized wavelets by using metaplectic structure. We consider the wavelet approach to the calculations of the Melnikov functions in the theory of homoclinic chaos in the perturbed Hamiltonian systems in part III B and for calculation of Arnold–Weinstein curves (closed loops) in the Floer variational approach in part III C. In parts III D and III E we consider applications of a very useful fast wavelet transform technique (part III F) to calculations in a symplectic scale of spaces and to quasiclassical evolution dynamics. This method gives maximally sparse representation of a (differential) operator that allows us to take into account a contribution from each level of resolution. In part IV A we consider symplectic and Lagrangian structures for the case of discretization of flows by corresponding maps; and in part IV B construction of corresponding solutions by applications of a generalized wavelet approach which is based on generalization of multiresolution analysis for the case of maps.

II SEMIDIRECT PRODUCT, DYNAMICS, WAVELET REPRESENTATION

A Semidirect Product

Relativity groups such as Euclidean, Galilean, or Poincare groups are particular cases of semidirect product construction, which is very useful and of simple general construction in the group theory [9]. We may consider as a basic example in the Euclidean group $SE(3) = SO(3) \bowtie \mathbf{R}^3$, the semidirect product of rotations and translations. In the general case we have $S = G \bowtie V$, where group G (Lie group or automorphisms group) acts on a vector space V and on its dual V^*. Let V be a vector space and G is the Lie group, which acts on the left by linear maps on V (G also acts on the left on its dual space V^*). The semidirect product $S = G \bowtie V$ is the Cartesian product $S = G \times V$ with group multiplication

$$(g_1, v_1)(g_2, v_2) = (g_1 g_2, v_1 + g_1 v_2), \tag{1}$$

where the action of $g \in G$ on $v \in V$ is denoted as gv. Of course, we can consider the corresponding definitions both in the case of the right actions and in the case when G is a group of automorphisms of the vector space V. As we shall explain below, both cases — Lie groups and automorphisms groups — are important for us.

So, the Lie algebra of S is the semidirect product of Lie algebra, $s = \mathcal{G} \bowtie V$ with brackets

$$[(\xi_1, v_1), (\xi_2, v_2)] = ([\xi_1, \xi_2], \xi_1 v_2 - \xi_2 v_1), \tag{2}$$

where the induced action of $\mathcal{G}$ by concatenation is denoted as $\xi_1 v_2$. Also we need expression for adjoint and coadjoint actions for semidirect products. Let $(g, v) \in S = G \times V$, $(\xi, u) \in s = \mathcal{G} \times V$, $(\mu, a) \in s^* = \mathcal{G}^* \times V^*$, $g\xi = Ad_g \xi$, $g\mu = Ad^*_{g^{-1}}\mu$, ga denoting the induced left action of g on a (the left action of G on V inducing a left action on V^* — the inverse of the transpose of the action on V), $\rho_v : \mathcal{G} \to V$ a linear map given by $\rho_v(\xi) = \xi v$, $\rho_v^* : V^* \to \mathcal{G}^*$ its dual. Then these actions are given by simple concatenation:

$$(g, v)(\xi, u) = (g\xi, gu - (g\xi)v), \tag{3}$$
$$(g, v)(\mu, a) = (g\mu + \rho_v^*(ga), ga) \quad .$$

Below we use the following notation: $\rho_v^* a = v \diamond a \in \mathcal{G}^*$ for $a \in V^*$, which is a bilinear operation in v and a. So, we have the coadjoint action:

$$(g, v)(\mu, a) = (g\mu + v \diamond (ga), ga). \tag{4}$$

Using concatenation notation for Lie algebra actions, we have an alternative definition of $v \diamond a \in \mathcal{G}^*$. For all $v \in V$, $a \in V^*$, $\eta \in \mathcal{G}$ we have

$$< \eta a, v > = - < v \diamond a, \eta > \quad . \tag{5}$$

71

B   The Lie-Poisson Equations and Semiproduct Structure

Below we consider the manifestation of the semiproduct structure of the symmetry group on a dynamical level. Let F, G be real valued functions on the dual space $\mathcal{G}^*$, $\mu \in \mathcal{G}^*$. The functional derivative of F at μ is the unique element $\delta F / \delta \mu \in \mathcal{G}$:

$$\lim_{\epsilon \to 0} \frac{1}{\epsilon} [F(\mu + \epsilon \delta \mu) - F(\mu)] = < \delta \mu, \frac{\delta F}{\delta \mu} > \quad , \tag{6}$$

for all $\delta \mu \in \mathcal{G}^*$, $<, >$ is pairing between $\mathcal{G}^*$ and $\mathcal{G}$.

Define the ($\pm$) Lie-Poisson brackets by

$$\{F, G\}_{\pm}(\mu) = \pm < \mu, [\frac{\delta F}{\delta \mu}, \frac{\delta G}{\delta \mu}] > \quad . \tag{7}$$

The Lie-Poisson equations, determined by

$$\dot{F} = \{F, H\} \quad , \tag{8}$$

read intrinsically

$$\dot{\mu} = \mp ad^*_{\partial H / \partial \mu} \mu. \tag{9}$$

For the left representation of G on V $\pm$ Lie-Poisson bracket of two functions $f, k : s^* \to \mathbf{R}$ is given by

$$\{f, k\}_{\pm}(\mu, a) = \pm < \mu, [\frac{\delta f}{\delta \mu}, \frac{\delta k}{\delta \mu}] > \pm < a, \frac{\delta f}{\delta \mu} \frac{\delta k}{\delta a} - \frac{\delta k}{\delta \mu} \frac{\delta f}{\delta a} >, \tag{10}$$

where $\delta f / \delta \mu \in \mathcal{G}$, $\delta f / \delta a \in V$ are the functional derivatives of f (6). The Hamiltonian vector field of $h : s^* \in \mathbf{R}$ has the expression

$$X_h(\mu, a) = \mp (ad^*_{\delta h / \delta \mu} \mu - \frac{\delta h}{\delta a} \diamond a, -\frac{\delta h}{\delta \mu} a). \tag{11}$$

Thus, Hamiltonian equations on the dual of a semidirect product are [9]:

$$\dot{\mu} = \mp ad^*_{\delta h / \delta \mu} \mu \pm \frac{\delta h}{\delta a} \diamond a \tag{12}$$

$$\dot{a} = \pm \frac{\delta h}{\delta \mu} a \quad .$$

So, we can see the explicit difference between Poisson brackets (7) and (10) and the equations of motion (9) and (12), which come from the semiproduct structure.

C Reduction of Dynamics on Semiproduct

There is the technique for reducing dynamics that is associated with the geometry of the semidirect product reduction theorem[9]. Let us have a Hamiltonian on T^*G that is invariant under the isotropy G_{a_0} for $a_0 \in V^*$. The semidirect product reduction theorem states that reduction of T^*G by G_{a_0} gives reduced spaces that are simplectically diffeomorphic to coadjoint orbits in the dual of the Lie algebra of the semidirect product $(\mathcal{G} \bowtie V)^*$. If one reduces the semidirect group product $S = G \bowtie V$ in two stages, first by V and then by G one recovers this semidirect product reduction theorem. Thus, let $S = G \bowtie V$, choose $\sigma = (\mu, a) \in \mathcal{G}^* \times V^*$ and reduce T^*S by the action of S at σ, giving the co-adjoint orbit $\mathcal{O}_\sigma$ through $\sigma \in S^*$. There is a symplectic diffeomorphism between $\mathcal{O}_\sigma$ and the reduced space obtained by reducing T^*G by the subgroup G_a (the isotropy of G for its action on V^* at the point $a \in V^*$) at the point $\mu|\mathcal{G}_a$, where $\mathcal{G}_a$ is the Lie algebra of G_a.

Then we have the following procedure.

1. We start with a Hamiltonian H_{a_0} on T^*G that depends parametrically on a variable $a_0 \in V^*$.

2. The Hamiltonian regarded as a map: $T^*G \times V^* \to \mathbf{R}$ is assumed to be an invariant on T^*G under the action of G on $T^*G \times V^*$.

3. Condition 2 is equivalent to the invariance of the function H defined on $T^*S = T^*G \times V \times V^*$ extended to be constant in the variable V under the action of the semidirect product.

4. By the semidirect product reduction theorem, the dynamics of H_{a_0} reduced by G_{a_0}, the isotropy group of a_0, is simplectically equivalent to Lie-Poisson dynamic on $s^* = \mathcal{G}^* \times V^*$.

5. This Lie-Poisson dynamics is given by equations (12) for the function $h(\mu, a) = H(\alpha_g, g^{-1}a)$, where $\mu = g^{-1}\alpha_g$.

D Lagrangian Theory, the Euler-Poincare Equations, Variational Approach on Semiproduct

Now we make a consideration based on [9], the Lagrangian side of a theory. This approach is based on variational principles with symmetry and is not dependent on a Hamiltonian formulation, although it is demonstrated in [9] that this purely Lagrangian formulation is equivalent to the Hamiltonian formulation on duals of the semidirect product (the corresponding Legendre transformation is a diffeomorphism).

We consider the case of the left representation and the left invariant Lagrangians (ℓ and L), which depend in addition on another parameter $a \in V^*$

(dynamical parameter), where V is the representational space for the Lie group G, and L has an invariance property related to both arguments. It should be noted that the resulting equations of motion, the Euler-Poincare equations, are not the Euler-Poincare equations for the semidirect product Lie algebra $\mathcal{G} \bowtie V^*$ or $\mathcal{G} \bowtie V$.

So, we have the following:

1. There is a left presentation of Lie group G on the vector space V and G acting in the natural way on the left on $TG \times V^* : h(v_g, a) = (hv_g, ha)$.

2. The function $L : TG \times V^* \in \mathbf{R}$ is the left G-invariant.

3. Let $a_0 \in V^*$, Lagrangian $L_{a_0} : TG \rightarrow \mathbf{R}$, $L_{a_0}(v_g) = L(v_g, a_0)$. L_{a_0} is the left invariant under the lift to TG of the left action of G_{a_0} on G, where G_{a_0} is the isotropy group of a_0.

4. Left G-invariance of L permits us to define

$$\ell : \mathcal{G} \times V^* \rightarrow \mathbf{R} \tag{13}$$

by

$$\ell(g^{-1}v_g, g^{-1}a_0) = L(v_g, a_0). \tag{14}$$

This relation defines for any $\ell : \mathcal{G} \times V^* \rightarrow \mathbf{R}$ the left G-invariant function $L : TG \times V^* \rightarrow \mathbf{R}$.

5. For a curve $g(t) \in G$ let

$$\xi(t) := g(t)^{-1}\dot{g}(t) \quad . \tag{15}$$

and define the curve $a(t)$ as the unique solution to the following linear differential equation with time dependent coefficients

$$\dot{a}(t) = -\xi(t)a(t), \tag{16}$$

with initial condition $a(0) = a_0$. The solution can be written as $a(t) = g(t)^{-1}a_0$.

Then we have four equivalent descriptions of the corresponding dynamics:

1. If a_0 is fixed, then Hamilton's variational principle

$$\delta \int_{t_1}^{t_2} L_{a_0}(g(t), \dot{g}(t)) \mathrm{d}t = 0 \tag{17}$$

holds for variations $\delta g(t)$ of $g(t)$ vanishing at the endpoints.

2. $g(t)$ satisfies the Euler-Lagrange equations for L_{a_0} on G.

3. The constrained variational principle

$$\delta \int_{t_1}^{t_2} \ell(\xi(t), a(t)) \mathrm{d}t = 0 \qquad (18)$$

holds on $\mathcal{G} \times V^*$, using variations of ξ and a of the form $\delta\xi = \dot{\eta} + [\xi, \eta]$, $\delta a = -\eta a$, where $\eta(t) \in \mathcal{G}$ vanishes at the endpoints.

4. The Euler-Poincare equations hold on $\mathcal{G} \times V^*$

$$\frac{\mathrm{d}}{\mathrm{d}t} \frac{\delta\ell}{\delta\xi} = ad_\xi^* \frac{\delta\ell}{\delta\xi} + \frac{\delta\ell}{\delta a} \diamond a \qquad (19)$$

So, we may apply our wavelet methods either on the level of the variational formulation (17) or on the level of the Euler-Poincare equations (19).

E   Continuous Wavelet Transform

Now we need take into account the Hamiltonian or Lagrangian structures related to systems (12) or (19). Therefore, we need to consider generalized wavelets, which allow us to consider the corresponding structures instead of compactly supported wavelet representation from paper [1].

In wavelet analysis the following three concepts are used now: 1) a square integrable representation U of a group G, 2) coherent states (CS) over G, and 3) the wavelet transform associated to U. We consider now their unification [10], [11].

Let G be a locally compact group and U_a a strongly continuous, irreducible, unitary representation of G on Hilbert space $\mathcal{H}$. Let H be a closed subgroup of G, $X = G/H$ with (quasi) invariant measure ν and $\sigma : X = G/H \to G$ a Borel section in a principal bundle $G \to G/H$. Then we say that U is a square integrable $mod(H, \sigma)$ if there exists a non-zero vector $\eta \in \mathcal{H}$ so that

$$0 < \int_X | < U(\sigma(x))\eta|\Phi > |^2 \mathrm{d}\nu(x) =< \Phi|A_\sigma\Phi > < \infty, \quad \forall \Phi \in \mathcal{H} \quad . \qquad (20)$$

Given such a vector $\eta \in \mathcal{H}$ called admissible for (U, σ), we define the family of (covariant) coherent states or wavelets, indexed by points $x \in X$, as the orbit of η under G, though the representation U and the section σ [10], [11]

$$S_\sigma = \eta_{\sigma(x)} = U(\sigma(x))\eta | x \in X \quad . \qquad (21)$$

So, coherent states or wavelets are simply the elements of the orbit under U of a fixed vector η in representational space. We have the following fundamental properties:

75

1. Overcompleteness:
   The set S_σ is total in $\mathcal{H} : (S_\sigma)^\perp = 0$.

2. Resolution property:
   the square integrability condition (20) may be represented as a resolution relation:

$$\int_X |\eta_\sigma(x) > < \eta_{\sigma(x)}| \mathrm{d}\nu(x) = A_\sigma, \tag{22}$$

where A_σ is a bounded, positive operator with a densely defined inverse. Define the linear map

$$W_\eta : \mathcal{H} \to L^2(X, \mathrm{d}\nu), (W_\eta \Phi)(x) = < \eta_{\sigma(x)} | \Phi > \quad . \tag{23}$$

Then the range H_η of W_η is complete with respect to the scalar product $< \Phi | \Psi >_\eta = < \Phi | W_\eta A_\sigma^{-1} W_\eta^{-1} \Psi >$ and W_η is the unitary operator from $\mathcal{H}$ onto $\mathcal{H}_\eta$. W_η is a Continuous Wavelet Transform (CWT).

3. Reproducing kernel:
   The orthogonal projection from $L^2(X, \mathrm{d}\nu)$ onto $\mathcal{H}_\eta$ is an integral operator K_σ, and H_η is a reproducing kernel Hilbert space of functions:

$$\Phi(x) = \int_X K_\sigma(x, y) \Phi(y) \mathrm{d}\nu(y), \quad \forall \Phi \in \mathcal{H}_\eta. \tag{24}$$

The kernel is given explicitly by $K_\sigma(x, y) = < \eta_{\sigma(x)} | A_\sigma^{-1} \eta_{\sigma(y)} >$, if $\eta_{\sigma(y)} \in D(A_\sigma^{-1})$, $\forall y \in X$. So, the function $\Phi \in L^2(X, \mathrm{d}\nu)$ is a wavelet transform (WT) if it satisfies this reproducing relation.

4. Reconstruction formula.
   The WT W_η may be inverted on its range by the adjoint operator, $W_\eta^{-1} = W_\eta^*$ on $\mathcal{H}_\eta$ to obtain for $\eta_{\sigma(x)} \in D(A_\sigma^{-1})$, $\forall x \in X$,

$$W_\eta^{-1} \Phi = \int_X \Phi(x) A_\sigma^{-1} \eta_{\sigma(x)} \mathrm{d}\nu(x), \quad \Phi \in \mathcal{H}_\eta. \tag{25}$$

This is inverse WT.

If A_σ^{-1} is bounded, then S_σ is called a frame; if $A_\sigma = \lambda I$, then S_σ is called a tight frame. These two cases are a generalization of a simple case, when S_σ is an (ortho)basis.

The most simple cases of this construction are:
1. $H = \{e\}$. This is the standard construction of WT over a locally compact group. It should be noted that the square integrability of U is equivalent to U belonging to the discrete series. The most simple example is related to the affine $(ax + b)$ group and yields the usual one-dimensional wavelet analysis

$$[\pi(b,a)f](x) = \frac{1}{\sqrt{a}}f\left(\frac{x-b}{a}\right). \qquad (26)$$

For $G = SIM(2) = \mathbf{R}^2 \rtimes (\mathbf{R}_*^+ \times SO(2))$, the similitude group of the plane, we have the corresponding two-dimensional wavelets.

2. $H = H_\eta$, the isotropy (up to a phase) subgroup of η: this is the case of the Gilmore-Perelomov CS. Some cases of group G are:

a) Semisimple groups, such as SU(N), SU(N|M), SU(p,q), Sp(N,$\mathbf{R}$).

b) the Weyl-Heisenberg group G_{WH} which leads to the Gabor functions, i.e., canonical (oscillator) coherent states associated with windowed Fourier transform or Gabor transform (see also part III A):

$$[\pi(q,p,\varphi)f](x) = \exp(i\mu(\varphi - p(x-q)))f(x-q) \quad . \qquad (27)$$

In this case, H is the center of G_{WH}. In both cases the time-frequency plane corresponds to the phase space of group representation.

c) The similitude group SIM(n) of $\mathbf{R}^n (n \geq 3)$: for $H = SO(n-1)$ we have the axisymmetric n-dimensional wavelets.

d) Also we have the case of a bigger group, containing both affine and Weyl-Heisenberg group, which interpolate between affine wavelet analysis and windowed Fourier analysis: affine Weyl–Heisenberg group [11].

e) Relativity groups. In a nonrelativistic setup, the natural kinematical group is the (extended) Galilean group. Also we may adds independent space and time dilations and obtain an affine Galilean group. If we restrict the dilations by the relation $a_0 = a^2$, where a_0, a are the time and space dilation, we obtain the Galilei-Schrödinger group, invariance group of both Schrödinger and heat equations. We consider these examples in the next section. In the same way we may consider as a kinematical group the Poincare group. When $a_0 = a$ we have affine Poincare or Weyl-Poincare group. We consider a useful generalization of that affinization construction for the case of a hidden metaplectic structure in section III A.

But the usual representation is not square–integrable and must be modified: restriction of the representation to a suitable quotient space of the group (the associated phase space in our case) restores square – integrability: $G \longrightarrow$ homogeneous space.

Also, we have a more general approach which allows consideration of wavelets corresponding to more general groups and representations [12], [13].

Our goal is to apply these results to problems of Hamiltonian dynamics and as a consequence we need to take into account the symplectic nature of our dynamical problem. Also, the symplectic and wavelet structures must be consistent (this must resemble the symplectic or Lie-Poisson integrator theory). We use the point of view of geometric quantization theory (orbit method) instead of harmonic analysis. Because of this we can consider (a) – (e) analogously.

F   Bases for Solutions

We consider an important particular case of the affine relativity group (relativity group combined with dilations) — affine Galilei group in n-dimensions. So, we have a combination of the Galilei group with independent space and time dilations: $G_{aff} = G_m \bowtie D_2$, where $D_2 = (\mathbf{R}_*^+)^2 \simeq \mathbf{R}^2$, G_m is the extended Galilei group corresponding to mass parameter $m > 0$ (G_{aff} is a non-central extension of $G \bowtie D_2$ by $\mathbf{R}$, where G is usual Galilei group). The generic element of G_{aff} is $g = (\Phi, b_0, b; v; R, a_0, a)$, where $\Phi \in \mathbf{R}$ is the extension parameter in G_m, $b_0 \in \mathbf{R}$, $b \in \mathbf{R}^n$ are the time and space translations, $v \in \mathbf{R}^n$ is the boost parameter, $R \in SO(n)$ is a rotation and $a_0, a \in \mathbf{R}_*^+$ are time and space dilations. The actions of g on space-time is then $x \mapsto aRx + a_0 vt + b$, $t \mapsto a_0 t + b_0$, where $x = (x_1, x_2, ..., x_n)$. The group law is

$$gg' = (\Phi + \frac{a^2}{a_0}\Phi' + avRb' + \frac{1}{2}a_0 v^2 b_0', b_0 + a_0 b_0', b + aRb' + a_0 v b_0';$$
$$v + \frac{a}{a_0}Rv', RR'; a_a a_0', aa') \quad . \tag{28}$$

It should be noted that D_2 acts nontrivially on G_m. Space-time wavelets associated to G_{aff} correspond to unitary irreducible representation of spin zero. It may be obtained via orbit method. The Hilbert space is $\mathcal{H} = L^2(\mathbf{R}^n \times \mathbf{R}, dk d\omega)$, $k = (k_1, ..., k_n)$, where $\mathbf{R}^n \times \mathbf{R}$ may be identified with the usual Minkowski space, and we have for representation:

$$(U(g)\Psi)(k, \omega) = \sqrt{a_0 a^n}\exp i(m\Phi + kb - \omega b_0)\Psi(k', \omega'), \tag{29}$$

with $k' = aR^{-1}(k + mv)$, $\omega' = a_0(\omega - kv - \frac{1}{2}mv^2)$, and $m' = (a^2/a_0)m$. Mass m is a coordinate in the dual of the Lie algebra and these relations are a part of the coadjoint action of G_{aff}. This representation is unitary and irreducible but not square integrable. So, we need to consider reduction to the corresponding quotients $X = G/H$. We consider the case in which H={phase changes Φ and space dilations a}. Then the space $X = G/H$ is parametrized by points $\bar{x} = (b_0, b; v; R; a_0)$. There is a dense set of vectors $\eta \in \mathcal{H}$ admissible mod(H, σ_β), where σ_β is the corresponding section. We have a two-parameter family of functions β(dilations): $\beta(\bar{x}) = (\mu_0 + \lambda_0 a_0)^{1/2}$, $\lambda_0, \mu_0 \in \mathbf{R}$. Then any admissible vector η generates a tight frame of Galilean wavelets

$$\eta_{\beta(\bar{x})}(k, \omega) = \sqrt{a_0(\mu_0 + \lambda_0 a_0)^{n/2}}e^{i(kb - \omega b_0)}\eta(k', \omega'), \tag{30}$$

with $k' = (\mu_0 + \lambda_0 a)^{1/2}R^{-1}(k + mv)$, $\omega' = a_0(\omega - kv - mv^2/2)$. The simplest examples of admissible vectors (corresponding to the usual Galilei case) are the Gaussian vector: $\eta(k) \sim \exp(-k^2/2mu)$ and the binomial vector: $\eta(k) \sim (1 + k^2/2mu)^{-\alpha/2}$, $\alpha > 1/2$, where u is a kind of internal energy. When we impose the relation $a_0 = a^2$, then we have the restriction to the Galilei-Schrödinger group $G_s = G_m \bowtie D_s$, where D_s is the one-dimensional subgroup of D_2. G_s is a natural invariance group of both the Schrödinger equation and the heat equation. The restriction to G_s of representation (29) splits into the direct sum of two irreducible ones, $U = U_+ \oplus U_-$, corresponding to the decomposition $L^2(\mathbf{R}^n \times \mathbf{R}, \mathrm{d}k\mathrm{d}\omega) = \mathcal{H}_+ \oplus \mathcal{H}_-$, where

$$\mathcal{H}_\pm = L^2(D_\pm, \mathrm{d}k\mathrm{d}\omega) \tag{31}$$
$$= \{\psi \in L^2(\mathbf{R}^n \times \mathbf{R}, \mathrm{d}k\mathrm{d}\omega), \quad \psi(k,\omega) = 0 \quad for \quad \omega + k^2/2m = 0\} \quad .$$

These two subspaces are the analogues of the usual Hardy spaces on $\mathbf{R}$, i.e. the subspaces of (anti)progressive wavelets (see also below, part III A). The two representation $U_\pm$ are square integrable modulo the center. There is a dense set of admissible vectors η, and each of them generates a set of CS of the Gilmore-Perelomov type. Typical wavelets of this kind are:
the Schrödinger-Marr wavelet,

$$\eta(x,t) = (i\partial_t + \frac{\triangle}{2m})\mathrm{e}^{-(x^2+t^2)/2} \quad , \tag{32}$$

the Schrödinger-Cauchy wavelet,

$$\psi(x,t) = (i\partial_t + \frac{\triangle}{2m})\frac{1}{(t+i)\prod_{j=1}^{n}(x_j+i)} \quad . \tag{33}$$

So, in the same way, we can construct invariant bases with an explicit manifestation of the underlying symmetry for solving Hamiltonian (12) or Lagrangian (19) equations.

III   SYMPLECTIC STRUCTURES, QUANTIZATION AND FAST WAVELET TRANSFORM

A   Metaplectic Group and Representations

Let $Sp(n)$ be symplectic group, $Mp(n)$ be its unique two- fold covering – metaplectic group [14]. Let V be a symplectic vector space with symplectic form (,), then $R \oplus V$ is nilpotent Lie algebra - Heisenberg algebra:

$$[R, V] = 0, \quad [v, w] = (v, w) \in R, \quad [V, V] = R.$$

$Sp(V)$ is a group of automorphisms of Heisenberg algebra.

Let N be a group with Lie algebra $R \oplus V$, i.e. Heisenberg group. By Stone–von Neumann theorem Heisenberg group has unique irreducible unitary representation in which $1 \mapsto i$. Let us also consider the projective representation of simplectic group $Sp(V)$: $U_{g_1} U_{g_2} = c(g_1, g_2) \cdot U_{g_1 g_2}$, where c is a map: $Sp(V) \times Sp(V) \to S^1$, i.e. c is S^1-cocycle.

But this representation is unitary representation of universal covering, i.e. metaplectic group $Mp(V)$. We give this representation without Stone-von Neumann theorem. Consider a new group $F = N' \bowtie Mp(V)$, $\bowtie$ is semidirect product (we consider instead of $N = R \oplus V$ the $N' = S^1 \times V$, $S^1 = (R/2\pi Z)$). Let V^* be dual to V, $G(V^*)$ be automorphism group of V^*. Then F is subgroup of $G(V^*)$, which consists of elements, which acts on V^* by affine transformations.

This is the key point!

Let $q_1, ..., q_n; p_1, ..., p_n$ be symplectic basis in V, $\alpha = pdq = \sum p_i dq_i$ and $d\alpha$ be symplectic form on V^*. Let M be fixed affine polarization, then for $a \in F$ the map $a \mapsto \Theta_a$ gives unitary representation of G: $\Theta_a : H(M) \to H(M)$

Explicitly we have for representation of N on H(M):

$$(\Theta_q f)^*(x) = e^{-iqx} f(x), \quad \Theta_p f(x) = f(x - p)$$

The representation of N on H(M) is irreducible. Let A_q, A_p be infinitesimal operators of this representation

$$A_q = \lim_{t \to 0} \frac{1}{t}[\Theta_{-tq} - I], \quad A_p = \lim_{t \to 0} \frac{1}{t}[\Theta_{-tp} - I],$$

then $\quad A_q f(x) = i(qx)f(x), \quad A_p f(x) = \sum p_j \frac{\partial f}{\partial x_j}(x)$

Now we give the representation of infinitesimal basic elements. Lie algebra of the group F is the algebra of all (nonhomogeneous) quadratic polynomials of (p,q) relatively Poisson bracket (PB). The basis of this algebra consists of elements $1, q_1, ..., q_n, p_1, ..., p_n, q_i q_j, q_i p_j, p_i p_j, \quad i, j = 1, ..., n, \quad i \le j$,

$$PB \text{ is } \{f, g\} = \sum \frac{\partial f}{\partial p_j} \frac{\partial g}{\partial q_i} - \frac{\partial f}{\partial q_i} \frac{\partial g}{\partial p_i} \quad \text{and} \quad \{1, g\} = 0 \quad \text{for all } g,$$

$\{p_i, q_j\} = \delta_{ij}, \quad \{p_i q_j, q_k\} = \delta_{ik} q_j, \quad \{p_i q_j, p_k\} = -\delta_{jk} p_i, \quad \{p_i p_j, p_k\} = 0,$
$\{p_i p_j, q_k\} = \delta_{ik} p_j + \delta_{jk} p_i, \quad \{q_i q_j, q_k\} = 0, \{q_i q_j, p_k\} = -\delta_{ik} q_j - \delta_{jk} q_i$

so, we have the representation of basic elements $f \mapsto A_f : 1 \mapsto i, q_k \mapsto ix_k,$

$$p_l \mapsto \frac{\delta}{\delta x^l}, p_i q_j \mapsto x^i \frac{\partial}{\partial x^j} + \frac{1}{2} \delta_{ij}, \quad p_k p_l \mapsto \frac{1}{i} \frac{\partial^k}{\partial x^k \partial x^l}, q_k q_l \mapsto ix^k x^l$$

This gives the structure of the Poisson manifolds to representation of any (nilpotent) algebra or in other words to continuous wavelet transform.

The Segal-Bargman Representation. Let $z = 1/\sqrt{2} \cdot (p - iq)$, $\bar{z} = 1/\sqrt{2} \cdot (p + iq)$, $p = (p_1, ..., p_n)$, F_n is the space of holomorphic functions of n complex variables with $(f, f) < \infty$, where

$$(f, g) = (2\pi)^{-n} \int f(z)\overline{g(z)}e^{-|z|^2} dp dq$$

Consider a map $U : H \to F_n$, where H is with real polarization, F_n is with complex polarization, then we have

$$(U\Psi)(a) = \int A(a, q)\Psi(q)dq, \qquad \text{where} \quad A(a, q) = \pi^{-n/4}e^{-1/2(a^2+q^2)+\sqrt{2}aq}$$

i.e. the Bargmann formula produce wavelets. We also have the representation of Heisenberg algebra on F_n :

$$U\frac{\partial}{\partial q_j}U^{-1} = \frac{1}{\sqrt{2}}\left(z_j - \frac{\partial}{\partial z_j}\right), \qquad Uq_jU^{-1} = -\frac{i}{\sqrt{2}}\left(z_j + \frac{\partial}{\partial z_j}\right)$$

and also : $\omega = d\beta = dp \wedge dq$, where $\beta = i\bar{z}dz$.

Orbital Theory for Wavelets. Let coadjoint action be $< g \cdot f, Y > = < f, Ad(g)^{-1}Y >$, where $<,>$ is pairing $g \in G$, $f \in g^*$, $Y \in \mathcal{G}$. The orbit is $\mathcal{O}_f = G \cdot f \equiv G/G(f)$. Also, let A=A(M) be algebra of functions, V(M) is A-module of vector fields, A^p is A-module of p-forms. Vector fields on orbit is

$$\sigma(\mathcal{O}, X)_f(\phi) = \frac{d}{dt}(\phi(\exp tXf))\Big|_{t=0}$$

where $\phi \in A(\mathcal{O})$, $f \in \mathcal{O}$. Then $\mathcal{O}_f$ are homogeneous symplectic manifolds with 2-form $\Omega(\sigma(\mathcal{O}, X)_f, \sigma(\mathcal{O}, Y)_f) = < f, [X, Y] >$, and $d\Omega = 0$. PB on $\mathcal{O}$ have the next form $\{\Psi_1, \Psi_2\} = p(\Psi_1)\Psi_2$ where p is $A^1(\mathcal{O}) \to V(\mathcal{O})$ with definition $\Omega(p(\alpha), X) = i(X)\alpha$. Here $\Psi_1, \Psi_2 \in A(\mathcal{O})$ and $A(\mathcal{O})$ is Lie algebra with bracket $\{,\}$. Now let N be a Heisenberg group. Consider adjoint and coadjoint representations in some particular case. $N = (z, t) \in C \times R, z = p + iq$; compositions in N are $(z, t) \cdot (z', t') = (z+z', t+t'+B(z, z'))$, where $B(z, z') = pq' - qp'$. Inverse element is $(-t, -z)$. Lie algebra n of N is $(\zeta, \tau) \in C \times R$ with bracket $[(\zeta, \tau), (\zeta', \tau')] = (0, B(\zeta, \zeta'))$. Centre is $\bar{z} \in n$ and generated by $(0,1)$; Z is a subgroup $\exp \bar{z}$. Adjoint representation N on n is given by formula $Ad(z, t)(\zeta, \tau) = (\zeta, \tau + B(z, \zeta))$ Coadjoint: for $f \in n^*$, $g = (z, t)$, $(g \cdot f)(\zeta, \zeta) = f(\zeta, \tau) - B(z, \zeta)f(0, 1)$ then orbits for which $f|_{\bar{z}} \neq 0$ are plane in n^* given by equation $f(0, 1) = \mu$. If $X = (\zeta, 0)$, $Y = (\zeta', 0)$, $X, Y \in n$ then symplectic structure is

$$\Omega(\sigma(\mathcal{O}, X)_f, \sigma(\mathcal{O}, Y)_f) = < f, [X, Y] > = f(0, B(\zeta, \zeta'))\mu B(\zeta, \zeta')$$

Also we have for orbit $\mathcal{O}_\mu = N/Z$ and $\mathcal{O}_\mu$ is Hamiltonian G-space.

According to this approach we can construct by using methods of geometric quantization theory many "symplectic wavelet constructions" with corresponding symplectic or Poisson structure on it. Very useful particular spline–wavelet basis with uniform exponential control on stratified and nilpotent Lie groups was considered in [13].

B Applications to Melnikov Functions Approach

We give now some point of application of wavelet methods from the preceding parts to the Melnikov approach in the theory of homoclinic chaos in perturbed Hamiltonian systems for examples from [1].

In the Hamiltonian form we have:

$$\dot{x} = J \cdot \nabla H(x) + \varepsilon g(x, \Theta), \quad \dot{\Theta} = \omega, \quad (x, \Theta) \in R^n \times T^m,$$

for $\varepsilon = 0$ we have:

$$\dot{x} = J \cdot \nabla H(x), \quad \dot{\Theta} = \omega \quad . \tag{34}$$

For $\varepsilon = 0$ we have homoclinic orbit $\bar{x}_0(t)$ to the hyperbolic fixed point x_0. For $\varepsilon \neq 0$ we have the normally hyperbolic invariant torus T_ε and condition on transversal intersection of stable and unstable manifolds $W^s(T_\varepsilon)$ and $W^u(T_\varepsilon)$ in terms of Melnikov functions $M(\Theta)$ for $\bar{x}_0(t)$:

$$M(\Theta) = \int_{-\infty}^{\infty} \nabla H(\bar{x}_0(t)) \wedge g(\bar{x}_0(t), \omega t + \Theta) dt$$

. This condition has the next form:

$$M(\Theta_0) = 0, \quad \sum_{j=1}^{2} \omega_j \frac{\partial}{\partial \Theta_j} M(\Theta_0) \neq 0 \quad .$$

According to the approach of Birkhoff-Smale-Wiggins, we determined the region in parameter space in which we can observe the chaotic behaviour [4]. If we cannot solve equations (34) explicitly in time, then we use the wavelet approach from paper [1] for the computations of homoclinic (heteroclinic) loops as the wavelet solutions of system (34). For computations of quasiperiodic Melnikov functions,

$$M^{m/n}(t_0) = \int_0^{mT} DH(x_\alpha(t)) \wedge g(x_\alpha(t), t + t_0) dt$$

. we used the periodization of wavelet construction from paper [1]. We also used the symplectic Melnikov function approach in which we have:

$$M_i(z) = \lim_{j \to \infty} \int_{-T_j^*}^{T_j} \{h_i, \hat{h}\}_{\Psi(t,z)} dt$$

$$d_i(z, \varepsilon) = h_i(z_\varepsilon^u) - h_i(z_\varepsilon^s) = \varepsilon M_i(z) + O(\varepsilon^2)$$

where $\{,\}$ is the Poisson bracket and $d_i(z, \varepsilon)$ is the Melnikov distance. So, we need symplectic invariant wavelet expressions for Poisson brackets. The computations are produced according to invariant calculation of Poisson brackets, which is based on consideration of part III A and on operator representation from part III F (see below).

C   Floer Approach for Closed Loops

Now we consider the generalization of the wavelet variational approach to the symplectic invariant calculation of closed loops in Hamiltonian systems [15]. As we demonstrated in [4] we have the parametrization of our solution by some reduced algebraical problem; but in contrast to the cases from paper [1], where the solution is parametrized by construction based on the scalar refinement equation, in the symplectic case we have parametrization of the solution by matrix problems – Quadratic Mirror Filters equations. Now we consider a different approach.

Let (M, ω) be a compact symplectic manifold of dimension $2n$, ω being a closed 2-form (nondegenerate) on M which induces an isomorphism $T^*M \to TM$. Thus every smooth time-dependent Hamiltonian $H : \mathbf{R} \times M \to \mathbf{R}$ corresponds to a time-dependent Hamiltonian vector field $X_H : \mathbf{R} \times M \to TM$ defined by

$$\omega(X_H(t, x), \xi) = -d_x H(t, x)\xi \tag{35}$$

for $\xi \in T_x M$. Let H (and X_H) be periodic in time: $H(t + T, x) = H(t, x)$ and consider the corresponding Hamiltonian differential equation on M:

$$\dot{x}(t) = X_H(t, x(t)) \quad . \tag{36}$$

The solutions $x(t)$ of (36) determine a 1-parameter family of diffeomorphisms $\psi_t \in \mathrm{Diff}(M)$ satisfying $\psi_t(x(0)) = x(t)$. These diffeomorphisms are symplectic: $\omega = \psi_t^* \omega$. Let $L = L_T M$ be the space of contractible loops in M which are represented by smooth curves $\gamma : \mathbf{R} \to M$, satisfying $\gamma(t + T) = \gamma(t)$. Then the contractible T-periodic solutions of (36) can be characterized as the critical points of the functional $S = S_T : L \to \mathbf{R}$:

$$S_T(\gamma) = -\int_D u^* \omega + \int_0^T H(t, \gamma(t)) dt, \tag{37}$$

where $D \subset \mathbf{C}$ is a closed unit disc and $u : D \to M$ is a smooth function, which on boundary agrees with γ, i.e., $u(\exp\{2\pi i\Theta\}) = \gamma(\Theta T)$. Because $[\omega]$, the cohomology class of ω, vanishes, then $S_T(\gamma)$ is independent of choice of u. Tangent space $T_\gamma L$ is the space of vector fields $\xi \in C^\infty(\gamma^* TM)$ along γ satisfying $\xi(t + T) = \xi(t)$. Then we have for the 1-form $\mathrm{d}f : TL \to \mathbf{R}$

$$\mathrm{d}S_T(\gamma)\xi = \int_0^T (\omega(\dot{\gamma}, \xi) + \mathrm{d}H(t, \gamma)\xi) dt \quad , \tag{38}$$

and the critical points of S are contractible loops in L which satisfy the Hamiltonian equation (36). Thus the critical points are precisely the required T-periodic solution of (36).

To describe the gradient of S we choose a on an almost complex structure on M which is compatible with ω. This is an endomorphism $J \in C^\infty(\text{End}(TM))$ satisfying $J^2 = -I$ so that

$$g(\xi, \eta) = \omega(\xi, J(x)\eta), \qquad \xi, \eta \in T_x M \tag{39}$$

defines a Riemannian metric on M. The Hamiltonian vector field is then represented by $X_H(t, x) = J(x)\nabla H(t, x)$, where ∇ denotes the gradient w.r.t. the x-variable using the metric (39). Moreover the gradient of S w.r.t. the induced metric on L is given by

$$\text{grad}S(\gamma) = J(\gamma)\dot{\gamma} + \nabla H(t, \gamma), \qquad \gamma \in L \quad . \tag{40}$$

Studying the critical points of S is confronted with the well-known difficulty that the variational integral is neither bounded from below nor from above. Moreover, at every possible critical point the Hessian of f has an infinite dimensional positive and an infinite dimensional negative subspace, so the standard Morse theory is not applicable. The additional problem is that the gradient vector field on the loop space L,

$$\frac{\mathrm{d}}{\mathrm{d}s}\gamma = -\text{grad}f(\gamma) \quad , \tag{41}$$

does not define a well-posed Cauchy problem. But Floer [15] found a way to analyse the space $\mathcal{M}$ of bounded solutions consisting of the critical points, together with their connecting orbits. He used a combination of a variational approach and Gromov's elliptic technique. A gradient flow line of f is a smooth solution $u : \mathbf{R} \to M$ of the partial differential equation

$$\frac{\partial u}{\partial s} + J(u)\frac{\partial u}{\partial t} + \nabla H(t, u) = 0, \tag{42}$$

which satisfies $u(s, t + T) = u(s, t)$. The key point is to consider (42) not as the flow on the loop space but as an elliptic boundary value problem. It should be noted that (42) is a generalization of an equation for Gromov's pseudoholomorphic curves (corresponding to the case $\nabla H = 0$ in (42)). Let $\mathcal{M}_T = \mathcal{M}_T(H, J)$, the space of bounded solutions of (42), i.e. the space of smooth functions $u : \mathbf{C}/iT\mathbf{Z} \to M$, which are contractible, solve equation (42) and have finite energy flow:

$$\Phi_T(u) = \frac{1}{2}\int_{-\infty}^{\infty}\int_0^T (|\frac{\partial u}{\partial s}|^2 + |\frac{\partial u}{\partial t} - X_H(t, u)|^2)\mathrm{d}t\mathrm{d}s \quad < \infty. \tag{43}$$

For every $u \in M_T$ there exists a pair x, y of contractible T-periodic solutions of (36), so that u is a connecting orbit from y to x:

$$\lim_{s \to -\infty} u(s, t) = y(t), \qquad \lim_{s \to +\infty} = x(t) \quad . \tag{44}$$

Then the approach from [1], which we may apply either on the level of standard boundary problem (42) or on the level of variational approach (43), and representation of operators (in our case, J and ∇) according to part III F (see below) lead us to wavelet representation of closed loops.

D   Quasiclassical Evolution

Let us consider classical and quantum dynamics in phase space $\Omega = R^{2m}$, with coordinates (x, ξ) and generated by Hamiltonian $\mathcal{H}(x, \xi) \in C^\infty(\Omega; R)$. If $\Phi_t^{\mathcal{H}} : \Omega \longrightarrow \Omega$ is (classical) flow, then time evolution of any bounded classical observable or symbol $b(x, \xi) \in C^\infty(\Omega, R)$ is given by $b_t(x, \xi) = b(\Phi_t^{\mathcal{H}}(x, \xi))$. Let $H = Op^W(\mathcal{H})$ and $B = Op^W(b)$ be the self-adjoint operators or quantum observables in $L^2(R^n)$, representing the Weyl quantization of the symbols $\mathcal{H}, b$ [14]

$$(Bu)(x) = \frac{1}{(2\pi\hbar)^n} \int_{R^{2n}} b\left(\frac{x+y}{2}, \xi\right) \cdot e^{i<(x-y),\xi>/\hbar} u(y) dy d\xi,$$

where $u \in S(R^n)$ and $B_t = e^{iHt/\hbar} B e^{-iHt/\hbar}$ is the Heisenberg observable or quantum evolution of the observable B under unitary group generated by H. B_t solves the Heisenberg equation of motion

$$\dot{B}_t = \frac{i}{\hbar}[H, B_t].$$

Let $b_t(x, \xi; \hbar)$ be a symbol of B_t; then we have the following equation for it:

$$\dot{b}_t = \{\mathcal{H}, b_t\}_M, \tag{45}$$

with initial condition $b_0(x, \xi, \hbar) = b(x, \xi)$. Here $\{f, g\}_M(x, \xi)$ are the Moyal brackets of the observables $f, g \in C^\infty(R^{2n})$, $\{f, g\}_M(x, \xi) = f \sharp g - g \sharp f$, where $f \sharp g$ is the symbol of the operator product and is presented by the composition of the symbols f, g,

$$(f \sharp g)(x, \xi) = \frac{1}{(2\pi\hbar)^{n/2}} \int_{R^{4n}} e^{-i<r,\rho>/\hbar + i<\omega,\tau>/\hbar} \cdot f(x+\omega, \rho+\xi) \cdot$$
$$g(x+r, \tau+\xi) d\rho d\tau dr d\omega.$$

For our problems it is useful that $\{f, g\}_M$ admits the formal expansion in powers of $\hbar$:

$$\{f, g\}_M(x, \xi) \sim \{f, g\} + 2^{-j} \sum_{|\alpha+\beta|=j\geq 1} (-1)^{|\beta|} \cdot (\partial_\xi^\alpha f D_x^\beta g) \cdot (\partial_\xi^\beta g D_x^\alpha f),$$

where $\alpha = (\alpha_1, \ldots, \alpha_n)$ is a multi-index, $|\alpha| = \alpha_1 + \ldots + \alpha_n$, $D_x = -i\hbar\partial_x$. So, evolution (45) for symbol $b_t(x, \xi; \hbar)$ is

$$\dot{b}_t = \{\mathcal{H}, b_t\} + \frac{1}{2^j} \sum_{|\alpha|+|\beta|=j\geq 1} (-1)^{|\beta|} \cdot \hbar^j (\partial_\xi^\alpha \mathcal{H} D_x^\beta b_t) \cdot (\partial_\xi^\beta b_t D_x^\alpha \mathcal{H}). \tag{46}$$

At $\hbar = 0$ this equation transforms to the classical Liouville equation

$$\dot{b}_t = \{\mathcal{H}, b_t\}. \tag{47}$$

Equation (46) plays a key role in many quantum (semiclassical) problems. We note only the problem of relation between quantum and classical evolutions or how long the evolution of the quantum observables is determined by the corresponding classical one [14]. Our approach to the solution of systems (46), (47) is based on our technique from [1]-[7] and the very useful linear parametrization for differential operators which we present in section III F.

E   SYMPLECTIC HILBERT SCALES VIA WAVELETS

We can solve many important dynamical problems so that KAM perturbations, spread of energy to higher modes, weak turbulence, growth of solutions of Hamiltonian equations only if we consider scales of spaces instead of one functional space. For the Hamiltonian system and its perturbations for which we need take into account the underlying symplectic structure, we need to consider symplectic scales of spaces. So, if $\dot{u}(t) = J\nabla K(u(t))$ is the Hamiltonian equation, we need a wavelet description of the symplectic or quasicomplex structure on the level of functional spaces. It is very important that, according to [16], Hilbert basis is in the same time a Darboux basis to the corresponding symplectic structure. We need to provide the Hilbert scale $\{Z_s\}$ with the symplectic structure [16], [17]. All we need is the following. J is a linear operator, $J : Z_\infty \to Z_\infty$, $J(Z_\infty) = Z_\infty$, where $Z_\infty = \cap Z_s$. J determines an isomorphism of the scale $\{Z_s\}$ of the order $d_J \geq 0$. The operator J with domain of definition Z_∞ is antisymmetric in Z: $< Jz_1, z_2 >_Z = - < z_1, Jz_2 >_Z$, $z_1, z_2 \in Z_\infty$. Then the triple

$$\{Z, \{Z_s | s \in R\}, \quad \alpha = < \bar{J}dz, dz >\}$$

is the symplectic Hilbert scale. So, we may consider any dynamical Hamiltonian problem on a functional level. As an example, for the KdV equation we have

$$Z_s = \{u(x) \in H^s(T^1) | \int_0^{2\pi} u(x)\mathrm{d}x = 0\}, \, s \in R, \quad J = \partial/\partial x \quad ;$$

J is isomorphism of the scale of order one, and $\bar{J} = -(J)^{-1}$ is isomorphism of order -1. According to [18], general functional spaces and scales of spaces such as Holder–Zygmund, Triebel–Lizorkin and Sobolev can be characterized through wavelet coefficients or wavelet transforms. As a rule, the faster the wavelet coefficients decay, the more the analyzed function is regular [18]. An example most important to us is the scale of Sobolev spaces. Let $H_k(R^n)$ be the Hilbert space of all distributions with finite norm

$$\|s\|^2_{H_k(R^n)} = \int \mathrm{d}\xi (1 + |\xi|^2)^{k/2} |\hat{s}(\xi)|^2.$$

Let us consider the wavelet transform

$$W_g f(b, a) = \int_{R^n} \mathrm{dx} \frac{1}{\mathrm{a^n}} \bar{\mathrm{g}} \left(\frac{\mathrm{x} - \mathrm{b}}{\mathrm{a}} \right) \mathrm{f(x)},$$

$b \in R^n$, $a > 0$, w.r.t. analyzing wavelet g, which is strictly admissible, i.e.

$$C_{g,g} = \int_0^\infty \frac{\mathrm{da}}{a} |\hat{g}(\bar{a}k)|^2 < \infty.$$

Then there is a $c \geq 1$ so that

$$c^{-1} \|s\|^2_{H_k(R^n)} \leq \int_{H^n} \frac{\mathrm{dbda}}{a} (1 + a^{-2\gamma}) |W_g s(b, a)|^2 \leq c \|s\|^2_{H_k(R^n)}.$$

This shows that localization of the wavelet coefficients at a small scale is linked to local regularity.

So, we need representation for a differential operator (J in our case) in the wavelet basis. We consider it in the next section.

F   FAST WAVELET TRANSFORM FOR DIFFERENTIAL OPERATORS

Let us consider multiresolution representation $\ldots \subset V_2 \subset V_1 \subset V_0 \subset V_{-1} \subset V_{-2} \ldots$ (see our other paper from this Proceedings for details of wavelet machinery). Let T be an operator $T : L^2(R) \to L^2(R)$, with the kernel $K(x, y)$ and $P_j : L^2(R) \to V_j$ ($j \in Z$) be the projection operators on the subspace V_j corresponding to j level of resolution:

$$(P_j f)(x) = \sum_k < f, \varphi_{j,k} > \varphi_{j,k}(x).$$

Let $Q_j = P_{j-1} - P_j$ be the projection operator on the subspace W_j; then we have the following "microscopic or telescopic" representation of operator T which takes into account contributions from each level of resolution from different scales, starting with the coarsest and ending to the finest scales:

$$T = \sum_{j \in Z} (Q_j T Q_j + Q_j T P_j + P_j T Q_j).$$

We remember that this is the result of the presence of the affine group inside this construction. The non-standard form of operator representation [19] is a representation of an operator T as a chain of triples $T = \{A_j, B_j, \Gamma_j\}_{j \in Z}$, acting on the subspaces V_j and W_j:

$$A_j : W_j \to W_j, B_j : V_j \to W_j, \Gamma_j : W_j \to V_j,$$

where operators $\{A_j, B_j, \Gamma_j\}_{j \in Z}$ are defined as

$$A_j = Q_j T Q_j, \quad B_j = Q_j T P_j, \quad \Gamma_j = P_j T Q_j.$$

The operator T admits a recursive definition via

$$T_j = ($$

$A_{j+1} B_{j+1}$
$\Gamma_{j+1} T_{j+1},$
where $T_j = P_j T P_j$ and T_j works on $V_j : V_j \rightarrow V_j$. It should be noted that operator A_j describes interaction on the scale j independently from other scales, operators B_j, Γ_j describe interaction between the scale j and all coarser scales, the operator T_j is an "averaged" version of T_{j-1}.

The operators A_j, B_j, Γ_j, T_j are represented by matrices $\alpha^j, \beta^j, \gamma^j, s^j$

$$\alpha^j_{k,k'} = \int \int K(x,y) \psi_{j,k}(x) \psi_{j,k'}(y) \mathrm{dxdy}$$

$$\beta^j_{k,k'} = \int \int K(x,y) \psi_{j,k}(x) \varphi_{j,k'}(y) \mathrm{dxdy} \qquad (48)$$

$$\gamma^j_{k,k'} = \int \int K(x,y) \varphi_{j,k}(x) \psi_{j,k'}(y) \mathrm{dxdy}$$

$$s^j_{k,k'} = \int \int K(x,y) \varphi_{j,k}(x) \varphi_{j,k'}(y) \mathrm{dxdy} \quad .$$

We may compute the non-standard representations of operator d/dx in the wavelet bases by solving a small system of linear algebraical equations. So, we have for objects (48)

$$\alpha^j_{i,\ell} = 2^{-j} \int \psi(2^{-j}x - i) \psi'(2^{-j} - \ell) 2^{-j} \mathrm{dx} = 2^{-j} \alpha_{i-\ell}$$

$$\beta^j_{i,\ell} = 2^{-j} \int \psi(2^{-j}x - i) \varphi'(2^{-j}x - \ell) 2^{-j} \mathrm{dx} = 2^{-j} \beta_{i-\ell}$$

$$\gamma^j_{i,\ell} = 2^{-j} \int \varphi(2^{-j}x - i) \psi'(2^{-j}x - \ell) 2^{-j} \mathrm{dx} = 2^{-j} \gamma_{i-\ell},$$

where

$$\alpha_\ell = \int \psi(x - \ell) \frac{\mathrm{d}}{\mathrm{dx}} \psi(x) \mathrm{dx}$$

$$\beta_\ell = \int \psi(x - \ell) \frac{\mathrm{d}}{\mathrm{dx}} \varphi(x) \mathrm{dx}$$

$$\gamma_\ell = \int \varphi(x - \ell) \frac{\mathrm{d}}{\mathrm{dx}} \psi(x) \mathrm{dx} \quad ;$$

then by using refinement equations,

$$\varphi(x) = \sqrt{2} \sum_{k=0}^{L-1} h_k \varphi(2x - k),$$

$$\psi(x) = \sqrt{2} \sum_{k=0}^{L-1} g_k \varphi(2x - k),$$

$g_k = (-1)^k h_{L-k-1}, \quad k = 0, \ldots, L-1$ we have in terms of filters (h_k, g_k):

$$\alpha_j = 2 \sum_{k=0}^{L-1} \sum_{k'=0}^{L-1} g_k g_{k'} r_{2i+k-k'},$$

$$\beta_j = 2 \sum_{k=0}^{L-1} \sum_{k'=0}^{L-1} g_k h_{k'} r_{2i+k-k'},$$

$$\gamma_i = 2 \sum_{k=0}^{L-1} \sum_{k'=0}^{L-1} h_k g_{k'} r_{2i+k-k'},$$

where

$$r_\ell = \int \varphi(x - \ell) \frac{d}{dx} \varphi(x) dx, \ell \in Z.$$

Therefore, the representation of d/dx is completely determined by the coefficients r_ℓ or by representation of d/dx only on the subspace V_0. The coefficients $r_\ell, \ell \in Z$ satisfy the following system of linear algebraical equations:

$$r_\ell = 2 \left[r_{2l} + \frac{1}{2} \sum_{k=1}^{L/2} a_{2k-1} (r_{2\ell-2k+1} + r_{2\ell+2k-1}) \right] ,$$

and $\sum_\ell \ell r_\ell = -1$, where $a_{2k-1} = 2 \sum_{i=0}^{L-2k} h_i h_{i+2k-1}, \ k = 1, \ldots, L/2$ are the autocorrelation coefficients of the filter H. If a number of vanishing moments $M \geq 2$ then this linear system of equations has a unique solution with a finite number of non-zero r_ℓ, $r_\ell \neq 0$ for $-L + 2 \leq \ell \leq L - 2, r_\ell = -r_{-\ell}$. For the representation of operator d^n/dx^n we have the similar reduced linear system of equations. Then finally we have for the action of operator $T_j(T_j : V_j \rightarrow V_j)$ on sufficiently smooth function f:

$$(T_j f)(x) = \sum_{k \in Z} \left(2^{-j} \sum_\ell r_\ell f_{j,k-\ell} \right) \varphi_{j,k}(x),$$

where $\varphi_{j,k}(x) = 2^{-j/2} \varphi(2^{-j} x - k)$ is the wavelet basis and

$$f_{j,k-1} = 2^{-j/2} \int f(x) \varphi(2^{-j} x - k + \ell) dx$$

are wavelet coefficients. So, we have simple linear parametrization of the matrix representation of our differential operator in the wavelet basis and of the action of this operator on the arbitrary vector in our functional space. Then we may use such representation in all preceding sections.

IV MAPS AND WAVELET STRUCTURES

A   Veselov-Marsden Discretization

Discrete variational principles lead to evolution dynamics analogous to the Euler-Lagrange equations [9]. If Q is a configuration space, then a discrete Lagrangian is a map $L : Q \times Q \rightarrow \mathbf{R}$. usually L is obtained by approximating the given Lagrangian. For $N \in N_+$ the action sum is the map $S : Q^{N+1} \to \mathbf{R}$ defined by

$$S = \sum_{k=0}^{N-1} L(q_{k+1}, q_k), \qquad (49)$$

where $q_k \in Q$, $k \geq 0$. The action sum is the discrete analog of the action integral in the continuous case. Extremizing S over $q_1, ..., q_{N-1}$ with fixing q_0, q_N, we have the discrete Euler-Lagrange equations (DEL):

$$D_2 L(q_{k+1}, q_k) + D_1(q_k, q_{q-1}) = 0, \qquad (50)$$

for $k = 1, ..., N - 1$.
    Let

$$\Phi : Q \times Q \to Q \times Q \quad , \qquad (51)$$

and

$$\Phi(q_k, q_{k-1}) = (q_{k+1}, q_k) \qquad (52)$$

is a discrete function (map), then we have for DEL:

$$D_2 L \circ \Phi + D_1 L = 0 \quad , \qquad (53)$$

or in coordinates q^i on Q we have DEL

$$\frac{\partial L}{\partial q_k^i} \circ \Phi(q_{k+1}, q_k) + \frac{\partial L}{\partial q_{k+1}^i}(q_{k+1}, q_k) = 0. \qquad (54)$$

It is very important that the map Φ exactly preserve the symplectic form ω:

$$\omega = \frac{\partial^2 L}{\partial q_k^i \partial q_{k+1}^j}(q_{k+1}, q_k) \mathrm{d}q_k^i \wedge \mathrm{d}q_{k+1}^j \quad . \qquad (55)$$

B   Generalized Wavelet Approach

Our approach to the solutions of Equation (54) is based on applications of general and very efficient methods developed by A. Harten [20], who produced a "General Framework" for multiresolution representation of discrete data. It is based on consideration of basic operators, decimation and prediction, which connect adjacent resolution levels. These operators are constructed from two basic blocks: the discretization and reconstruction operators. The former obtains discrete information from given continuous functions (flows); and the latter produces an approximation to those functions, from discrete values, in the same function space to which the original function belongs. A "new scale" is defined as the information on a given resolution level which cannot be predicted from discrete information at lower levels. If the discretization and reconstruction are local operators, the concept of "new scale" is also local. The scale coefficients are directly related to the prediction errors, and thus to the reconstruction procedure. If scale coefficients are small at a certain location on a given scale, it means that the reconstruction procedure on that scale gives a proper approximation of the original function at that particular location. This approach may be considered as some generalization of standard wavelet analysis approach. It allows the consideration of multiresolution decomposition when the usual approach is impossible (δ-functions case).

Let F be a linear space of mappings

$$F \subset \{f | f : X \to Y\}, \tag{56}$$

where X, Y are linear spaces. Let also D_k be a linear operator

$$D_k : f \to \{v^k\}, \quad v^k = D_k f, \quad v^k = \{v_i^k\}, \quad v_i^k \in Y. \tag{57}$$

This sequence corresponds to k level discretization of X. Let

$$D_k(F) = V^k = \text{span}\{\eta_i^k\} \tag{58}$$

and the coordinates of $v^k \in V^k$ in this basis are $\hat{v}^k = \{\hat{v}_i^k\}$, $\hat{v}^k \in S^k$:

$$v^k = \sum_i \hat{v}_i^k \eta_i^k, \tag{59}$$

D_k is a discretization operator. The main goal is to design a multiresolution scheme (MR) [20] that applies to all sequences $s \in S^L$, but corresponds for those sequences $\hat{v}^L \in S^L$, which are obtained by the discretization (56).

Since D_k maps F onto V^k then for any $v^k \subset V^k$ there is at least one f in F so that $D_k f = v^k$. Such correspondence from $f \in F$ to $v^k \in V^k$ is reconstruction and the corresponding operator is the reconstruction operator R_k:

$$R_k : V_k \to F, \qquad D_k R_k = I_k, \tag{60}$$

where I_k is the identity operator in V^k (R^k is right inverse of D^k in V^k).

Given a sequence of discretization $\{D_k\}$ and sequence of the corresponding reconstruction operators $\{R_k\}$, we define the operators D_k^{k-1} and P_{k-1}^k

$$D_k^{k-1} = D_{k-1} R_k : V_k \to V_{k-1} \tag{61}$$
$$P_{k-1}^k = D_k R_{k-1} : V_{k-1} \to V_k \quad .$$

If the set D_k in nested [20], then

$$D_k^{k-1} P_{k-1}^k = I_{k-1} \quad , \tag{62}$$

and we have for any $f \in F$ and any $p \in F$ for which the reconstruction R_{k-1} is exact:

$$D_k^{k-1}(D_k f) = D_{k-1} f \tag{63}$$
$$P_{k-1}^k(D_{k-1} p) = D_k p \quad .$$

Let us consider any $v^L \in V^L$, Then there is $f \in F$ so that

$$v^L = D_L f, \tag{64}$$

and it follows from (63) that the process of successive decimation [20]

$$v^{k-1} = D_k^{k-1} v^k, \qquad k = L, ..., 1 \tag{65}$$

yields for all k

$$v^k = D_k f \quad . \tag{66}$$

Thus the problem of prediction, which is associated with the corresponding MR scheme, can be stated as a problem of approximation: knowing $D_{k-1} f$, $f \in F$, find a "good approximation" for $D_k f$. It is very important that each space V^L has a multiresolution basis

$$\bar{B}_M = \{\bar{\phi}_i^{0,L}\}_i, \{\{\bar{\psi}_j^{k,L}\}_j\}_{k=1}^L \quad , \tag{67}$$

and that any $v^L \in V^L$ can be written as

$$v^L = \sum_i \hat{v}_i^0 \bar{\phi}_i^{0,L} + \sum_{k=1}^L \sum_j d_j^k \bar{\psi}_j^{k,L}, \tag{68}$$

where $\{d_j^k\}$ are the k scale coefficients of the associated MR, and $\{\hat{v}_i^0\}$ is defined by (59) with $k = 0$. If $\{D_k\}$ is a nested sequence of discretization [20] and $\{R_k\}$ is any corresponding sequence of linear reconstruction operators, then we have from (68) for $v^L = D_L f$ applying R_L:

$$R_L D_L f = \sum_i \hat{f}_i^0 \phi_i^{0,L} + \sum_{k=1}^L \sum_j d_j^k \psi_j^{k,L}, \tag{69}$$

where

$$\phi_i^{0,L} = R_L \bar{\phi}_i^{0,L} \in F, \quad \psi_j^{k,L} = R_L \bar{\psi}_j^{k,L} \in F, \quad D_0 f = \sum \hat{f}_i^0 \eta_i^0. \tag{70}$$

When $L \to \infty$, we have sufficient conditions which ensure that the limiting process $L \to \infty$ in (69, 70) yields a multiresolution basis for F. Then, according to (67), (68) we have a very useful representation for solutions of equations (54) or for different map construction in the forms which are counterparts for discrete (different) cases of construction from paper [1].

ACKNOWLEDGEMENTS

We would like to thank Professors M. Cornacchia, C. Pellegrini, L. Palumbo, Mrs. M. Laraneta, J. Kono, and G. Nanula for hospitality, help, support before and during the Arcidosso meeting, and all participants for interesting discussions. We are very grateful to Professor M. Cornacchia, Mrs. J. Kono and M. Laraneta, because without their permanent encouragement this paper would not have been written.

REFERENCES

1. Fedorova, A.N., Zeitlin, M.G., and Parsa, Z., 'Variational Approach in Wavelet Framework to Polynomial Approximations of Nonlinear Accelerator Problems', this volume.
2. Fedorova, A.N., and Zeitlin, M.G., 'Wavelets in Optimization and Approximations', *Math. and Comp. in Simulation*, **46**, 527-534 (1998).
3. Fedorova, A.N., and Zeitlin, M.G., 'Wavelet Approach to Polynomial Mechanical Problems', New Applications of Nonlinear and Chaotic Dynamics in Mechanics, Kluwer, 101-108, (1998).
4. Fedorova, A.N., and Zeitlin, M.G., 'Wavelet Approach to Mechanical Problems. Symplectic Group, Symplectic Topology and Symplectic Scales', New Applications of Nonlinear and Chaotic Dynamics in Mechanics, Kluwer, 31-40, 1998.
5. Fedorova, A.N., and Zeitlin, M.G., 'Nonlinear Dynamics of Accelerator via Wavelet Approach', AIP Conf. Proc., vol. 405, ed. Z. Parsa, 87-102, 1997, Los Alamos preprint, physics/9710035.
6. Fedorova, A.N., Zeitlin, M.G, and Parsa, Z., 'Wavelet Approach to Accelerator Problems', parts 1-3, Proc. PAC97 vol. **2**, 1502-1504, 1505-1507, 1508-1510, IEEE, 1998.
7. Fedorova, A.N., Zeitlin, M.G, and Parsa, Z., 'Nonlinear Effects in Accelerator Physics: from Scale to Scale via Wavelets', 'Wavelet Approach to Hamiltonian, Chaotic and Quantum Calculations in Accelerator Physics', Proc. EPAC'98, 930-932, 933-935, Institute of Physics, 1998.
8. Dragt, A.J., Lectures on Nonlinear Dynamics, CTP, 1996.
9. Marsden, J.E., Park City Lectures on Mechanics, Dynamics and Symmetry, CALTECH, 1998.
10. Antoine, J.-P., UCL, 96-08, 95-02.
11. Kalisa, C., and Torresani, B., CPT-92 P.2811 Marseille, 1992.
12. Kawazoe, T.: Proc. Japan Acad. **71** Ser. A, 1995, p. 154.
13. Lemarie P.G.: Proc. Int. Math. Congr., Satellite Symp., 1991, p. 154.
14. G.B. Folland, 'Harmonic Analysis in Phase Space', Princeton, 1989.
15. Hofer, E., Zehnder, E., *Symplectic Topology*: Birkhauser, 1994.
16. Kuksin, S., 'Nearly Integrable Hamiltonian Systems', Springer, 1993.
17. Bourgain, J., IMRN, 6, 277-304, 1996.
18. Holschneider, M., CPT-96/P3344, Marseille.
19. Beylkin, G., Coifman, R. R., and Rokhlin, V., Comm. Pure and Appl. Math, **44**, 141-183, 1991.
20. Harten, A., *SIAM J. Numer. Anal.*, **31**, 1191-1218, 1994.

A Synthetic Geometrical Approach To Betatronic Motion: Some Remarks

Paolo Freguglia*

*Dipartimento di Scienze,Università di Chieti-Pescara, Viale Pindaro,42
I - 65127 PESCARA, Italy
*Domus Galilaena, Via S.Maria, 26   I - 56126 PISA, Italy

Abstract. This contribution examines some themes for an easy computer approach to betatronic motion. A synthetic geometrical model of this motion is presented. I shall take the starting-point for my proposal from geometrical optical simulation.

1. Optical Path in a Circular Crown: Complex Numbers

In a circular crown we can consider a certain finite number of (sections of) instruments of refraction. For instance we can consider (sections) of optical prisms like instruments of refraction. At first we consider a *fixed in advance optical path* $\wp$ (succession of optical beams) which is established by n made firm optical instruments which are distributed symmetrically in a circular crown. This fixed optical path is a vectorial closed polygon (see Fig.1 for $n = 8$). If these optical instruments undergo some little positional (by rotations) variations, these influence the refraction and hence these perturb the optical fixed before-hand optical path. The above mentioned variations are performed by the angles α_K which depend respectively on optical instruments of refraction. These angles α_k are added or subtracted by the fixed angle δ (see Figure 2). In this case we will call

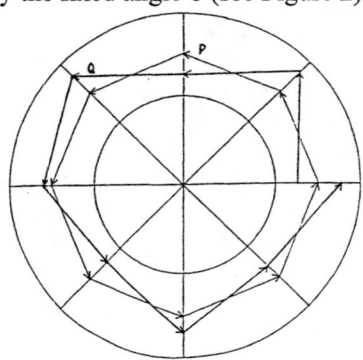

FIGURE 1.

CP468, *Nonlinear and Collective Phenomena in Beam Physics–1998 Workshop,*
edited by S. Chattopadhyay, M. Cornacchia, and C. Pellegrini

this new optical path, which is an open vectorial polygon, *real optical path 2* relating to ℘ The optical paths are determined by vertexes of their polygons.

In what follows we leave out of consideration the connected to insertion of lenses problems (to focus). These problems can be examined afterwards. It is easy to demonstrate the following:

Prop.1: *If are given*:
   n : *number* ($\in$ **N**) *which divides* 2π
   $\{\alpha_k\}$: *finite succession* (n *elements*) *of refraction angles relating to* n *optical instruments*,

then the vertexes of the fixed (polygonal) optical path ℘ *in a circular crown is determined by the solutions* $z_k \in$ **C** *of the following algebraic equation*:

$$z^n + b = 0 \tag{1.1}$$

where (see Fig.2) $z_k = {}^n\sqrt{(-b)} = |OA_0| \exp i\theta_k$, and $|OA_0| = |OA_1| = ... = |OA_n|$,
with $\theta_k = 2k\pi/n$ *and* $k = 0, 1, ..., n-1$, *while the real optical path 2 relating to* ℘ *determined by the solutions* $z_k \in$ **C** *of the algebraic equation*:

$$z^n + a_{(n)} = 0 \tag{1.2}$$

where $z_k = {}^n\sqrt{(-a_{(n)})} = |OT_k| \exp i\theta_k$ *with* $\theta_k = 2k\pi/n$ *and* $k = 0,1..., n-1$ *and*

$$|OT_{k\_} = [\_OT_{k-1}| \sin(((n-2)\_/2n) + \alpha_k)] / [\sin(((n-2)\pi/2n) - \alpha_k)] \tag{1.3}$$

or $|OT_k| = [|OT_{k-1}| \sin(((n-2)\pi/2n) - \alpha_k)] / [\sin(((n-2)\pi/2n) + \alpha_k)$

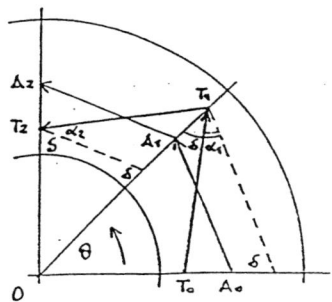

FIGURE 2.

We can call *space of phases for* $\mathcal{2}$ the Euclidean space $\mathbf{R}^2$ where we have the following phase space co-ordinates:

$$x_k \equiv (x_{1k}, x_{2k}), \quad \text{where: } x_{1k} = |OT_k| \quad \text{and} \quad x_{2k} = \alpha_k \quad (1.4)$$

with $k = 0, 1, ..., n - 1$.

In this space it is possible to introduce the *transfer map* $M^{(k)}$ which is a function that transforms the phase space co-ordinates x_{k-1} to x_k according to equation (2.2).

2. Optical Path in a Torus: Quaternions

In three-dimensional case a torus (doughnut) corresponds to the circular crown (two-dimensional case), and the circular crown can be seen as an horizontal section of a torus. Every vector v_k belonging to $\mathcal{p}$ (reference path) with the corresponding vector u_k belonging $\mathcal{2}$ (real path) lie from time to time on generally different plane (see Fig.3). In this case also we consider the vertexes of vectorial polygons. Hence we can utilise the quaternions:

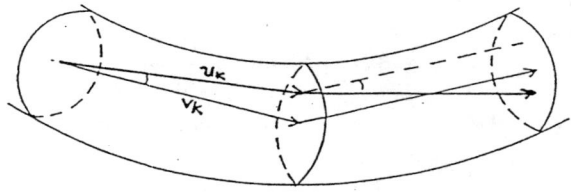

FIGURE 3.

$$q_k = a_k + b_k \mathbf{i} + c_k \mathbf{j} + d_k \mathbf{k} = Sq_k + Vq_k \qquad (2.1)$$

with $a_k, b_k, c_k, d_k \in \mathbf{R}$ and $\mathbf{i}, \mathbf{j}$ and $\mathbf{k}$ are versors, where $a_k = Sq_k$ (*scalar part* of quaternion q_k), $b_k \mathbf{i} + c_k \mathbf{j} + d_k \mathbf{k} = Vq_k$ (*vectorial part* of quaternion q_k) and $k = 0, 1, ..., n-1$. We remember that (see [2]) $\mathbf{Q}$ (where $q_k \in \mathbf{Q}$) is a (right) vectorial 2-dimensional space on $\mathbf{C}$. The basis of $\mathbf{Q}$ is given by two elements 1 and j. Hence elements of $\mathbf{Q}$ have the form $q = 1z + jw = z + jw$, where $z, w \in \mathbf{C}$. Moreover the value of expression (2.1) depend on the optical instrument of refraction. We know that if $v_k \in \mathcal{p}$ and $u_k \in \mathcal{2}$ then:

$$\forall k \qquad q_k = - u_k \cdot v_k + u_k \times v_k \qquad k = 0,1,..., n-1 \qquad (2.2)$$

It is well known that:

$$S_{qk} = |u_k| / |v_k| \qquad (2.3)$$

and

$$V_{qk} \text{ determines three angles: } (\alpha_k, \beta_k, \gamma_k) \qquad (2.4)$$

where α_k is the angle between the two vectors u_k and v_k. β_k and γ_k determine the plane where u_k and v_k lie.

We have the following:

Prop.2: $\{q_k\}$ *determines a manifold of vectors-points in* $\mathbf{R}^4$

It can be interesting to study the geometry of $\{q_k\}$ in $\mathbf{R}^4$ in connection with the positions in $\mathbf{R}^3$ of p and of $\mathcal{Q}$.

Finally we have the:

Prop.3: *If* $\mathcal{P} \equiv \{v_k\}$ *and* $\{q_k\}$ *with* k = 0, 1, ..., n - 1 *are given then we can determine the vertexes of* $\mathcal{Q} \equiv \{u_k\}$ *with* k = 0, 1, ..., n - 1.

As it is well known (see [2]), if we define in **Q** the product:

$$qr = (z + jw)(z_1 + jw_1) = (zz_1 - w^*w_1) + j(z^*w_1 + wz_1)$$

where $r = z_1 + jw_1$ is an other element of **Q**, and w^*, z^* are the conjugates of w and z, then **Q** is a ring with division. **C** is subring of **Q** and $xq = qx$ for every $q \in \mathbf{Q}$ and $x \in \mathbf{R}$.

In this case the *phase space* is a $\mathbf{R}^4$ and the phase space co-ordinates, according to (2.3) and (2.4), are:

$$x1_k = S_{qk} \ , \ x2_k = \alpha_k \ , \ x3_k = \beta_k \ , \ x4_k = \gamma_k \qquad (2.5)$$

3. The Betatronic Motion: Analogies

If we consider the well known equations pertinent to betatronic motion written like difference equations (either for the linear case or for nonlinear case), that is:

$$x_{k+2} - 2x_{k+1} + (1 - (1 - k_{1k}\, \rho^2_k\,)/\, \rho^2_k\,)x_k = F(x^n_{\ k}) \text{ or } = 0 \qquad (3.1.1)$$

$$(3.1)$$

$$y_{k+2} - 2y_{k+1} + (1 - k_{1k})y_k = G(x^n_{\ k}) \text{ or } = 0 \qquad (3.1.2)$$

with $n \geq 2$ (in the case of circular crown we have only the first (3.1.1) of the (3.1)) then we observe analogies with the optical examined situations. This fact is well known, but on our opinion the geometrical methods which we have proposed are interesting. Indeed the equation (3.1.1), in the nonlinear case, can be written in the in the following form:

$$p_{k+1} - p_k + ((1 - k_{1k}\, \rho^2_k\,)/\, \rho^2_k\,)x_k = F(x^n_{\ k}) \qquad (3.2.1)$$

$$(3.2)$$

$$x_{k+1} - x_k - p_k = 0 \qquad (3.2.2)$$

where x_k is on the horizontal axe perpendicular to the orbit, k is the curvilinear co-ordinate and $x_{k+1} - x_k = p_k$ is the momentum. Analogously for the (3.1.2), where y_k lies on the vertical axe perpendicular to the orbit, and for the linear cases. We have the following correspondences with the optical simulation, as regards with the space of phase:

- *circular crown case*:

we have a two dimensional phase space where:

$$x_k \quad \text{corresponds to } |OT_k| \text{ (position)}$$

$$p_k \quad \text{corresponds to } \alpha_k \quad \text{(velocity)}$$

- *torus case*:

we have a four-dimensional phase space:

x_{1k}, x_{2k} (positions) correspond to suitable projections, on the co-ordinate planes, of the two vectors u_k and v_k. These projections depend on Sq_k and α_k.

p_{1k}, p_{2k} (momenta) correspond to suitable projections, on the co-ordinate planes, of difference of two vectors u_k and v_k. These projections depend on β_k and γ_k.

4. The Linear Case

When we consider the linear case, for the optical path we can have (see Fig.4) the following successions:

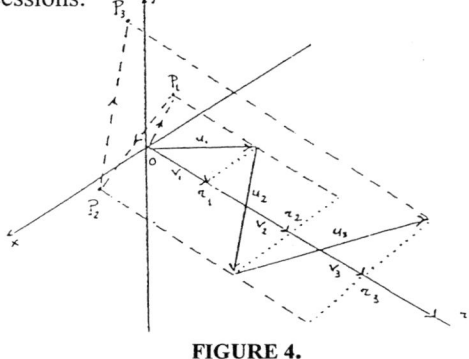

FIGURE 4.

$$0 \quad 1 \quad 2 \quad 3 \quad 4 \quad 5 \quad ....$$

$$\{v_k\} \equiv \{r_k\}: \quad 0 \quad r_1 \quad r_2 \quad r_3 \quad r_4 \quad r_5 \quad ....$$

$$\{P_k\}: P_0 \quad P_1 \quad P_2 \quad P_3 \quad P_4 \quad P_5 \quad ....$$

where $P_k \equiv (x_k, y_k)$. Therefore we have:

Prop.4: *If the succession $\{r_k\}$ determines $\wp \equiv \{r_k\}$ and the succession $\{P_k\}$ is given, then the succession $\mathcal{Z} \equiv \{q_k\}$ is determined.*

Now if for the betatronic motion we consider, for example, the case of <u>the linear accelerator with everywhere constant magnetic field</u>, the equations (3.1) become:

$$x_{k+2} - 2x_{k+1} + (1 - \rho)x_k = 0 \qquad (4.1.1)$$
$$\qquad\qquad (4.1)$$
$$y_{k+2} - 2y_{k+1} + (1 - \rho)y_k = 0 \qquad (4.1.2)$$

where ρ is constant. For $\rho = 1$, for instance, the (4.1.1), and analogously the (4.1.2), becomes:

$$x_{k+2} - x_{k+1} - (x_{k+1} - x_k) = - x_k \qquad (4.2.1)$$

If we suppose: $x_{k+1} - x_k = p_k$, then we can write the (4.2.1) so:

$$x_{k+1} - x_k = p_k$$
$$\qquad\qquad (4.3.1)$$
$$p_{k+1} - p_k = - x_k$$

99

by multiplying the second equation of (4.3.1) by i and afterward by adding member to member the equations (4.3.1) we have:

$$x_{k+1} + ip_{k+1} - (x_k + ip_k) = -i(x_k + ip_k) \qquad (4.4.1)$$

If we suppose $x_k + ip_k = z_k$ then we can write the (4.4.1) so:

$$z_{k+1} - z_k = -iz_k \qquad (4.5.1)$$

Analogously for the (4.1.2), if we suppose $y_{k+1} - y_k = r_k$ and $y_k + ir_k = w_k$, we have:

$$w_{k+1} - w_k = -iw_k \qquad (4.5.2)$$

The system of equations (4.5.1) and (4.5.2) gives in $\mathbf{Q}$ a difference equations of quaternions. In fact by multiplying the (4.5.2) by j and by adding member to member the equations (4.5.1) and (4.5.2), we have:

$$q_{k+1} - q_k = -iq_k \qquad (4.6)$$

which can be seen as a vectorial difference equation in $\mathbf{R}^4$ (where $i \equiv \mathbf{i}$ and $j \equiv \mathbf{j}$ and $\mathbf{k} = \mathbf{ij}$).

5. ACKNOWLEDGMENTS

I should like to thank Armando Bazzani (Univ.Bologna, INFN) and Giorgio Turchetti (Univ.Bologna, INFN); with them I have discussed the themes of the present paper.

6. REFERENCES

1. Bazzani, A., Todesco, E., Turchetti, G., Servizi, G., *A normal form approach to the theory of nonlinear betatronic motion*, CERN, Geneva, 1994.
2. Mac Lane, S., Birkhoff, G., *Algebra*, Mac Millan Comp., New York, 1967.
3. Mickens, R.E., *Difference Equations, Theory and Applications*, VNR, New York, 1990.
4. Tait, P.G., *Traité élémentaire des quaternions,* (French tr. by G.Plarr), Gouthier-Villars, Paris, 1882.

Beam tracking simulation for an rf gun apparatus in the SPring-8

Akihiko Mizuno, Tsutomu Taniuchi, Shinsuke Suzuki, Kenichi Yanagida, Hiroshi Abe, Takao Asaka, Hideaki Yokomizo, and Hirofumi Hanaki

Japan Synchrotron Radiation Research Institute,(SPring-8)
Mikazuki, Sayo, Hyogo 679-5198, Japan

Abstract. A photo cathode rf gun has been studied in the SPring-8 Linac to obtain a lower emittance beam. In order to perform a comparison with the beam characteristics, a beam tracking simulation code for an rf gun test apparatus has also been developed. In this code, the electric and magnetic forces between all particles are calculated. Accordingly, a lot of time is required, but a high accuracy can be expected in the calculations in comparison with other codes like PIC.

In this paper, we describe the above rf gun test apparatus in the SPring-8, the simulation code, and present some calculation results such as the emittance of the emitted beam compared with MAFIA's data.

INTRODUCTION

In the SPring-8, we are developing a photo cathode rf gun for the conventional injector of the linac, and also for future applications such as single pass FEL based on the SASE. For the SASE, in particular, a lower emittance of around several πmm·mrad and a high peak current of around several hundred amperes are required simultaneously. In order to achieve these values, some calculation codes able to predict the beam characteristics as accuracy as possible are needed. Ready-made tracking codes, however, are not considered to be enough, because they include a lot of assumptions. Therefore, we are developing our own simulation code which is suitable for our rf gun system and includes as few assumptions as possible. As the first step for the above photo cathode rf gun system, we have decided to design the rf cavity in a very simple manner to make an easy comparison with the simulation code.

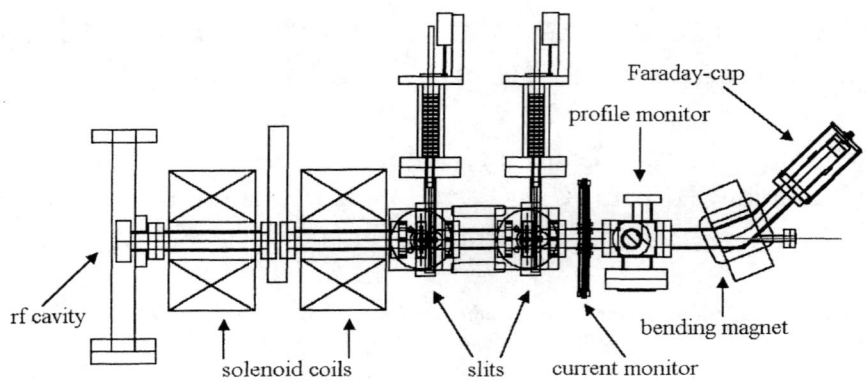

FIGURE 1. The rf gun test apparatus in the SPring-8 Linac

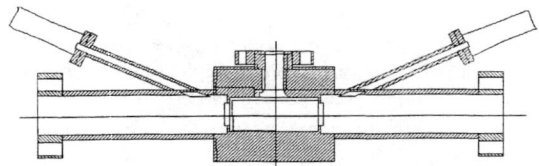

FIGURE 2. Cross section schematic of an rf cavity. There are two windows for laser injection. The incident angle is 24 degrees.

THE TEST APPARATUS

Figure 1 shows the test apparatus employed for the rf gun experiments. Two slits and a current monitor are used for emittance measurement, a bending magnet and a Faraday-cup are used for energy analysis. The distance from the cathode to the Faraday-cup is about 2 meters.

The rf cavity [1] is shown in Figure 2. The reason for choosing a single cell cavity is that the field distribution is simpler than that for a multi cell cavity and it is easier to compare with simulations. For the same reason, the bottom side of the cavity in Figure 2, which is made of copper, is used as the cathode. The accelerating gap was designed by MAFIA so that the emittance becomes minimum. This cavity has two coupling ports to improve the field symmetry and shorten the filling time. The rf from the right hand side port travels to the opposite port and is fed to a dummy load.

One of the purposes of these experiments is to measure the emittance of the emitted beam. The emittance just after the rf cavity, however, is unable to be measured because of beam expansion caused by space charge effects. It is therefore measured by the two slits mounted after the two solenoid coils. Consequently however, the

2D-emittance is varied by the solenoid fields. If we discuss the characteristics of the rf cavity we also have to discuss the solenoid field effects.

Experiments for this test apparatus are scheduled to start from autumn 1998.

THE BEAM TRACKING CODE

The beam tracking code is designed to be very simple. Because of recent progress concerning the CPU speed, we try to calculate all space charge effects for each electron. However, we assume that the charge and mass for each electron are larger than for real electrons. If we assume uniform motion for the electrons, the electric and magnetic fields at point A caused by electron B are expressed as follows;

$$\mathbf{E}_A = \frac{1}{4\pi\epsilon_0\gamma^2} \frac{-e\mathbf{r}}{\left[|\mathbf{r}|^2 - \frac{|\mathbf{v}_B \times \mathbf{r}|^2}{c^2}\right]^{3/2}} \qquad \mathbf{B}_A = \frac{1}{c^2}\mathbf{v}_B \times \mathbf{E}_A \qquad (1)$$

where $\mathbf{r}$ is the vector from B to A, $\mathbf{v}_B$ is the velocity of the electron B, and γ is the relative factor of electron B. These fields act on electron A as follows;

$$\mathbf{F}_A = -e\left(\mathbf{E}_A + \mathbf{v}_A \times \mathbf{B}_A\right) \qquad (2)$$

The equation of motion for each electron is derived to the following equation and becomes adaptable for the Runge-Kutta method.

$$-e\left(\mathbf{v} \times \mathbf{B} + \mathbf{E}\right) = m_0 \frac{d\left(\gamma\mathbf{v}\right)}{dt} \quad \Longrightarrow \quad \frac{d\mathbf{v}}{dt} = -\frac{e}{\gamma m_0}\left(\mathbf{v} \times \mathbf{B} + \mathbf{E} - \frac{(\mathbf{v} \cdot \mathbf{E})}{c^2}\mathbf{v}\right)$$

$$(3)$$

The electromagnetic fields in the cavity are calculated using MAFIA and are included in the tracking code. The magnetic fields of the two solenoid coils are also calculated in our code. We only consider the above three effects as fields.

In a typical calculation, the number of electrons is set to 5000. Using a CPU of 333MHz on a DEC Alpha 21164, the elapsed time for 0.1 meter tracking is about two hours, with a time step of 1ps.

SIMULATION

Beam tracking simulations in the rf cavity were carried out using both MAFIA and our tracking code.

At first, using MAFIA, the beam characteristics were surveyed by changing an initial rf phase at which the top of the bunch was on the cathode. It was found that the emittance became minimum at an initial rf phase of 30 degrees. The left side of Figure 3 shows the calculated normalized emittance using MAFIA, as a function of the distance from the cathode. In the figure, the exit of the rf cavity is 0.075m

TABLE 1. Parameters used in MAFIA's calculation

Charge per Bunch	[nC]	1
Transverse Profile		Uniform
Spot Size on the Cathode	[mm]	1
Longitudinal Profile		Uniform
Bunch Length	[ps]	10
Initial Emittance	[πmm·mrad]	0
Initial rf Phase	[degree]	30
The Number of Macro Particles		10000

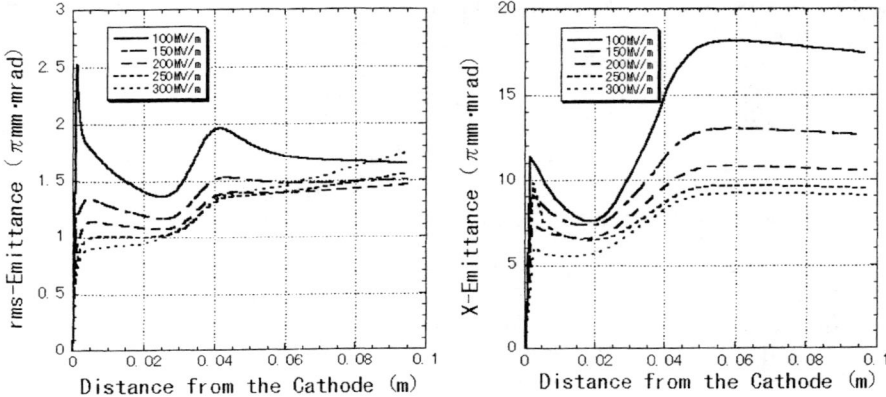

FIGURE 3. Calculated emittance in the rf cavity, as a function of the distance from the cathode. The left hand side graph is calculated by MAFIA, and the right hand side graph is calculated by our tracking code.

from the cathode. The calculations were carried out by changing the maximum field gradient on the cathode. The parameters used in this calculation are shown in Table 1.

The calculations using our tracking code were done under as identical conditions as possible. Table 2 lists the parameters. The major differences between the parameters are with the initial beam transverse profile and initial rf phase. In MAFIA, the profile is uniform, but in our code, it is a Gaussian distribution. For the initial rf phase, it is optimized in our code, around 45 degrees is preferable.

The results are shown on the right hand side of Figure 3, where the normalized rms emittance is defined by $\epsilon_x = \langle \gamma \rangle \langle \beta \rangle \sqrt{\langle x^2 \rangle \langle x'^2 \rangle - \langle x \cdot x' \rangle^2}$, and is the total bunch emittance.

Comparing the two simulation results, the magnitudes of the emittance do not agree. Several reasons are considered. In our code, the effects of image forces by the

TABLE 2. Parameters used in our tracking code

Charge per Bunch	[nC]	1
Transverse Profile		Gaussian
Spot Size on the Cathode	[mm]	$\phi 0.5$ (1σ)
Longitudinal Profile		Uniform
Bunch Length	[ps]	10
Initial Emittance	[πmm·mrad]	0
Initial rf Phase	[degree]	45
The Number of Particles		5000
Charge of One Particle	[C]	2×10^{-13}
Time Step of Calculation	[ps]	1

cathode surface are not included. We assume that the electron motion is uniform motion, because this simplifies field calculations using Liénard-Wiechert potentials. In actuality, however, the electron motion in the cavity is not uniform.

The difference is also considered to be caused from the accuracy of MAFIA. In MAFIA's data, an optimum field gradient of about 200MV/m can be seen. In our code, however, the optimum gradient can not be found. The reduction of the emittance becomes saturated as the gradient gets higher. It is considered to be more reasonable.

Figure 4 shows the emittance calculated by our code, from the cathode to the Faraday-cup. The change of the emittance corresponds to the distribution of the

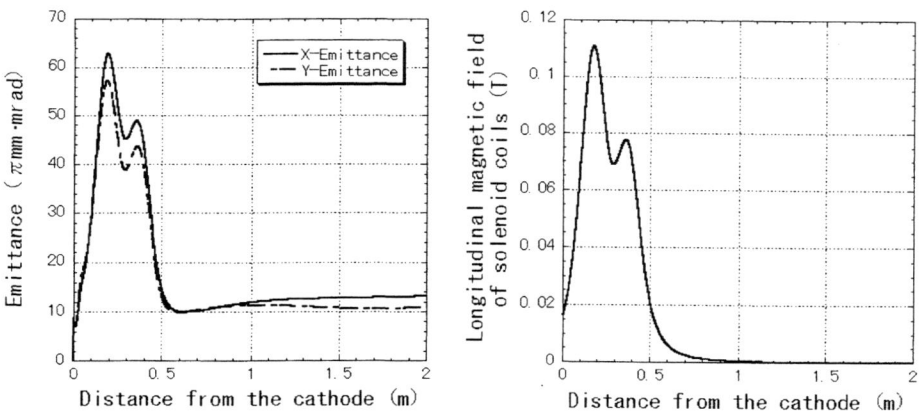

FIGURE 4. Calculated emittance from the cathode to the Faraday-cup with the solenoid coils included. The field of the solenoid coils is shown on the right hand side. In this case, the maximum field gradient on the cathode is 150MV/m. The other parameters are the same as in Table 2.

105

longitudinal field of the solenoid coils. This calculation is important because we can measure the emittance only after the solenoid coils, that is, the position of the slits. This calculation includes the optimum solenoid fields, which are selected so as not to enlarge the beam radius after the solenoids. Accordingly, the value of the emittance at the exit of the rf cavity does not grow even after the solenoids.

CONCLUSION

A beam tracking code has been developed for an rf gun test apparatus in the Spring-8 Linac. In this code, we try to calculate all space charge effects for each electron. Therefore a CPU time of about two hours is needed for tracking in a cavity, but the code can be used for precise calculation purposes. The rms emittance of the beam which is extracted from the rf cavity is around 10 πmm·mrad and is larger than MAFIA's result. Several reasons are considered for the difference. Experiments are scheduled to start from autumn 1998. One of the purposes of these experiments is to compare the beam characteristics with simulations.

REFERENCES

1. Taniuchi T. et al., *Proc. of the 11th Symp. on Accelerator Science and Technology* SPring-8, Harima, JAPAN 203 (1997).

Normal Form Approaches and Resonance Analysis of LHC Models

Yannis Papaphilippou and Frank Schmidt

CERN, CH-1211 Geneva 23, Switzerland

Abstract. In order to understand the dynamic aperture limitations in LHC optics versions 4 and 5 at injection energy (450 GeV), a thorough resonance analysis is performed through Lie perturbation methods and Normal Form construction. In this respect, a simple numerical tool has been developed, the Graphical Representation of Resonances (GRR), allowing the evaluation and graphical representation of the resonance strengths and detuning, up to a desired order. The resonance analysis performed by means of GRR enabled us to understand the effect of the large errors in some special quadrupoles of LHC optics version 5. We were also able to identify and minimise the resonance which was correlated with the drop of the dynamic aperture, following the introduction of a large skew octupole bias in the LHC optics version 5, using the "target" error table. As shown by subsequent tracking studies, the proposed correction procedures led to a considerable improvement of the dynamic aperture of the studied LHC models.

INTRODUCTION

A crucial point in the design of hadron colliders is the long term beam stability determined by the dynamic aperture (DA). The straightforward procedure to calculate the DA of a given accelerator model is performed through element by element tracking of particles. Considering the fact that the usual injection time for accelerators like the LHC corresponds to 10^7 particle turns, tracking simulations are limited by the lack of available computer power for following the full particle orbits. A more serious problem comes from the fact that tracking cannot provide any understanding regarding the resonance structure in the phase space of the system. What seems to be of high interest is to recover the reasons triggering these resonances thereby limiting the DA.

The necessary insight regarding the system's non-linear dynamics can be given by applying the methods of high order perturbation theory [1–3]. These approaches can be conducted by using an explicit form of the system's Poincaré map. This map is usually computed through a Taylor expansion around the 1-periodic orbit of the accelerator beam, with the assistance of Lie algebraic tools. In general, the map is written in variables which are not close to the invariants of motion. The

CP468, *Nonlinear and Collective Phenomena in Beam Physics–1998 Workshop*,
edited by S. Chattopadhyay, M. Cornacchia, and C. Pellegrini

construction of Normal Forms consists in transforming the original variables of the map to a new set of variables, which are close to the invariants of motion, in order for the resulting map to have a simpler form. The generating function, through which this symplectic transformation is performed, is computed by means of a perturbative scheme using as parameter the distance to the origin. This function contains the information regarding the distortion of the phase space due to non-linearities and the influence of resonances on the dynamics of the system.

On the other hand, what could be of great interest is an indication about the resonance strengths for specific initial conditions, especially close to the parts of the phase space where the beam is lost. This is a key point for coping with the non-linear dynamics of an accelerator, taking into account that particular multi-polar magnet errors have a specific contribution to particular resonance strengths. Thus, the resonance strengths establish a sort of quality factor [6], which can determine whether the specifications proposed for the design of an accelerator lattice are optimal with respect to the phase space regularity of the model. Through a resonance analysis of this kind, one can associate the physical characteristics of the accelerator to the DA of the model, find the limits and provide the necessary cures, ensuring the long-term stability of the beam. In this respect, we developed a simple numerical tool, the Graphical Representation of Resonances (GRR), which uses as input the output of standard Lie algebraic numerical codes [4] and allows the evaluation and graphical representation of the resonance strengths and detuning, up to a given order (see also [5,6]).

In this article, we present the impact of the use of GRR in the ongoing studies for understanding the dynamics of the LHC optics versions 4 and 5. The article is organised as follows: in Sect. I, we outline the basic ideas regarding Normal Form analysis and the GRR tool. In the next section, we present examples of the application of this resonance analysis in order to understand the dynamics of LHC models. In the last section, the principal results are summarised together with some objectives for future studies.

I GRAPHICAL REPRESENTATION OF RESONANCES

A Resonance Strength

By neglecting the weak coupling between the vertical and horizontal motion and the longitudinal one, the system is restricted to the 4D phase space. In order to construct the Normal Form $\mathcal{U}$ of a Poincaré map $\mathcal{M}$ representing the successive intersections of a kicked accelerator beam at a fixed position of the path variable s, one performs a symplectic transformation expressed by the functional equation

$$\mathcal{U} = \Phi^{-1} \circ \mathcal{M} \circ \Phi \ ,$$

which transforms the variables of the original map $\mathbf{z} = (z_x, z_x^*, z_y, z_y^*)$ to a new set $\zeta = (\zeta_x, \zeta_x^*, \zeta_y, \zeta_y^*)$, where the $(*)$ denotes complex conjugate variables. The original variables are usually the complexified version of the Courant-Snyder coordinates

$$z_{x,y} = \sqrt{2\,J_{x,y}}\,e^{-i(\varphi_{x,y}+\varphi_{x0,y0})}\,, \tag{1}$$

where $J_{x,y}$ and $\varphi_{x,y}$ are the action-angle variables of the integrable problem, in the linear case. The actions are a function of the horizontal and vertical emittance

$$\epsilon_x = 2\,J_x = n_\sigma(\epsilon_n/\gamma)^{1/2}\cos\phi \quad \text{and} \quad \epsilon_y = 2\,J_y = n_\sigma(\epsilon_n\gamma)^{1/2}\sin\phi\,, \tag{2}$$

where the normalisation factor n_σ determines the amplitude of the particle in terms of the rms beam size σ, ϵ_n is the normalised emittance, γ the energy factor and ϕ is the amplitude ratio ($\tan\phi = \epsilon_y/\epsilon_x$).

On the other hand, the new variables are expressed

$$\zeta_{x,y} = \sqrt{2\,I_{x,y}}\,e^{-i(\psi_{x,y}+\psi_{x0,y0})}\,, \tag{3}$$

i.e. as a function of the non-liner invariants of motion $I_{x,y}$ and their conjugate phases $\psi_{x,y}$.

The transformation Φ, as well as the maps, may be represented using the Lie formalism, e.g.:

$$\Phi = e^{:F:}\,, \tag{4}$$

which denotes a series of Poisson brackets operations. In general, the generating function F is represented by a sum of homogeneous polynomials of the new variables. By taking the inverse transformation Φ^{-1}, the generating function F can be expressed as a function of the old variables

$$F = \sum_{jklm} f_{jklm}\,z_x^{\,j}\,z_x^{*\,k}\,z_y^{\,l}\,z_y^{*\,m} \quad \text{with} \quad f_{jklm}\in\mathbb{C} \ \text{ and } \ j,k,l,m\in\mathbb{N}\,. \tag{5}$$

Inserting the expressions of the $\mathbf{z}$ variables in the series (5), one obtains:

$$F = \sum_{jklm} f_{jklm}\,(\epsilon_x)^{\frac{j+k}{2}}\,(\epsilon_y)^{\frac{l+m}{2}}\,e^{-i\varphi_{jklm}}\,, \tag{6}$$

where the phase term variable is $\varphi_{jklm} = (j-k)(\varphi_x + \varphi_{x0}) + (l-m)(\varphi_y + \varphi_{y0})$. The expression (6) is usually computed order by order by means of Lie algebraic numerical tools, as the DaLie code [4], which is used in the present study.

The infinite series (6) is not convergent by construction. Furthermore, the number of terms grows very sharply with the order $n = j+k+l+m$ [1–3]. In practice, one takes a truncated expression of the series (6), and in our case, we usually found sufficient to carry out the calculation of the perturbing series up to 12th order which corresponds to an 11th order map.

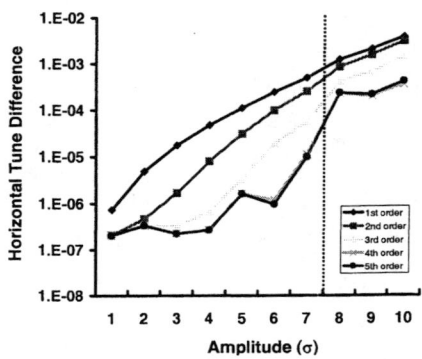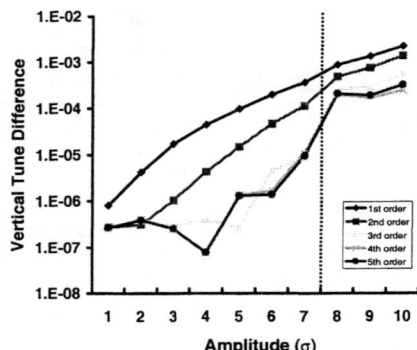

FIGURE 1. *Difference between the tune values calculated with two methods, for a specific realisation of the magnet errors (seed 25) of LHC optics version 5 "nominal" table, for $\phi = 15°$ and for different particle amplitudes.*

The expression of the phase φ_{jklm} shows that each term of the series (6) corresponds to a resonance condition of the form

$$(j - k)\nu_x + (l - m)\nu_y + c = 0 \quad \text{with} \quad c \in \mathbb{Z} \,,$$

with ν_x and ν_y representing the frequencies of motion. These resonances are related with the non-linear dynamical behaviour of the system. The norm of the coefficients $|f_{jklm}|$ provides an indication about the strength of the corresponding resonances, these last being associated with the multi-polar magnetic field errors. In fact, multipole errors of a certain order n_m will have a contribution to the coefficients f_{jklm} of order $n \geq n_m$. Hence, one could infer that the importance of a specific resonance regarding the dynamics of the system can be revealed by the norm of the corresponding coefficients. However, the series terms associated with the same resonance $(a, b) = (j - k, l - m)$ of order n will also appear in higher orders $n + n'$, with n' a non-zero pair number. Thus, a more precise estimation of the resonance strength in a given position of the phase space can be efficiently computed only by including the contribution of these higher order terms. To do this, one may fix the phase at an arbitrary value, e.g. $\varphi_{jklm} = 0$, without loss of generality. The strength of a specific resonance (a, b) is given by the norm of

$$F_{(a,b)} = \sum_{\substack{jklm \\ j+k+l+m \leq n \\ j-k=a\,,\,l-m=b}} f_{jklm} \left(\epsilon_x\right)^{\frac{j+k}{2}} \left(\epsilon_y\right)^{\frac{l+m}{2}} \,, \tag{7}$$

where the contribution of all terms up to an order n is taken into account.

110

B   Tune Shift

In the same way, one can also compute the tune shift due to the non-linearities of the system. In fact, in the non-resonant case, the variables $\zeta'_{x,y}$ after one conjugation of the map $\mathcal{U}$ are:

$$\zeta'_{x,y} = e^{-i\Omega_{x,y}(\rho_x,\rho_y)}\zeta_{x,y} \ , \tag{8}$$

where $\Omega_{x,y}$ are the non-linear frequencies associated with the horizontal and vertical motion and $\rho_{x,y} = 2I_{x,y} = \zeta_{x,y}\zeta^*_{x,y}$ are the generalisation of the vertical and horizontal emittance in the non-linear case. Taking into account that the tunes are

$$\Omega_{x,y} = \frac{\partial \, \overline{H}}{\partial \rho_{x,y}} \tag{9}$$

where $\overline{H}$ is the new Hamiltonian associated with the map, the non-linear tunes can be written as a series of homogeneous polynomials

$$\Omega_{x,y}(\rho_x,\rho_y) = \sum_{\substack{jklm \\ j+k+l+m\leq[(n-1)/2] \\ j-k=0\,,\,l-m=0}} \omega_{jklm}\,(\rho_x)^{\frac{j+k}{2}}\,(\rho_y)^{\frac{l+m}{2}} \ , \tag{10}$$

where the (j,k,l,m) are such that the phase dependence vanishes, leaving only the amplitude dependent terms in the polynomial. In order to evaluate the series (10), the knowledge of the non-linear invariants $\rho_{x,y} = 2I_{x,y}$ is necessary. These last can be computed through the inverse transformation $\Phi^{-1}(\mathbf{z})$. Hence, the non-linear invariant can be expressed as a function of the linear ones $\epsilon_{x,y}$ and the non-linear tunes can be calculated by the series (10) written in these latter variables. Note also that the maximum detuning order is now the integer part of the expression $(n-1)/2$, e.g. the terms of an 11th order map will contribute up to a detuning order of 5.

The precision of the calculation using the Normal Form machinery can be checked by comparing the tunes computed through the series expansion (10) with the values provided by a direct application of a Laskar type frequency analysis [7] (we use the the SUSSIX code [8]) of tracking data generated by SIXTRACK [9] for several particle orbits, in the case of the "nominal" error table of LHC optics version 5. We generated 10 particle orbits, the amplitude of which varies from $n_s = 1\sigma$ to 10σ and the amplitude ratio is kept fixed $\phi = 15°$. In Fig.1, we present graphically the difference between the horizontal (left) and vertical (right) tune given by the two computation methods for a specific realisation of the magnet errors ("seed" number 25) . The different graphs correspond to different orders in the calculation of the tune (from 1st to 5th order) and the horizontal axis corresponds to different amplitudes measured in σ. The vertical axis represents the tune difference in logarithmic scale and the vertical doted line denotes the barrier after which the particles begin to be chaotic. It is apparent that, at least up to the point where the particles are

111

Amplitude

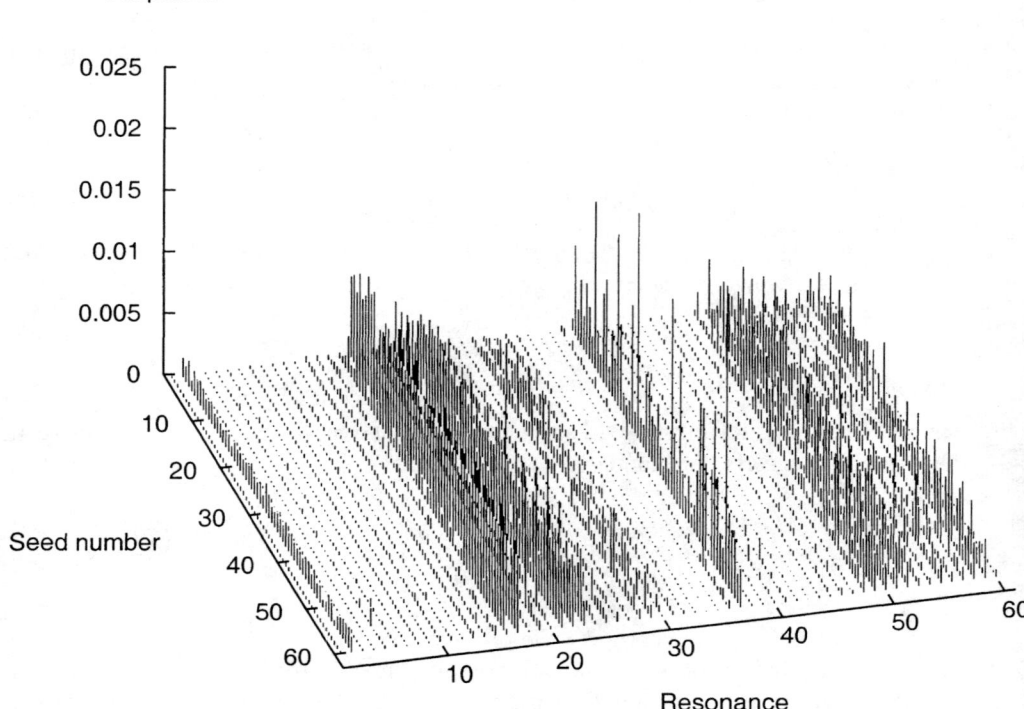

FIGURE 2. *Norm of the 7th order resonance coefficients f_{jklm} of the generating function (6), for 60 seeds of LHC optics versions 5 using the "nominal" error table.*

stable and after adding the 4th order contribution, the two approaches converge to the same value, with a precision of 10^{-5}. Let us just remark that this result is only valid for a regular orbit evolving on a KAM torus. For a chaotic orbit, the calculation becomes meaningless as the tune is not constant.

II   APPLICATION TO THE LHC

A   Limitations in the Dynamic Apperture of LHC Optics Versions 4 and 5

A global picture of the resonance strengths given by the generating function F is represented graphically in the 3D plot of Fig. 2. This graph represents the norm of the 7th order resonance coefficients f_{jklm} with $n = 7$, for LHC optics version 5, using the "nominal" error table and for 60 seeds. Each number on the horizontal axis corresponds to a different 7th order resonance (60 in total - see Table 1). The first 7 series of spikes represent the amplitudes of the normal resonances and the next 23 the normal sub-resonances. The remaining 30 correspond to the 7th order skew resonances and sub-resonances.

The importance of the (7,0) resonance for predominantly horizontal motion was reported in previous studies [10,11]. Nevertheless, this is not at all visible in this picture, i.e. the first series of spikes is quite small as compared with certain sub-resonances and skew resonances, in the middle of the graph. This is indeed due to the fact that the subresonance strengths mask the importance of the (7,0) resonance. On the other hand, the higher order series terms contributing to a specific resonance are not included in the coefficients f_{jklm}.

Through Eq.(7), we were able to evaluate the strength $\left|F_{(a,b)}\right|$ of the 7th order resonances up to 12th order, for an amplitude ratio of 15° at 8σ, which is close to the minimum DA of this LHC model. A graphical representation of these resonances (14 in total) is given in Fig. 3. The depicted points correspond to the average value of the resonance driving terms over the 60 random realisations of the magnet errors and the error bars are equal to one standard deviation. The three different lines represent resonance strengths the computation of which is conducted up to 3

TABLE 1. Correspondence of numbers labelling the horizontal axis of Fig. 2 with the 7th order resonances

| | resonances | | sub-resonances | | | | | |
| | 7th order | | 5th order | | 3th order | | 1st order | |
	(j,k,l,m)	N_o	(j,k,l,m)	N_o	(j,k,l,m)	N_o	(j,k,l,m)	N_o
	(7,0,0,0)	1	(6,1,0,0)	8	(5,2,0,0)	18	(4,3,0,0)	27
	(5,0,2,0)	2	(5,0,1,1)	9	(4,1,1,1)	19	(3,2,1,1)	28
	(3,0,4,0)	3	(4,1,2,0)	10	(3,0,2,2)	20	(2,1,2,2)	29
Normal	(1,0,6,0)	4	(3,0,3,1)	11	(3,2,2,0)	21	(1,0,3,3)	30
	(5,0,0,2)	5	(2,1,4,0)	12	(2,1,3,1)	22		
	(3,0,0,4)	6	(1,0,5,1)	13	(1,0,4,2)	23		
	(1,0,0,6)	7	(4,1,0,2)	14	(3,2,0,2)	24		
			(3,0,1,3)	15	(2,1,1,3)	25		
			(2,1,0,4)	16	(1,0,2,4)	26		
			(1,0,1,5)	17				
	(6,0,1,0)	31	(5,1,1,0)	38	(4,2,1,0)	48	(3,3,1,0)	57
	(4,0,3,0)	32	(4,0,2,1)	39	(3,1,2,1)	49	(2,2,2,1)	58
	(2,0,5,0)	33	(3,1,3,0)	40	(2,0,3,2)	50	(1,1,3,2)	59
	(0,0,7,0)	34	(2,0,4,1)	41	(2,2,3,0)	51	(0,0,4,3)	60
Skew	(6,0,0,1)	35	(1,1,5,0)	42	(1,1,4,1)	52		
	(4,0,0,3)	36	(0,0,6,1)	43	(0,0,5,2)	53		
	(2,0,0,5)	37	(5,1,0,1)	44	(4,2,0,1)	54		
			(4,0,1,2)	45	(3,1,1,2)	55		
			(3,1,0,3)	46	(2,0,2,3)	56		
			(2,0,1,4)	47				

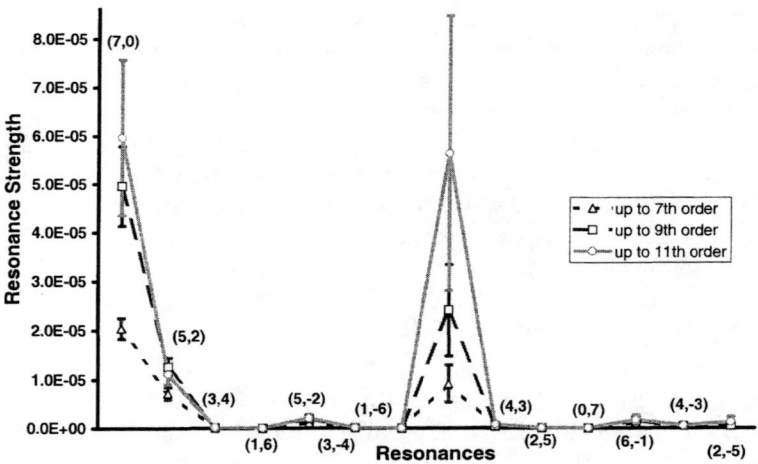

FIGURE 3. *Average value and standard deviation of the 7th order resonance strength at 8σ and $\phi = 15°$ for LHC optics version 5 "nominal" error table, over 60 seeds. The three different graphs represent resonance driving terms computed up to 3 different orders.*

different orders (7th, 9th and 11th). It is now clear that the (7,0) resonance has a quite big influence in the dynamics of the system. Moreover, the contribution of the higher orders to this resonance are quite important, a fact which could have not been revealed by simply checking the coefficients of the function F (Eq. 6). On the other hand, this graph also shows that the strength of the (6,1) resonance is quite big. A further analysis of the detuning of the system reveals that at this part of the phase space, the particle tunes are close to this resonance.

By using GRR, we were able to understand the reason for this strong excitation of the (7,0) resonance, limiting the DA of LHC optics version 5 with respect to version 4, at least for motion close to the horizontal plane. It was found that the large multi-pole errors corresponding to some special type of quadrupoles ("warm" quadrupoles) on the two high-beta insertions (I.P.3 and I.P.7) of the LHC optics version 5 were in the heart of the problem [13]. In Fig. 4, we present the average (over 60 seeds) absolute value of the relative difference, that is the difference of the resonance strength weighted by the biggest one among the two, between the 12 most prominent resonances of LHC optics version 4 and 5. There is a 50% difference between the amplitudes of the (7,0) resonance in the two lattices. By switching off the errors in the "warm" quadrupoles, the two lattices become approximately identical. This effect is indeed due to the strong kick produced by the interplay of the important beta function values in these areas of the machine (of the order of 350 m in I.P. 3 and 600 m in I.P.7 [14]) with the strong multi-pole errors of the "warm" quadrupoles (especially the b_3 and b_7).

This resonance analysis study guided us in the correction of the LHC optics

TABLE 2. The effect of the multipole errors of warm quadrupoles on the dynamic aperture for different LHC optics versions 4 and 5 machines. The average and minimum DA over 60 seeds is shown, for $\phi = 15°$ and $45°$.

Phase	Type	DA (σ)	4	5 Nominal	5 Target
15°	Warm Quads switched ON	Average	10.0	9.1	10.4
		Minimum	8.5	7.4	8.6
	Warm Quads switched OFF	Average	10.7	11.6	12.4
		Minimum	9.6	10.3	11.3
45°	Warm Quads switched ON	Average	11.1	11.3	12.8
		Minimum	9.5	9.2	11.4
	Warm Quads switched OFF	Average	11.4	12.4	13.8
		Minimum	10.1	10.7	12.3

version 5 with an important average improvement of the DA. The results of the 6-dimensional tracking studies with SIXTRACK [9] are reported in Table 2. We give the minimum and average value of the DA for several cases of LHC optics version 4 and 5, with and without the errors in the "warm" quadrupoles. The "target" error table is the one proposed in order for the LHC to reach the target DA of 12σ so as to have a safety factor of 2 [11], considering the fact that the collimators will be positioned at 6σ. The tracking is conducted for two directions of the phase space (amplitude ratios of 15° and 45°) which, in our case, correspond roughly to the minimum and average values of the DA, over all phases [15]. For all cases and for both phase values, the average gain of the D.A. is of the order of 2σ. Especially for the "target" error table, we are able to reach the target DA of 12σ.

B Correction of the Effect of the Octupole Error Bias on the LHC Dipoles

A similar resonance analysis was followed in order to understand the drop of the DA in LHC optics version 5, using the "target" error table, after the inclusion of a large bias of the systematic per arc octupoles in the main dipoles of the model [16]. Indeed, the inclusion of these new realistic values for the octupole errors deteriorated the DA of the "target" table (see Table 3, "strong b_4", "strong a_4" and "strong b_4, a_4" cases).

The experience from LHC optics version 4 [10] has shown that the bias can be usually cancelled by erect and skew octupole spool pieces in half of the machine, each located at one end of the dipoles and powered in series (having one spool piece type in the outer channel corresponding to the 1st, 5th, 6th and 7th octant and

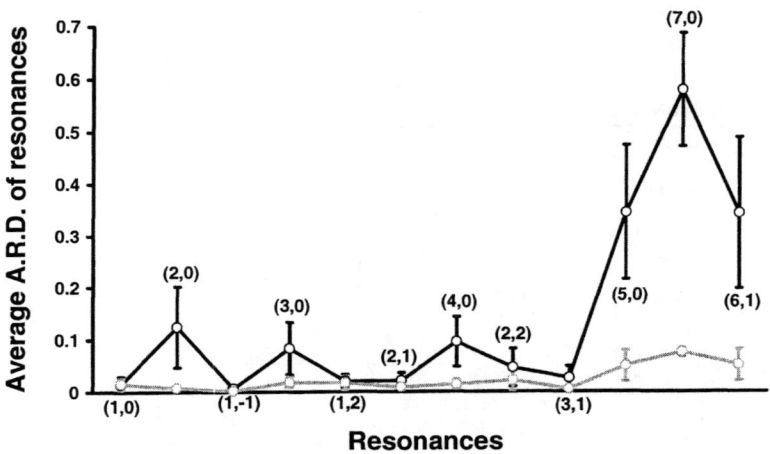

FIGURE 4. *Average absolute value of the resonance strengths relative difference at 8σ and 15°*
between LHC optics version 4 and 5 with and without the errors on the "warm" quadrupoles.

TABLE 3. The effect of realistic erect and skew octupolar errors and their
corrections on the dynamic aperture of LHC optics version 5 at injection.
The average and minimum DA over 60 seeds is shown, for $\phi = 15°$ and 45°.

Errors and Correctors	Phase [°]			
	15		45	
	Dynamic Aperture [σ]			
	Minimum	Average	Minimum	Average
Target error table	11.3	12.4	12.3	13.8
Strong b_4	9.6	12.2	10.8	13.4
$+b_4$ spool piece (SP)	11.8	12.6	12.0	13.7
Strong a_4	10.4	12.1	10.0	12.9
$+a_4$ SP	10.2	12.1	9.5	12.5
+ optimised a_4 SP	11.3	12.5	11.7	13.8
Strong b_4, a_4	10.1	12.0	10.0	12.6
$+b_4$ SP, a_4 SP	9.7	12.0	9.5	12.5
$+b_4$ SP, optimised a_4 SP	11.2	12.6	11.8	13.6

one in the inner channel corresponding to the remaining octants). The correction
of the bias of the erect octupole component with erect octupole spool pieces in half
of the machine fully restored the DA of the "target" error table (see Table 3, "$+b_4$
spool piece" case). Nevertheless, it was not sufficient to suppress the bias of the
skew octupole component following the same procedure (see Table 3, "$+a4$ SP"
and "$+b_4$ SP, a_4 SP" cases).

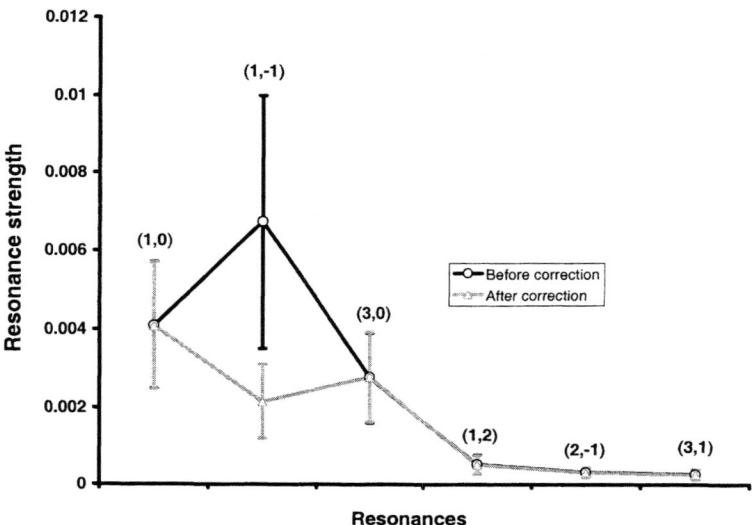

FIGURE 5. *Strength of resonances for LHC optics version 5 due to skew octupole errors before and after correction.*

A resonance analysis with GRR has revealed that the skew octupoles mainly excite the $(1, -1)$ resonance. In order to cancel the effect of the skew octupoles, a minimisation procedure was followed, by producing maps parameterised by the strength of the skew octupole spool pieces. The minimisation of the first order coefficients f_{1012}, f_{2101} of the $(1, -1)$ resonance with skew octupole spool pieces positioned in the outer channel of the LHC lattice and powered in series was very beneficial. This can be easily seen in Fig. 5 where we present the strength difference of some important resonances between the "target" error table and the tables with the large skew octupole bias before and after the correction. The resonances in this graph are evaluated at an amplitude of 8σ and for amplitude ratio of 15°. All points represent the average values over 60 seeds and the error bars correspond to the standard deviation. The effect of the correction of the $(1, -1)$ resonance is indeed very visible. The resonance strength are considerably reduced after the correction with the skew octupole spool pieces. In particular, the $(1, -1)$ resonance strength decreases by a factor of three so that there is approximately no difference left with the one of the "target" error table. This improvement was verified by the tracking studies, where the D.A. is fully restored after the correction with respect to the one of the initial "target" error table (Table 3, "optimised a_4 SP" and "+b_4 SP, optimised a_4 SP" cases).

III   CONCLUSIONS

We followed a resonance analysis procedure in order to understand the reasons which limit the DA of LHC optics version 4 and 5. For this, we used the standard Normal Form approaches of Hamiltonian perturbation theory assisted by semi-analytical numerical methods. For the accurate evaluation and representation of the resonance driving terms, we constructed a simple numerical tool, the Graphical Representation of Resonances. Using the GRR tool, we were able to identify the deteriorating effect of the large errors in some special quadrupoles with respect to the DA of LHC optics version 5. Further, we identified the $(1, -1)$ resonance which was correlated with the drop of the DA after the introduction of a large skew octupole bias in the dipoles of the LHC optics version 5 "target" error table. In order to recover the lost DA, we used octupole spool pieces which minimised the effect of this resonance. Particle tracking has shown that, by following the proposed correction schemes, the DA of the LHC models under study was considerably improved. We can thus be confident that this type of resonance analysis can be used as a guide in order to achieve an efficient correction of accelerator models. In the near future, the GRR tool will be appropriately standardised and documented in order to be accessible for anyone desiring to perform this type of resonance analysis in a lattice of interest.

REFERENCES

1. Berz, M. et al. *Part. Acc.*, **24**, 91 (1989).
2. Berz, M., *Part. Acc.*, **24**, 109 (1989).
3. Bazzani, A., et al., *CERN Yellow Report*, **94-02**, 1994.
4. Forest, E., "The DaLie Code" (unpublished), 1986.
5. Irwin, J., et al., 4th EPAC Conference, London, 1994.
6. Todesco, E., et al., *Comp. Phys. Commun.* **106**, 169 (1997).
7. Laskar, J. *Physica D*, **67**, 257 (1993).
8. Bartolini, R. and Schmidt, F., *Part.Acc.* **59**, 93 (1998).
9. Schmidt, F., *CERN SL report* **94–56**, 1994.
10. Böge, M., et al., EPAC'96, Barcelona, 1996.
11. Böge , M. and Schmidt, F., PAC'97, Vancouver, 1997.
12. Papaphilippou, Y. and Schmidt, F., *LHC project report* **235**, HEACC'98, Dubna, 1998.
13. Papaphilippou, Y. and Schmidt, F., *LHC project report*, in press (1999).
14. The LHC Study Group, *The Large Hadron Collider*, CERN/AC/95-05(LHC), 1995.
15. Böge , M., Schmidt, F. and Xu, G., *LHC project note* **154**, (1998).
16. Jin, L., Papaphilippou, Y. and Schmidt, F., *LHC project report* **253**, (1998).

MOMENTUM APERTURE OF THE ADVANCED LIGHT SOURCE

W. Decking and D. Robin

LBNL, Berkeley, CA, 94720, USA [1]

Abstract. This paper shows measurements of the momentum aperture of the Advanced Light Source (ALS) based on Touschek lifetime measurements. The measured data is compared with tracking simulations and a simple model for the apertures will help to explain the observed effects.

I INTRODUCTION

Particles in a storage ring oscillate about their design orbit. In some cases these oscillations can be irregular or unstable leading to particle losses. The storage ring design must provide a large region around the design orbit in which the oscillations are stable in order to insure good performance of the ring. The envelope of this region in the six-dimensional phase-space is called acceptance. Great effort is spent to increase the acceptance of a storage ring during the design and in operation. A large acceptance is important for two reasons: It facilitates injection into the storage ring and leads to high injection efficiencies, and it elongates the lifetime of the stored particles.

In this paper we describe our efforts to measure the acceptance and compare it with numerical and theoretical predictions. We primarily infer the acceptance from measurements of the Touschek lifetime.

The lifetime of a low energy, small emittance synchrotron radiation source like the ALS is usually limited by the momentum aperture of the ring. The momentum aperture limits the acceptance in the longitudinal direction. Large momentum apertures are reached by

- providing enough rf-voltage for a large bucket

- avoiding the degradation of the momentum aperture due to linear or non-linear synchro-betatron coupling where the particle exceeds the transverse aperture of the machine.

[1] This work was supported by the Director, Office of Energy Research, Office of Basic Energy Sciences, Materials Sciences Division, of the U.S. Department of Energy, under Contract No. DE-AC03-76SF00098.

CP468, *Nonlinear and Collective Phenomena in Beam Physics–1998 Workshop*,
edited by S. Chattopadhyay, M. Cornacchia, and C. Pellegrini

First we will define the several effects which limit the momentum aperture. Then we show measurements of the dependence of the momentum aperture of the ALS on different working points, chromaticities, and the addition of coupling and insertion devices. We introduce a simple model to explain the momentum dependent dynamic aperture and compare the experimental results and the model with tracking calculations.

II MOMENTUM APERTURE

The momentum aperture ε is the maximum relative momentum deviation $\delta = \frac{\Delta p}{p_0}$ from the design momentum p_0 a particle can experience without being lost. The momentum aperture is determined by two different processes: the height of the rf-bucket and the maximum stable transverse amplitudes.

The height of the rf-bucket (or rf-aperture) provided by the accelerating voltage in the cavity is

$$\varepsilon_{rf} \propto \pm\sqrt{\frac{V_{rf}}{\alpha h E}} , \tag{1}$$

where V_{rf} is the rf-voltage, α the momentum compaction factor, h the harmonic number, and E the nominal energy. Equation 1 is only valid for rf-voltages much higher than the energy loss per turn.

The transverse motion of particles is limited by the vacuum chamber aperture. Off-energy particles will oscillate about an off-energy orbit which may be closer to the vacuum chamber wall than the on-energy orbit. Vacuum chamber apertures as well as optical functions vary along the ring. Assuming linear particle motion the largest amplitude a particle can have without hitting the physical aperture is determined by the smallest phase space ellipse fitting through all apertures in the ring,

$$A_{phys,x}(\delta) = \min_{s\in[0,L]} \frac{(x_{vc.}(s) - \eta(s)\delta)^2}{\beta_x(s)} , \tag{2}$$

with η, β being the dispersion and β-function, and $x_{vc.}$ the vacuum chamber aperture. This is termed the physical aperture. Dispersion is usually only present in the horizontal plane, thus the vertical physical aperture is calculated by equation 2 without dispersion, leaving $A_{phys,y}$ momentum independent.

The particle motion can also reach the vacuum chamber aperture by nonlinear dynamic effects, i.e. the oscillation amplitudes are increased resonantly or chaoticaly until a particle gets lost. This limit is called dynamic aperture A_{dyn} and is obtained through tracking calculations or analytical methods. The transverse momentum dependent aperture is given by the minimum of the physical and the dynamic aperture.

When electrons scatter within a bunch, they may transfer enough momentum to be outside the momentum aperture of the storage ring. This effect is called Touschek effect and is proportional to the electron density within a bunch. After a momentum change due to a scattering within the bunch, the particle starts a betatron oscillation around the dispersion orbit. The induced linear invariant amplitude is

$$A_{ind,x}(s, \delta) = H(s)\delta^2 \,, \tag{3}$$

where $H(s) = \gamma(s)\eta(s)^2 + 2\alpha(s)\eta(s)\eta(s)' + \beta(s)\eta(s)'^2$. In the presence of linear coupling, the invariant amplitude couples in the vertical plane by $A_{ind,y}(s, \delta) = \kappa A_{ind,x}(s, \delta)$.

The induced amplitude should not exceed the maximum allowable transverse amplitude

$$H(s)\delta^2 \leq \min \left[A_{phys,x}(\delta), A_{dyn,x}(\delta), \frac{1}{\kappa}A_{phys,y}(\delta) \right] \,. \tag{4}$$

Solving this equation for the maximum δ around the ring gives a position dependent momentum aperture, which we call $\varepsilon_{trans}(s)$. The absolute momentum aperture is the smaller of ε_{rf} and $\varepsilon_{trans}(s)$ at any position in the ring.

Assuming a flat beam, and thus the main contribution of the velocity spread coming from horizontal motion (σ_x'), the Touschek lifetime becomes [1]:

$$\frac{1}{\tau_{tou}(s)} \propto \frac{1}{E^3} \frac{I_b}{V_b(s)\sigma_x'(s)} \frac{1}{\varepsilon(s)^2} f\left(\frac{\varepsilon(s)^2}{\sigma_x'(s)^2 E^2}\right) \,. \tag{5}$$

I_b is the bunch current and V_b the bunch volume. As bunch volume as well as momentum aperture vary around the ring, the Touschek lifetime has to be averaged around the ring [2].

Figure 1 shows a comparison of the maximum allowable apertures and the induced oscillation amplitudes at different locations in the ALS. The thin lines represent the induced amplitudes. In the straight section, where there is no dispersion, the induced amplitude is zero, that means the particle will only change its energy but not start a betatron oscillation. In the arc section with finite dispersion the induced amplitude shows the quadratic behavior as can be seen from equation 3. The thick lines represent the various apertures. The vertical lines show the rf-aperture with $\varepsilon_{rf} = 0.03$, corresponding to the presently available rf-voltage. The dashed-dotted lines show the physical apertures $A_{phys,x}$ and $A_{phys,y}$, where a

---

[2] A. Nadji et al [2] and others pointed out that the straight forward application of the above formulas leads to wrong results for large momentum deviations. Strong sextupoles lead to higher order dispersion which alters the off momentum closed orbit. In addition, the Twiss functions also vary with the beam energy, thus changing the $H(s)$ function into $H(\delta, s)$. However, at the ALS the linear, momentum independent calculations of the optical functions and the closed orbit vary only up to 5 % for momentum deviations of up to 5 %.

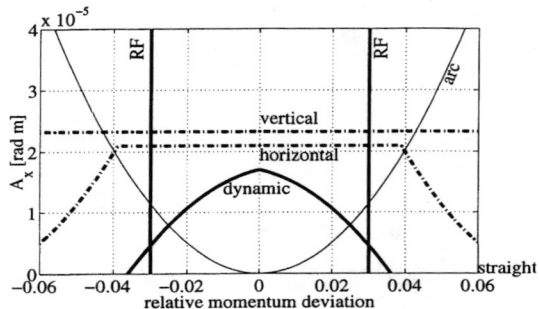

FIGURE 1. Contributions of the different aperture effects to the total momentum aperture. The thin lines are the invariant induced amplitudes at different positions in the ring. Dashed dotted lines represent the physical apertures, solid line is the dynamic aperture, and the vertical lines show the rf-aperture.

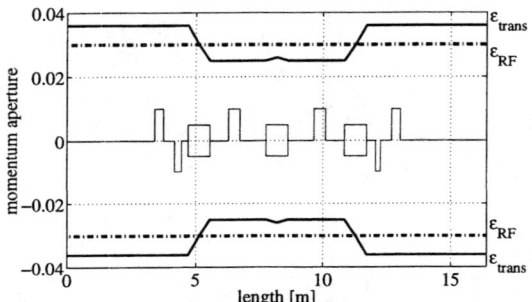

FIGURE 2. Momentum aperture along one cell of the ALS storage ring. Solid line is the momentum aperture from transverse limitations as derived from figure 1, dashed-dotted line the rf-aperture.

10 % coupling has been assumed to represent the vertical aperture as a limitation for horizontal motion. The solid line is a sketch of the dynamic aperture for a standard ALS lattice. We will discuss the underlying model in chapter IV. The momentum aperture is defined as the smallest crossing point of the thin and thick lines. As the induced amplitude varies around the ring so does the momentum aperture. Figure 2 shows the momentum aperture for one cell of the ALS. The solid line is the momentum aperture due to transverse limitations as derived from figure 1. The dashed-dotted line is the rf-aperture. In the straight sections the momentum aperture will be defined by the smallest of the rf-bucket height or the dynamic momentum aperture. In the arcs the momentum aperture will be defined by the dynamic momentum aperture, or for large coupling by the vertical physical aperture.

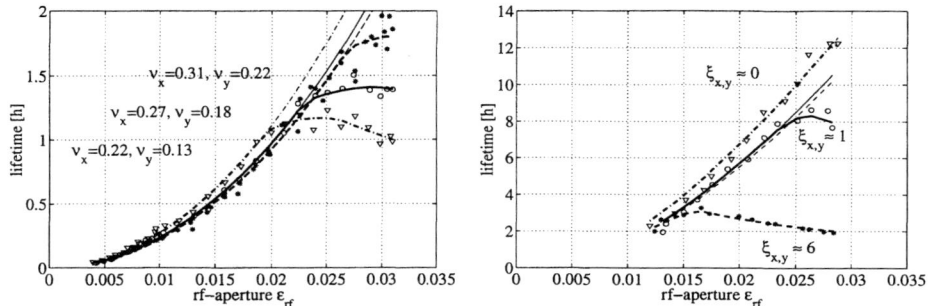

FIGURE 3. Beam lifetime as a function of the rf-aperture with three different working points (left) and three different chromaticities (right).

III MEASUREMENTS OF THE MOMENTUM APERTURE

Measurements of the Touschek lifetime as a function of the rf-voltage were conducted under different storage ring conditions. The synchrotron tune was measured simultaneously, from which one can calculate the bunch length and the rf-bucket height. To enhance the effect of the Touschek scattering over other lifetime effects a high current per bunch was filled in a few equidistantly spaced bunches. The low number of bunches avoids multi-bunch instabilities. The beam conditions thus were 1 mA/bunch and 8 bunches (out of 328) filled.

The data are fitted by applying equation 5 with the following fit parameters: Assuming an initial 1% coupling the bunch volume is corrected by a constant factor that takes into account any volume changes like variation of the coupling, instabilities, etc. The bunch volume is also adjusted according to the changing rf-voltage ($V_b \propto \sqrt{1/V_{rf}}$). The other parameters are the momentum apertures ε_{trans} in the straight section and in the arc. Thus the data can be fitted with just three parameters. The lines in figures 3 and 4 represent these fits, where the thin lines show the lifetime behavior if there would be no momentum aperture other than the rf-bucket.

Figure 3 left shows measurements of the lifetime with three different settings of the working point. The working point was moved along a line parallel to the $\nu_x - \nu_y$ coupling resonance. One can clearly see how the momentum aperture and the lifetime improves for higher working points. Figure 3 right displays three measurements with different chromaticities. The higher the chromaticity is set, the lower the momentum aperture.

Figure 4 left shows measurements with 1% coupling, while in figure 4 right the coupling was adjusted to $\approx 10\%$ with the help of skew quadrupoles. In addition a wiggler (2 T peak field, 0.16 m period length, and 3 m total length) was open (circles) or closed (stars) in both cases.

Table 1 summarizes the results for the measurements shown above as well as for

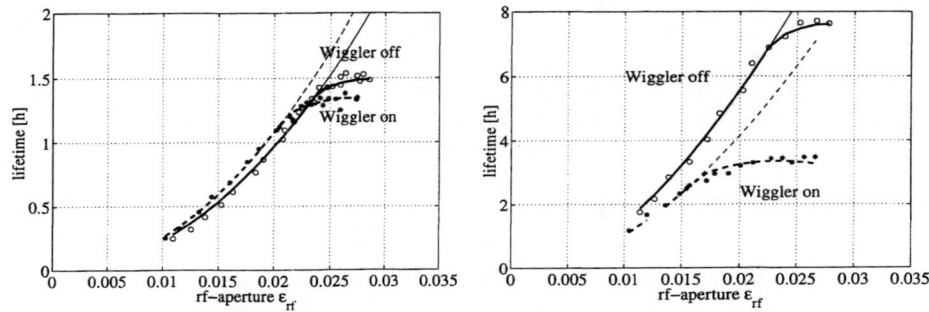

FIGURE 4. Beam lifetime as a function of the rf-aperture with no additional coupling (left) and 10 % coupling (right). In both cases the wiggler was open (circles) or closed (stars).

other measurements with different working points and chromaticities.

Let us first consider the cases with no wiggler. In these cases it appears as if the momentum aperture is determined by how close the working point is to the integer resonance $\nu_y = 8$. The momentum aperture can be increased by moving the tunes upwards along the coupling resonance $\nu_x - \nu_y$, thus increasing the distance to the integer. The momentum aperture decreases with increasing chromaticities according to the larger tune shift with energy in this case, which causes the tune to move more rapidly towards the integer. The introduction of coupling does not change the momentum aperture, which agrees with our measurements of the vacuum chamber apertures.

The cases with the wiggler on show a somewhat different behavior. The momentum aperture is slightly reduced for the low coupling case. Introducing coupling degrades the momentum aperture. Changing the tune upwards along the coupling resonance increases the momentum aperture again. The chromaticities have a similar effect as seen with no wiggler.

IV   SIMPLE MODEL FOR THE DYNAMIC APERTURE

A simple model of the dynamic limit could be as follows: Particles get lost when their tune satisfies a resonance condition. From knowing the tune shift terms with amplitude, $\frac{\partial \nu_y}{\partial A_x}$, $\frac{\partial \nu_y}{\partial A_y}$, and energy $\frac{\partial \nu_y}{\partial \delta}$, $\frac{\partial^2 \nu_y}{\partial \delta^2}$, one can estimate the tune shift $\Delta \nu_y$ due to momentum and transverse deviations:

$$\Delta \nu_y = \frac{\partial \nu_y}{\partial A_x} A_x + \frac{\partial \nu_y}{\partial A_y} \kappa A_x + \frac{\partial \nu_y}{\partial \delta} \delta + \frac{\partial^2 \nu_y}{\partial \delta^2} \delta^2 + \cdots , \qquad (6)$$

where linear coupling is treated as $A_y = \kappa A_x$. Setting $\Delta \nu_y$ as the distance to the closest 'deadly' resonance a momentum dependent dynamic aperture $A_{dyn,x}(\delta)$ can be calculated. In figure 5 left measurements of the tune shift with energy are shown

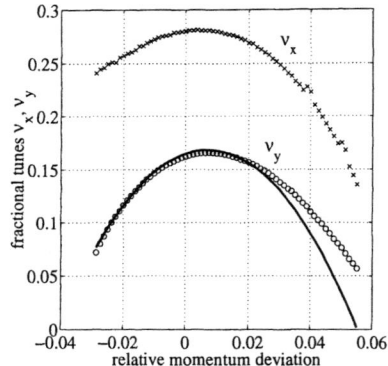 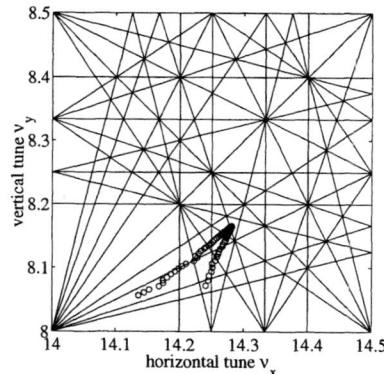

FIGURE 5. Measurements of the horizontal (crosses) and vertical (circles) tune shift with energy displayed together with model prediction for the vertical tune (solid line) on the left, and in the tune space on the right.

together with the model predictions for the vertical tune. The right side displays how the tunes are shifted in tune space.

The dynamic aperture drawn in figure 1 was computed using equation 6 with a tune shift $\Delta \nu_y = -0.14$.

V TRACKING CALCULATIONS

Tracking calculations have been performed to understand the experimental results and prove the accuracy of the model. Particles were tracked through the ALS lattice with a six-dimensional symplectic integrator[3]. The following errors and constraints were included in the model to simulate the realistic machine:

- Physical aperture borders were included in the tracking to prevent particle oscillations outside the realistic vacuum chamber. This is important because large amplitude particles may perform large, but stable oscillations which would be outside the physical aperture but not lead to a loss of the particle in the tracking.

- Linear field errors are simulated according to the optics measurements done at the ALS with the response-matrix fitting method [4]. This errors lead to a β-beat and thus a break in periodicity.

- Random skew quadrupole errors were distributed in all quadrupoles of the lattice and adjusted to obtain a 1 % coupling.

- The wiggler was simulated as a chain of hard-edge bending magnets obtaining the correct linear focusing and longitudinal dynamics properties.

---

[3] The tracking code TRACY2 was used.

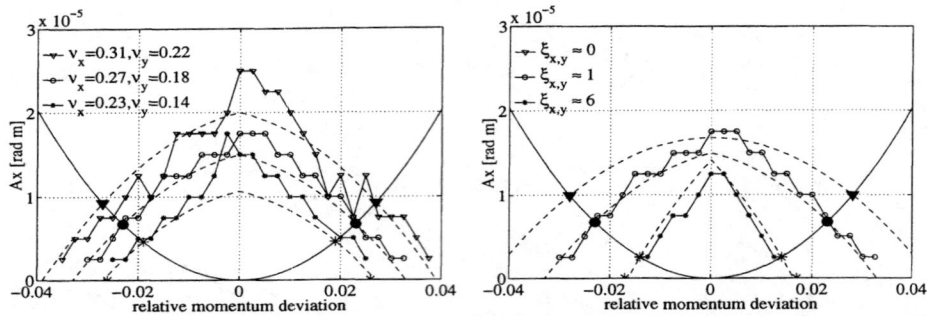

FIGURE 6. Maximum stable transverse emittance as a function of momentum deviation. Dashed lines show the dynamic aperture derived from a simple model. The left side is a case with three different working points, while the right side shows three different chromaticities.

Particles were launched off-energy and off-axis with respect to their off-energy orbit and tracked for 512 turns or until lost (damping times in the ALS are $\approx 10,000 - 20,000$ turns).

Results for three different tunes and three different chromaticities are shown in figure 6. Also included are the dynamic apertures obtained from the simple model (equation 6) assuming that the vertical tune is shifted towards the integer resonance. The tune shift is adjusted to match the momentum aperture measurements in the arc section of the ring. The thick symbols represent the measured momentum apertures, distributed along the curve showing the amplitude induced by Touschek scattering in the Arc. The agreement between measurement, tracking, and model is quite astonishing.

Figure 7 shows tracking results for different coupling and wiggler settings. In the left-hand side of figure 7 the tracking was done with the randomly distributed skew-quadrupoles leading to 1 % coupling. The wiggler was introduced as mentioned above. The agreement between measurements, tracking, and model again is very good. This situation changes if additional coupling is introduced into the tracking model by means of skew-quadrupole families distributed in the same way as in the ring. The transversal dynamic aperture for particles with smaller energy deviations is greatly reduced, while the dynamic aperture for large momentum deviations is roughly the same as in the low coupling case. The nominal vacuum chamber aperture is still large enough for the 10 % coupling case. This means that we either underestimate the linear coupling in the model, do not know the real vacuum chamber apertures well enough, or have a too simple picture of the influence of the skew quadrupoles.

The introduction of the wiggler in the highly coupled case shows at least qualitatively agreement between measurement and tracking. The momentum aperture is further reduced. We think this is due to the fact that the periodicity of the lattice is further destroyed by introducing a wiggler. This allows other resonances to be excited [5], thus changing the distance to the closest 'deadly' resonance. The effect

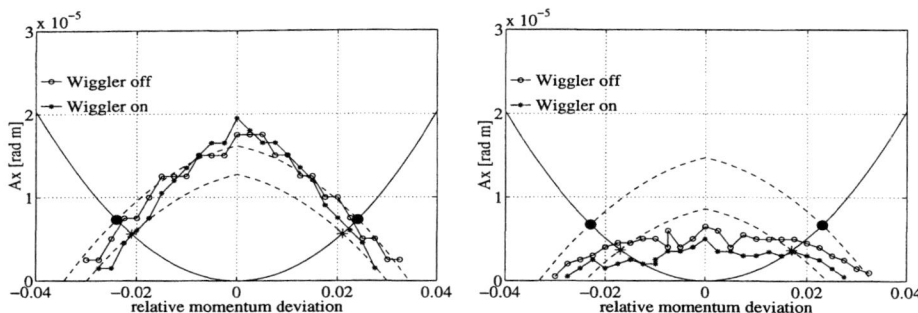

FIGURE 7. Maximum stable transverse emittance as a function of momentum deviation. Dashed lines show the dynamic aperture derived from a simple model. The left side is a case with low coupling, while the right side shows the case with additional coupling.

is less visible with small coupling because one still seems to need the coupling fields to excite large vertical oscillations which will hit the small gap vacuum chambers.

Table 1 summarizes the measurements and the tracking calculations for the cases mentioned throughout this text. It also shows if the tracking and the simple model agree qualitatively.

The reduction of the momentum aperture with wiggler and skew-quadrupoles is a disturbing effect for standard ALS operations. Usually the vertical beam size is enlarged by additional coupling to get a longer beam lifetime. To overcome this problem, the tune has been changed away from what we believe are the disturbing resonances. The momentum aperture (and thus the lifetime) improved, as can be seen in table 1. Further analysis, especially with the frequency analysis methods [6] should allow to predict which resonances are excited in the special cases.

VI CONCLUSION

Through measurements of the Touschek lifetime we are able to derive the momentum aperture of the Advanced Light Source. We find that the momentum aperture is not entirely determined by the rf-aperture (ε_{rf}). Also the physical apertures are large enough (for coupling up to 10%) that they do not limit the momentum aperture. The aperture is defined by dynamic effects, which are only partially described by our models. It is important to understand these dynamic limits to improve the lifetime in the ALS and accurately predict the lifetime (and dynamic aperture) in future machines.

TABLE 1. Measured and calculated momentum aperture in the straight and in the arc section for different machine conditions.

κ	$\xi_{x,y}$	ν_x	ν_y	measured ε [%] straight	arc	calculated[a] ε [%] straight	arc	agreement tracking-model
				Wiggler off				
0.01	1.0	0.31	0.22	>3.2	2.7	3.9	2.8	yes
0.01	1.0	0.27	0.18	>3.3	2.4	3.3	2.4	yes
0.01	1.0	0.22	0.13	2.6	2.0	2.8	1.9	yes
0.01	0.0	0.27	0.18	>2.8	2.8			
0.01	6.0	0.27	0.18	1.7	1.4	1.6	1.3	yes
0.10	1.0	0.27	0.18	>2.8	2.3	3.0	1.8	no
				Wiggler on				
0.01	1.0	0.27	0.18	>3.3	2.1	3.0	2.1	yes
0.10	1.0	0.27	0.18	>2.8	1.6	2.8	1.6	no
0.10	0.0	0.27	0.18	>2.8	1.7			
0.10	6.0	0.27	0.18	1.8	1.3			
0.10	1.0	0.31	0.22	>3.2	1.9	3.5	2.0	no

[a] Momentum deviation where dynamic aperture from tracking equals induced amplitude

REFERENCES

1. Bruck, *Accelerateurs Circulaires de Particles* (Presses Universitaires de France, Paris, 1966).
2. Nadji et al., in *Proc. Particle Accelerator Conference, Vancouver, 1997*, p. 1517.
3. Decking et al., in *Proc. European Particle Accelerator Conference, Stockholm, 1998*, p. 1262.
4. Robin et al., in *Proc. European Particle Accelerator Conference, Sitges, 1996*, p. 971.
5. Robin, Safranek, and Decking, submitted to *Phys. Rev. ST Accel. Beams*.
6. Laskar and Robin, *Particle Accelerators* **54**, (1996), p. 185-192.

Experimental Non Linear Beam Dynamics Studies with a Turn-by-Turn Phase Space Monitor at SPEAR

A.Terebilo*, C.Pellegrini*, M.Cornacchia**

*Department of Physics and Astronomy, UCLA 405 Hilgard Ave., Los Angeles, CA 90095
**Stanford Linear Accelerator Center,Stanford , California, 94309

Abstract. About 10 years ago the possibility of using turn-by-turn phase beam position monitors to gain insight into phase space dynamics of a single particle in a storage ring raised some interest among accelerator physicists. It was soon argued that Landau damping and collective effects would seriously complicate the interpretation of data. We have established that in SPEAR it is possible to lock a single bunch into a collective 'rigid body' mode. When in this mode the bunch will behave similar to a super particle of finite size with the charge equal to that of the bunch. In this paper we report on experiments that demonstrate this effect. We also numerically study the strong coupling limit, in which the transition to 'rigid body' motion occurs, for the two models proposed earlier. We present the experiments on non-linear resonance crossing and frequency map measurement in the 'super particle' framework.

INTRODUCTION

The motion of a single charged particle in a storage ring, including the effect of non-linear forces could, in principle, be studied by exciting oscillations around the equilibrium orbit with a fast kicker and measuring the transverse position at two points in the ring over many revolutions. This is equivalent to measuring a Poincaré section of the underlying dynamical system at time intervals naturally defined by the revolution period. This would be an ideal instrument that would provide direct insight into single particle dynamics in storage rings. It would also be an interesting non-linear system for scientists to study experimentally because it iterates much faster than non-linear circuits, systems with mechanical non-linear elements (springs) or celestial systems.

In practice, one cannot measure the position of a single particle. It is only possible to excite and measure the position of the center of mass (CM) of a bunch containing 10^9 - 10^{11} particles having different oscillation amplitudes and frequencies. This fact complicates the situation in two ways. Frequency spread within the bunch

CP468, *Nonlinear and Collective Phenomena in Beam Physics–1998 Workshop*,
edited by S. Chattopadhyay, M. Cornacchia, and C. Pellegrini

TABLE 1. Parameters of SPEAR electron storage ring

Energy	E_s	3.0 GeV
Revolution period	T	780 ns
Horizontal tune	v_x	7.17
Vertical tune	v_y	5.26
Horizontal chromaticity	ξ_x	0.8
Vertical chromaticity	ξ_y	1.1
Number of electrons in a bunch	N_b	(0.5-3) x 10^{10}
Bunch Length (mm)	l_b	10
Transverse bunch size (mm) At the BPM location	σ_x, σ_y	2.3 0.8
Radiation Damping Time	τ_x/T	11,000 turns
Momentum Compaction Factor	η_c	0.0013

leads to decoherence of initial oscillations which causes growth of the bunch size in the phase space [1][2]. Another problem arises from self-induced collective forces, which shift the frequency spectrum and introduce collective modes. However, under some conditions, the collective forces can be used to advantage to lock the bunch into a single CM dipole mode and make it behave similarly to a single particle with charge equal to that of the bunch. The growth rate of this mode can also be set to approximately or exactly compensate for radiation damping. This allows in principle to explore a region of the phase space for a much longer period of time, possibly long enough to apply standard analysis of non-linear time series, including extraction of invariant surfaces, Lyapunov exponents fractal dimension etc.

In this paper we report the observation of stable collective oscillation mode and the non-linear beam dynamics studies with it. The data was obtained in the electron storage ring SPEAR. The machine parameters relevant to this discussion are summarized in Table 1.

The transverse motion of each individual particle 'k' in the bunch can be described by the equation

$$\ddot{x}_k + \frac{1}{c\tau_x}\dot{x}_k + k_x(s)x_k = f(x_k, y_k, s) + \begin{bmatrix} Collective \\ Forces \end{bmatrix}$$

$$\ddot{y}_k + \frac{1}{c\tau_y}\dot{y}_k + k_y(s)y_k = g(x_k, y_k, s) + \begin{bmatrix} Collective \\ Forces \end{bmatrix} \tag{1}$$

The damping decrement terms are included to model radiation damping. Functions f and g result from x-y coupling and higher order terms in the Hamiltonian. To write the collective force terms explicitly one needs some knowledge of the impedance of the accelerator [5]. They contain fields from all particles at the location of the k-th particle generated at present (single turn) and all previous (multi-turn) revolutions. The dominant collective effect in SPEAR is the head-tail instability [4][5], a single turn effect with growth (damping) rate

$$\frac{1}{\tau_{x,y}}^{Head-Tail} = \frac{l_b \xi_{x,y} \cdot r_e \beta c N_b W_{\perp 0}(2l_b)}{(2\pi)^2 \gamma |\eta_c| v_{x,y}^2} \tag{2}$$

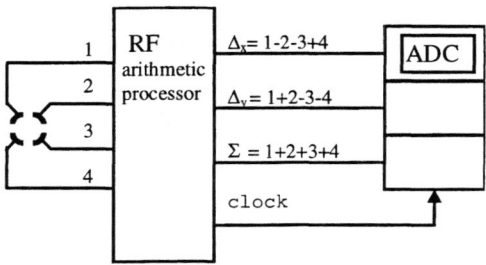

FIGURE 1. Data acquisition system block-diagram

where $W_{\perp 0}$ is the transverse wake function and r_e is the classical electron radius. Other terms in (2) are defined in Table 1. By adjusting the chromaticity or the bunch charge we can control the sign and the magnitude of the collective mode growth rate to make it larger or smaller than the radiation damping, thus obtaining damped, anti-damped or stable CM oscillations.

In the experiment we use horizontal and vertical beam kickers to produce large oscillations of the bunch compared to the bunch size. Under this condition we can write the position of an individual electron as $x_k=X_0+dx_k$, where X_0 is the center of mass motion produced by the kickers and dx_k is the amplitude spread related to the bunch transverse size. Since $X_0>>dx_k$, (1) can be rewritten approximately as an equation for the CM with the non-linear terms f and g evaluated at the CM position. This allows us to use the CM motion to study the non-linear beam dynamics.

The CM position is detected using Beam Position Monitors (BPMs), sets of button electrodes inserted in the vacuum pipe of an accelerator. The architecture of BPM data acquisition system is shown on Figure 1.

The raw button signals are linearly combined in the RF processor to produce signals proportional to the horizontal and vertical displacement, and a SUM signal proportional to the total bunch charge. The signals are sampled at the peak value by ADC modules and stored for up to 128K turns. The resolution of the single turn BPM measurements with time domain processing is determined by the geometry of the buttons, electronics noise and ADC bit resolution. At SPEAR we presently achieve 120 microns r.m.s. with 5mA in one bunch which degrades as 1/current.

EXPERIMENTAL RESULTS

The response to a horizontal kick when the chromaticity is set to its normal operating value $\xi_x=0.8$ is shown on Figure 2. In this plot the center of mass oscillation decays with damping time $\tau = T_0 \times 2000 \ turns$ (compare to radiation damping time $\tau_{rad} = T_0 \times 11,000 \ turns$) which indicates the presence of additional head–tail damping.

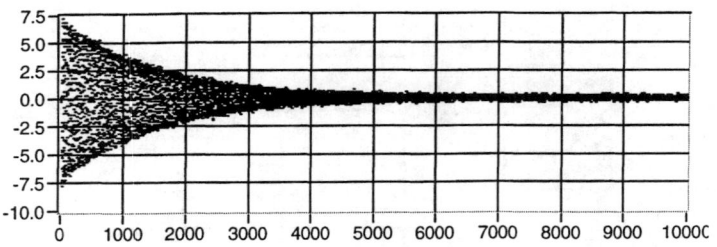

FIGURE 2. Horizontal position (mm) vs. turn number, $\xi_x = 0.8$

One indication that the motion remains coherent is the evolution of the horizontal tune. To demonstrate this, kicks of different amplitude were applied (Figure 3) and for each the instantaneous tune as a function of turn number and square amplitude was calculated using NAFF method [7]. The fact that the bunch traces the same detuning curve for all initial amplitudes with error no greater than the natural tune variation in SPEAR [6] indicates that the CM amplitude is the same as the amplitude of individual particles.

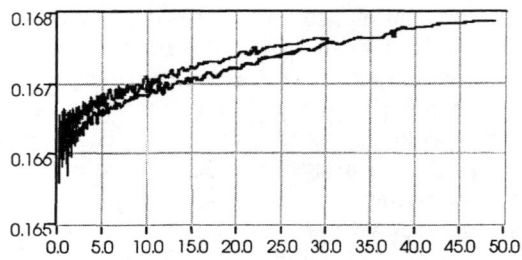

FIGURE 3. Horizontal tune versus square amplitude (mm^2) for three kicks with different initial amplitudes

It follows from (2) that one can balance the radiation damping with the head-tail growth. To demonstrate this we lowered the chromaticity to a negative value to obtain a quasi-steady mode with $\tau > T_0 \times 100,000\ turns$, as shown on Figure 4.

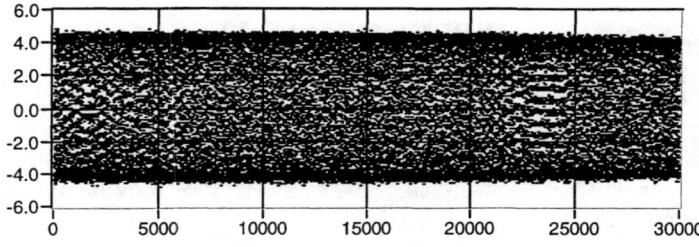

FIGURE 4. Horizontal position (mm) vs. turn number (Reduced chromaticity $\xi_x = -1.0$)

132

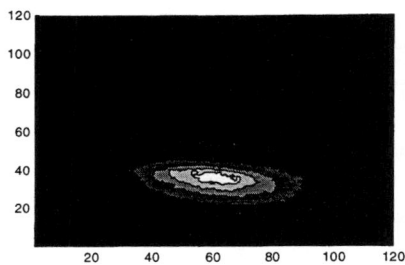

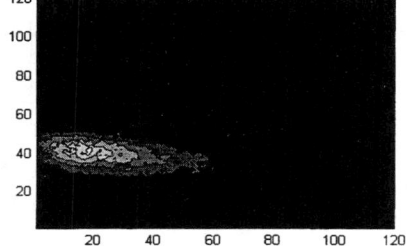

FIGURE 5. Transverse profile of unkicked beam

FIGURE 6. Transverse profile of a kicked beam after 30,000 turns

Snapshots of the transverse bunch density obtained by a gated CCD camera shown on Figures 5,6 further demonstrate, that the decoherence is not present and the bunch motion is that of a rigid body.

Knowledge that the oscillation measurements accurately capture the dynamics of a 'rigid body' mode justifies the approximation $X_0 >> dx_k$ made earlier and thus provides a powerful tool for studying non-linear dynamics in a storage ring.

With turn-by-turn systems, the dependence of tune on the amplitude is usually measured by extracting the tune immediately after many kicks of different amplitude, before decoherence sets in. Using the dipole mode with coherent particle motion in SPEAR, however, it is possible to measure tune shift with amplitude following a single kick. Figure 3 shows that the resulting detuning curve is the same as if it was measured using many kicks. This technique is faster and avoids errors introduced by slow variation of machine parameters, power supply ripple, or ground motion. By adjusting the chromaticity and thus the growth rate of the oscillation, one can observe particle loss in real time since the SUM signal is proportional to the instantaneous charge in the bunch. It can be seen from (2) that when some fraction of particles is lost the mode becomes damped. As shown in Figure 7 after the oscillation amplitude has grown to some critical value, enough particles are lost so that the response pattern changes to exponential damping.

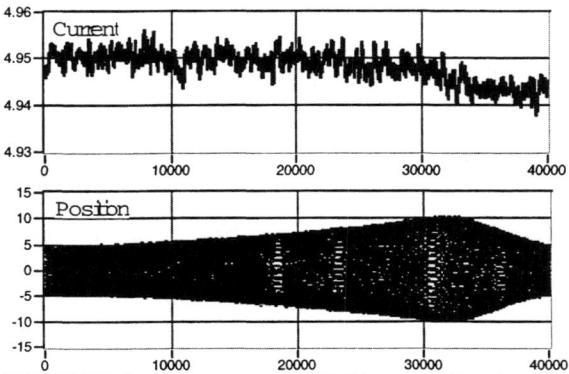

FIGURE 7. Instantaneous current (mA) and position (mm) versus turn number.

133

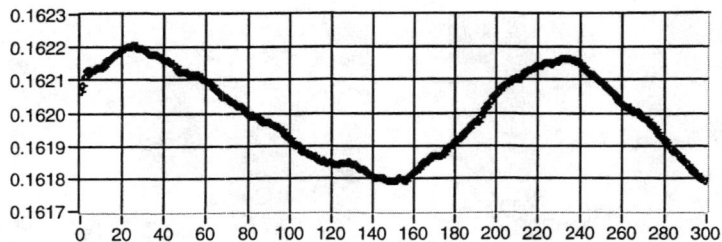

FIGURE 8. Fractional part of the horizontal tune versus turn number x 100

Tune variation due to the power supply ripple can be detected by measuring the tune at different phases with respect to 60 Hz line voltage [6]. In that case the individual measurements were performed with a few seconds time interval. Non ripple related factors such as ground motion and temperature drift affected the accuracy. Using the quasi-steady collective mode technique (Figure 8), ripple measurements take only about 30,000 revolutions, or 25 ms.

The collective mode can also be used to study the effect of non-linear resonances. When the beam is kicked the instantaneous tunes will deviate from the linear working point v_x^0, v_y^0 according to

$$v_x = v_x^0 + h_{xx}J_x + h_{xy}J_y + \dots$$
$$v_y = v_y^0 + h_{yy}J_y + h_{yx}J_x + \dots$$

(3)

where J_x, J_y are the action coordinates. The linear working point and the kick strength can be chosen so that the tune will start on one side of the resonance line and cross it as the amplitude changes. With the bunch locked in the dipole collective mode, it is possible to control the speed of the resonance crossing by changing the growth (damping) rate via current or chromaticity. Figure 10 shows the horizontal and vertical response to a horizontal kick near the resonance $2v_x - v_y = 9$. The combined radiation and head-tail damping is set to e-fold in about 1,300 turns. The horizontal oscillations coherently couple to the vertical plane.

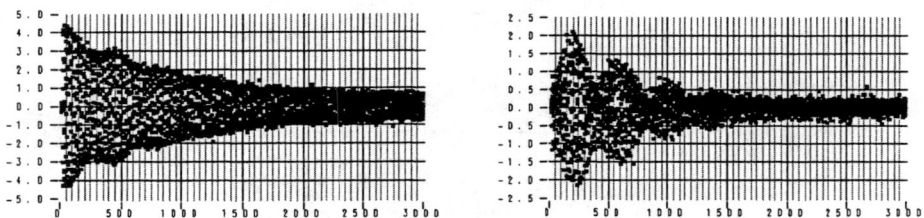

FIGURE 9. Horizontal and vertical position (mm) versus turn number. Fast resonance crossing. Horizontal motion coherently couples to vertical.

134

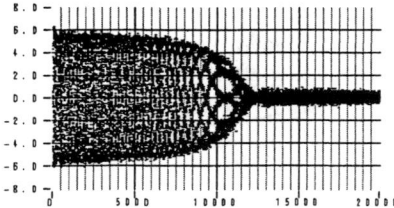

FIGURE 10. Horizontal position (mm) versus turn number. No coherent vertical oscillations are excited.

The response becomes qualitatively different with slower crossing (damping time ~30,000 turns). Figure 10 exhibits horizontal response under these conditions. In this case, no CM oscillations are observed in the vertical plane (not shown): slow crossing of a low order resonance effectively destroys the horizontal coherent mode.

The coherent mode is also useful to study higher order resonances in the frequency map [7]-[9]. We reported on the first proof-of-principle measurement of the frequency map in [10].

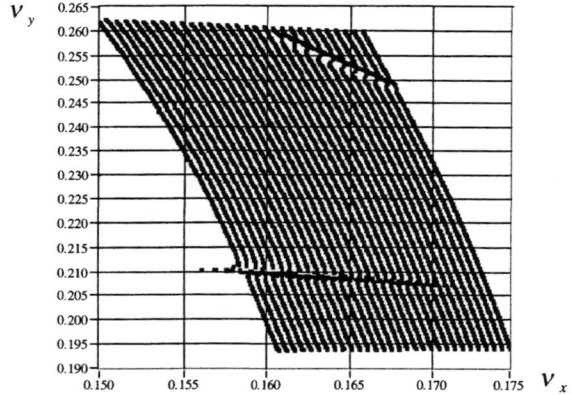

FIGURE 11. Frequency map of the SPEAR model. (Simulation)

The frequency map ('footprint') of SPE AR in Figure 11 was obtained by numerical tracking with initial conditions chosen on a uniform grid in J_x, J_y space. For each initial condition a single particle was tracked numerically for 2048 turns and the horizontal and vertical tune were computed using NAFF. The linear working point (ν_x = 7.166, ν_y = 5.26) is located in the upper right corner. The non-linear resonance lines act as attractors or repellers in the tune space. Four octupole magnets installed in SPEAR introduce a positive horizontal tune shift with horizontal amplitude [11]. To detect the presence of a resonance line we placed the linear working point slightly above it. The evolution of the tune (Figure 12) was then followed after applying a horizontal kick. In case (a), after the kick, the instantaneous tune starts on the other

horizontal tune

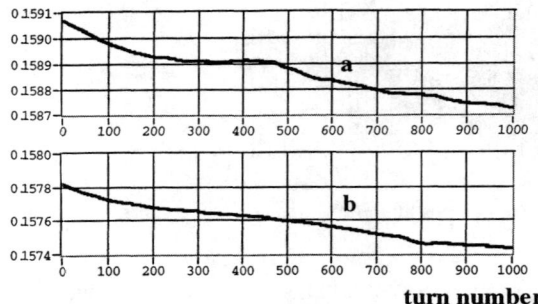

turn number

FIGURE 12. Experimental evidence of the $3v_x + 2v_y = 32$ resonance crossing. Horizontal tune vs. turn number for 2 different nearby linear working points set approximately to (a) (7.158; 5.261) and (b) (7.157; 5.261)

side of the resonance line due to the positive horizontal tune shift with amplitude. The tune then crosses the resonance as the amplitude decreases. The flat part of the upper graph is a manifestation of the resonance locking. In case (b) the, the linear tune is set below the line so it is not crossed.

MODELS

In the first approximation the bunch is modeled as a collection of independent oscillators with initial distribution in the phase space determined by the equilibrium bunch size and the kick parameters [1]. This approximation is valid for low enough current per bunch and chamber impedance. Several models were proposed [2][3], that include bunch interaction with accelerator environment. The first model assumes a large, but finite, set of coupled (with oversimplified coupling mechanism) linear oscillators. However simple and without rigorous assumptions, it gives an insight on how the transition between the Landau damped and 'rigid body' regimes happen. This linear system approach, though, will not predict any additional damping or anti damping. The second model we discuss describes coupling between the particles more physically through wake fields. It exhibits growth or damping of oscillations in addition to 'rigid body' locking.

Discrete Oscillator Model

Consider N coupled linear oscillators. The equations of motions are linear but the effect of non-linearity is accounted for in the distribution of frequencies $\rho(\omega)$. For each particle 'i',

136

$$y_i'' + \omega_i^2 y_i = \alpha \frac{1}{N} \sum y_i \qquad (4)$$

The coupling parameter α depends on the transverse impedance. In this simplified model every particle instantaneously sees the position of the center of mass of the bunch. This simplification is justified for short bunches. Otherwise one needs to add a second (longitudinal) dimension to the problem. The eigenvalue problem

$$(\Omega^2 - A) \times Y^{(k)} = \lambda \cdot Y^{(k)} \qquad (5)$$

where Ω - diagonal matrix with individual frequencies from the distribution $\rho(\omega)$
A - $N \times N$ matrix with all elements equal to α/N,

can be solved numerically for a reasonably large N. The solution satisfying the initial kick condition $Y'_i = Y'_0$ is

$$\overline{Y} = Y'_0 \sum_k^N \overline{Y}^{(k)} \left(\overline{Y}^{(k)} \cdot \overline{E} \right) \frac{1}{\omega^{(k)}} \sin \left(\omega^{(k)} t \right) \qquad (6)$$

$$\overline{E} = \begin{bmatrix} 1 \\ \dots \\ 1 \end{bmatrix}, \text{ unit vector} \qquad \omega^{(k)} = \sqrt{\lambda_k} \text{ frequency of the eigenmode}$$

Since BPMs detect only the position of the center of mass we need to sum over the components of the solution:

$$\langle Y \rangle_{C.M.} = Y'_0 \sum_k^N D^{(k)2} \frac{1}{\omega^{(k)}} \sin \left(\omega^{(k)} t \right) \qquad (7)$$
$$D^{(k)} = \left(\overline{Y}^{(k)} \cdot \overline{E} \right)$$

The contribution to the BPM signal of the 'k'-th eigenmode is proportional to the square of the sum of its components. It is easy to show that there is one eigenmode with frequency $\omega_{coh} \approx \left(1 - \alpha/2 \right) \cdot \omega_0$ where ω_0 is the center of $\rho(\omega)$. For $\alpha \gg \Delta$, characteristic width of $\rho(\omega)$, the contribution of this mode to the center of mass motion will be dominant.

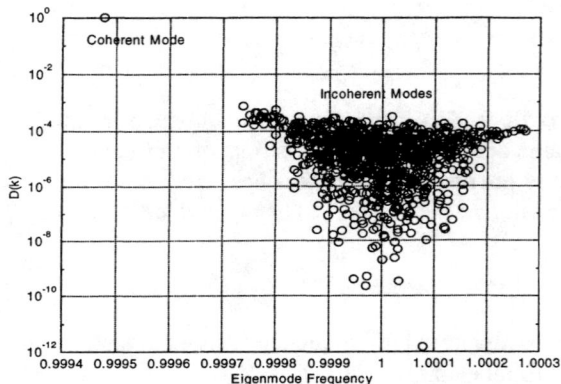

FIGURE 13. Strong coupling case. $N = 1000$, $\alpha = 0.001$, $\Delta = 0.0001$. Frequencies of the eigenmodes and their contribution to the centroid position as measured by BPMs.

Figure 13 shows the relative contribution to the center of mass of all modes for the case $\rho(\omega) \propto e^{\frac{-\omega^2}{\Delta^2}}$, $N = 1000$, $\alpha = 0.001$, $\Delta = 0.0001$. Figure 14 illustrates that only the coherent mode has significant projection on the unit vector and therefore contributes to the CM motion the most.

The centroid motion shown on Figure 15 can be well understood in terms of the dominant coherent eigenmode and incoherent modes. In this linear model the eigenmodes are uncoupled therefore the center of mass response will be determined primarily by the coherent mode. The position τ of the waist of the envelope reflects the

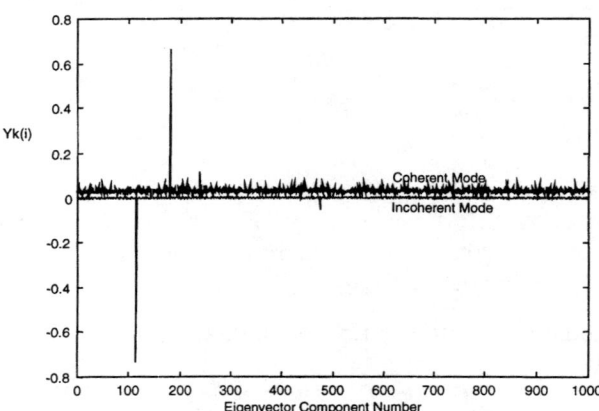

FIGURE 14. Components of the coherent and one of the incoherent eigenvectors

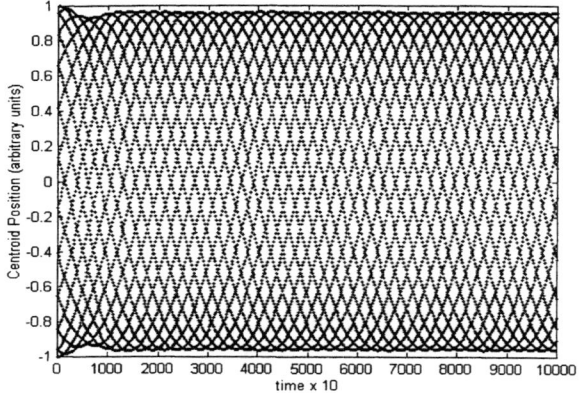

FIGURE 15. Strong coupling case. $N = 1000$, $\alpha = 0.001$, $\Delta = 0.0001$

condition $(\omega_{coh} - \omega_0)\tau = \pi/2$. This waist is well defined in all our experiments and can be used for instantaneous measurement of the coherent tune shift and impedance estimate.

In the case of weak coupling - small α there is no well defined coherent mode and the response of the centroid is Landau damped. Figures 16,17.

For intermediate cases the discrete oscillator model can be misleading producing repeating echoes with periodicity $T \sim N/\Delta\omega$.

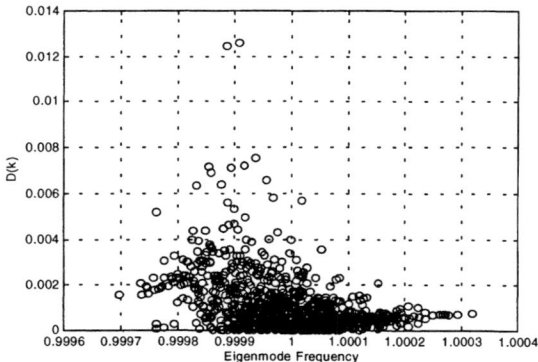

FIGURE 16. Weak coupling case. $N = 1000$, $\alpha = 0.0001$, $\Delta = 0.0001$. Frequencies of the eigenmodes and their contribution to the BPM signal. No well-defined coherent mode exists.

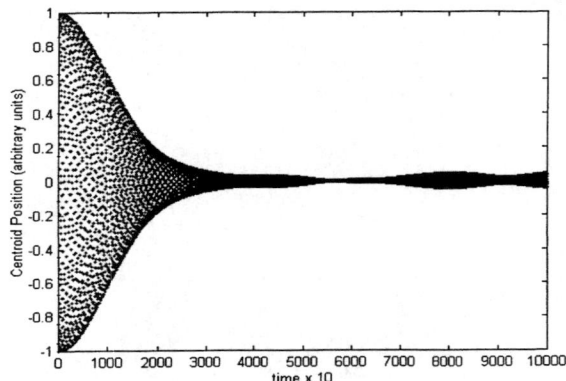

FIGURE 17. Weak coupling case. $N = 1000$, $\alpha = 0.0001$, $\Delta = 0.0001$ Center of mass oscillations are Landau damped.

Two Macro Particles with Continuous Distribution and Head-Tail Interaction

In this model [3] the bunch is divided into discrete head and tail parts executing longitudinal oscillations $z_1 = -z_2 = \hat{z} \sin(\omega_s s / c)$. Each has a frequency distribution $\rho(\omega)$ centered near $\omega_{1,2}$ which will change with the longitudinal phase $\omega_{1,2} = \omega_0 (1 + \xi \cdot z_{1,2} \omega_s / c\eta)$. ξ is the ring chromaticity.

The linearized equation of motion is

$$x''_{1,2} + \frac{\omega_{1,2}^2}{c^2} x_{1,2} = \varepsilon \cdot h_{2,1} \int x(s,\omega) \rho(\omega) d\omega \qquad (8)$$

where s is the longitudinal coordinate. ε is proportional to the wake field magnitude and the total bunch charge. h_i contains the shape of the wake and proper step-function to preserve causality. Equation 8 can be solved formally for the envelope of the oscillations by means of Laplace transforms:

$$\hat{X}(s) = \frac{1}{2i\pi} \hat{X}_0 \int_{\sigma - i\infty}^{\sigma + i\infty} \frac{k(p)}{1 - Rk(p)} e^{ps} dp \qquad (9)$$

where $k(p) = \int_0^\infty e^{-ps} ds \int e^{-i\Delta\omega s / c} \rho(\Delta\omega) d\Delta\omega$ is the Laplace transform of the Fourier image of the frequency distribution $\rho(\omega)$. $R = i(R_1 + iR_2 + iR_3\xi)$. R_I depends on $\Delta\omega_{coh}/\Delta\omega_0$ as parameter. The effect of imaginary impedance, real impedance and head-tail

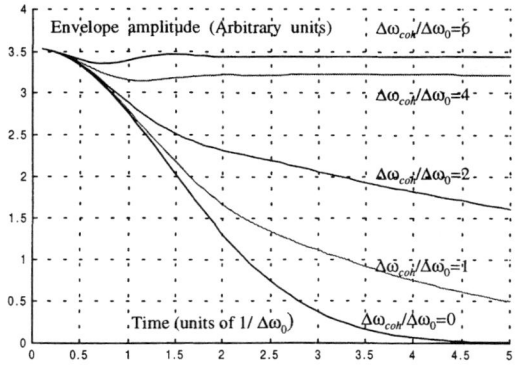

FIGURE 18. Envelope response. $R_2 = R_3 = 0$

interaction are separated in R_1, R_2, R_3 respectively. In case of zero wake-function all $R_i = 0$ and we recover the independent particles case [1]. Results of numerical integration of (8) for a Gaussian distribution of frequencies and different values of $\Delta\omega_{coh}/\Delta\omega_0$ are shown on Figure 18. As coupling between particles and therefore $\Delta\omega_{coh}/\Delta\omega_0$ increases the transition to a steady collective mode occurs.

Figure 19 shows that additional real impedance or head-tail effect on top of strong coupling result in exponential damping or anti damping of the centroid oscillation.

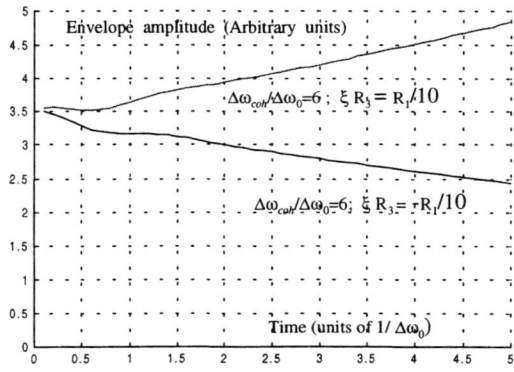

FIGURE 19. Envelope response. Additional head-tail effect leads to damping or anti damping

CONCLUSIONS AND SUGGESTIONS FOR FURTHER STUDIES

We have demonstrated that under normal operating current per bunch in SPEAR a single bunch behaves similarly to a macro particle. Using this macro particle as a phase space probe may lead to interesting studies such as measurements of frequency map and one turn map. We also plan to achieve an agreement between the experimental and numerical tracking on the turn-by-turn basis by treating the wake fields properly in our tracking codes.

REFERENCES

1.   Lee, S.Y., *Decoherence of Kicked Beams*, SSCL Report SSCL-N-749. (1991)
2.   Stupakov, G., Chao, A., SSCL Report SSCL-621 (1993).
3.   Stupakov, G., Chao, A., *Study of beam decoherence in the presence of head-tail instability using a two-particle model.* Proceedings of Particle Accelerators Conference 1995, Dallas, TX.
4.   Pellegrini, C., *Nuovo Cimento A* 64, 447 (1969).
5.   Chao, A., *Physics of Collective Beam Instabilities in High Energy Accelerators.* Wiley, New York, 1993.
6.   Terebilo, A., Pellegrini, C., Cornacchia, M., Corbett, J., Martin, D., *Measurement of the variation of machine parameters and the effect of the power supply ripple on the instantaneous tunes at SPEAR.* Proceedings of Particle Accelerators Conference 1997, Vancouver, Canada.
7.   Dumas, H., Laskar, J., *Phys. Rev. Lett.* **70** , 2975 (1993)
8.   Laskar, J., *Physica D* **67** 257 - 281 (1993).
9   Laskar, J., Robin, D., *Applications of frequency map analysis to the ALS*, Proceedings of the International Workshop on Single-Particle Effects in Large Hadron Colliders, Montreux, Switzerland (1995)
10.   Terebilo, A., Pellegrini, C., Cornacchia, M., Corbett, J., Martin, D., *Experimental non-linear beam dynamics studies at SPEAR,* Proceedings of Particle Accelerators Conference 1997, Vancouver, Canada.
11.   Tran, P., Pellegrini, C., Cornacchia, M., Lee, M., Corbett, W., *Measurement of the octupole induced tune shifts at SPEAR*, AIP Conference Proceedings Vol. **315**, 1994

Nonlinear Resonant Collimation for Future Linear Colliders

P.Emma, R.Helm, Y.Nosochkov, R.Pitthan, T.Raubenheimer,
K.Thompson, F.Zimmermann

Stanford Linear Accelerator Center, MS 26,
P.O.Box 4349, Stanford, CA 94309, USA

Abstract. We present a scheme for collimating large amplitude particles in the main linacs of a linear collider, by adding octupoles to the FODO lattice of the linac. With this scheme the requirements on downstream collimation can be greatly reduced or perhaps even eliminated. An analytic estimate of the amplitude at which particles are lost is made by calculating the separatrix of the fourth order resonance, and is in good agreement with the results of simulations. Simulations of particle distributions in the beam core and halo are presented, as well as alignment tolerances for the octupoles.

INTRODUCTION

Present designs for future linear colliders, such as the NLC [1] have dedicated collimation systems several kilometers in length. The collimation systems [1] are designed to serve two different functions: to protect all downstream systems against bunch trains which enter with large betatron excursions or large energy errors, and to remove the beam halo, which otherwise would cause unwanted background in the detector. An additional requirement is that in this scheme each betatron phase and each plane must be collimated twice.

It would of course be desirable to prevent or reduce the formation of the halo in the first place. However, there are many (probably unavoidable) sources of beam halo: (1) beam-gas Coulomb scattering, (2) beam-gas bremsstrahlung, (3) Compton scattering on thermal photons, (4) linac wakefields, (5) the sources, damping rings, and bunch compressors. In the Stanford Linear Collider (SLC) collimation of the beam before it enters the final focus and detector area was found to be essential, and it is expected that this will also be true for future linear colliders such as the NLC.

The NLC collimation system length of several kilometers is determined by the condition that spoilers and absorbers should survive the impact of an entire bunch train (nearly 10^{12} electrons). This requires a minimum spot size, in order that the collimator surface does not fracture or that the collimator does not melt somewhere

CP468, *Nonlinear and Collective Phenomena in Beam Physics–1998 Workshop,*
edited by S. Chattopadhyay, M. Cornacchia, and C. Pellegrini
1999 The American Institute of Physics 1-56396-862-2

inside its volume. For the NLC parameters, fracture and melting conditions give rise to about the same spot-size limit (roughly $10^6/\mu m^2$ for a copper absorber at 500 GeV [1]). While the surface fracture does not depend on the beam energy, the melting limit does, since the energy of an electromagnetic shower deposited per unit length increases in proportion to the beam energy. Therefore, the beam area at the absorbers must increase linearly with energy. Since, in addition, the emittances decrease inversely proportional to the energy, the beta functions must increase not linearly but quadratically. Assuming that the system length l scales in proportion to the maximum beta function at the absorbers, this results in a quadratic dependence [2], $l \propto \gamma^2$ ($\gamma \equiv E_{beam}/m_e c^2$). Counting both sides of the IP, the NLC collimation system is 5 km long. At 5 TeV the length of the collimation system could be 50 km. A schematic of such a conventional collimation system, as in the NLC design, is shown in Figure 1.

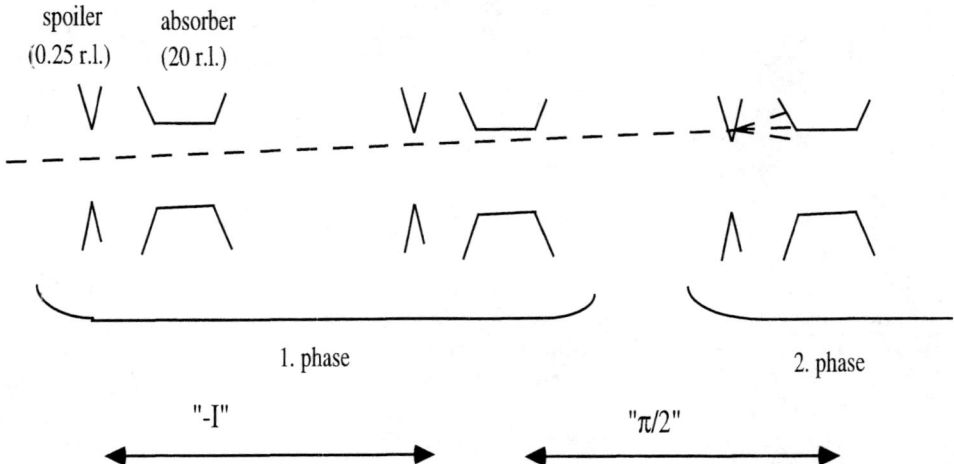

FIGURE 1. Schematic of a conventional collimation system, consisting of a series of spoilers and absorbers. The size of the spoilers and absorbers is approximately 1/4 and 20 radiation lengths (r.l.), respectively.

In this paper we present an alternative scheme for collimating large amplitude particles in the main linacs of a linear collider, by adding octupoles to the FODO lattice of the linac. The nonlinear fields of these octupoles are arranged in a configuration that resonantly destabilizes particles at large betatron amplitudes. The effective "dynamic aperture" of the linac can be controlled either through the octupole strength or through the phase advance per cell. Both halo particles and mis-steered beam pulses are dispersed by the nonlinear field, before they reach the end of the linac. Such a scheme could greatly reduce the requirements on a dedicated downstream collimation section, perhaps even eliminate it altogether.

The octupole magnets may be placed at every focusing quadrupole, where the horizontal beam size is largest, in order to remove horizontal beam tails. A similar beam line, with octupoles near the defocusing quadrupoles, may be employed for the vertical plane. Alternatively, an interleaved scheme which collimates both planes at the same time is conceivable.

In the NLC design, the required collimation depth is approximately 660 microns (corresponding to about $10\sigma_x$ at the defocusing quadrupoles). Note also that it may be acceptable to collimate further out by placing octupoles in front of the final focus quadrupole doublet; these have the effect of "folding in" residual beam tails before entering the doublet [3].

COLLIMATION IN ONE PLANE AT A TIME

We begin by considering the case of collimation in one plane at a time. As noted above, the octupoles are placed near every quadrupole that is focusing in this plane. First we give an analytic estimate of the collimating effect of such a system, and then we present simulations of the collimation, blow-up of the beam, and alignment tolerances.

Analytic estimate of collimation depth

For simplicity, in this section we describe the system by a smooth one-dimensional model and use a Hamiltonian approach to estimate the location of fourth-order separatrix introduced by the octupoles. A basic unit of the NLC octupole linac is assumed to consist of two FODO cells, with octupoles of effective integrated strength $k_1 = k_{oct,1}\beta^2$ and $k_2 = k_{oct,2}\beta^2$ (in the peculiar units of m^{-1}, where β is the beta function at the octupoles). Here $k_{oct,1}$, $k_{oct,2}$ are the conventional integrated octupole strengths in m^{-3}) near the two QF quadrupoles, i.e.

$$k_{oct} \equiv \frac{\partial^3 B_y/\partial x^3}{(B\rho)}\ell_{oct} \quad , \tag{1}$$

where ℓ_{oct} is the length of the octupole. This set of two FODO cells is then repeated periodically along the linac. The dynamics of this system is the same as for a storage ring with ring circumference equal to the length of the two FODO cells and with an effective tune equal to twice the phase advance per FODO cell divided by 2π. For the NLC [1], the effective tune is about 0.5, very appropriate for octupole collimation.

We choose as our canonical coordinates the action angle variables (I, ϕ) of the linear system. These are related to the physical transverse coordinates via

$$x(s) = \sqrt{2\beta(s)I}\cos\phi(s) \tag{2}$$

$$x'(s) = -\sqrt{2I/\beta(s)}\left(\sin\phi(s) + \alpha\cos\phi(s)\right), \tag{3}$$

where s is the position along this model linac. We will find it convenient to replace s by the azimuthal angle $\theta = 2\pi s/L$, where L denotes the circumference, and we will use θ as "time" variable.

The Hamiltonian describing this system then assumes the form

$$H(I,\phi,\theta) = QI + \frac{1}{4!}(2I)^2 \left[k_1 \cos^4 \phi \sum_{q=-\infty}^{\infty} \delta(\theta + q2\pi) \right.$$
$$\left. + k_2 \cos^4 \phi \sum_{q=-\infty}^{\infty} \delta(\theta + \pi + q2\pi) \right] \qquad (4)$$

where $\theta = 0$ is the location of the first octupole. We can expand the trigonometric functions and the delta functions, and then find

$$k_1 \cos^4 \phi \sum_{q=\infty}^{\infty} \delta(\theta + q2\pi) = \frac{k_1}{16}\left[e^{i4\phi} + e^{-i4\phi} + 4e^{i2\phi} + 4e^{-i2\phi} + 6 \right] \frac{1}{2\pi} \sum_p e^{i\theta p} \qquad (5)$$

$$k_2 \cos^4 \phi \sum_{q=-\infty}^{\infty} \delta(\theta + \pi + q2\pi) =$$
$$\frac{k_2}{16}\left[e^{i4\phi} + e^{-i4\phi} + 4e^{i2\phi} + 4e^{-i2\phi} + 6 \right] \frac{1}{2\pi} \sum_p e^{i(\theta+\pi)p} \qquad (6)$$

with $\sum_p$ a sum over all integer numbers p. There are infinitely many terms. Of relevance are only the resonant terms, which are not rapidly oscillating. For a linac phase advance of 90 degree per cell, we have $2\phi(\theta) \approx \theta$ (note that the "ring" comprises two FODO cells). Hence, we must keep the resonant terms $\pm(2\phi - \theta)$ and $\pm(4\phi - 2\theta)$, as well as the secular term.

If we choose $k_1 = -k_2$, the driving term of the $\pm(4\phi - 2\theta)$ resonance and the secular term cancel. Changing variables to $\psi = \phi - \theta/2$ and defining $\Delta Q = Q - 1/2$, the total Hamiltonian in the "rotating" frame is

$$\hat{H}(I,\psi)_{k_1=-k_2} = \Delta QI + \frac{1}{12\pi}I^2 \, k_1 \cos(2\psi) \qquad (7)$$

If, on the other hand, we use equal-sign octupoles, the Hamiltonian becomes:

$$\hat{H}(I,\psi)_{k_1=k_2} = \Delta QI + \frac{1}{48\pi}I^2 \, k_1 \cos(4\psi) + \frac{1}{16\pi}I^2 \, k_1 \qquad (8)$$

The odd-sign configuration of Eq. (7) is preferable, since the coefficient in front of the resonance driving term is four times larger and there is no amplitude-dependent tune, which could drive large-amplitude particles away from the resonance.

Let us then examine the odd-octupole configuration more closely. The Hamiltonian of Eq. (7), suggests that we might attempt to estimate the dynamic aperture limit A by setting the instantaneous tune where the particle spends most of its time $(\cos 2\psi \approx \pm 1)$ to the resonant tune of $1/2$:

$$\left[\frac{\partial \hat{H}}{\partial I}\right]_{I=A,\ \cos^2 2\psi=\pm1} = \frac{\partial}{\partial I}\left[\Delta Q I \pm \frac{1}{12\pi} I^2\, k_1\right]_{I=A} = 0 \qquad (9)$$

or

$$A \approx \left|\frac{6\pi \Delta Q}{k_1}\right| \quad , \qquad (10)$$

indicating a strong dependence on the tune difference ΔQ. Figure 2 compares this estimate with a simulation result for a two-dimensional map consisting of linear rotations and octupole kicks.

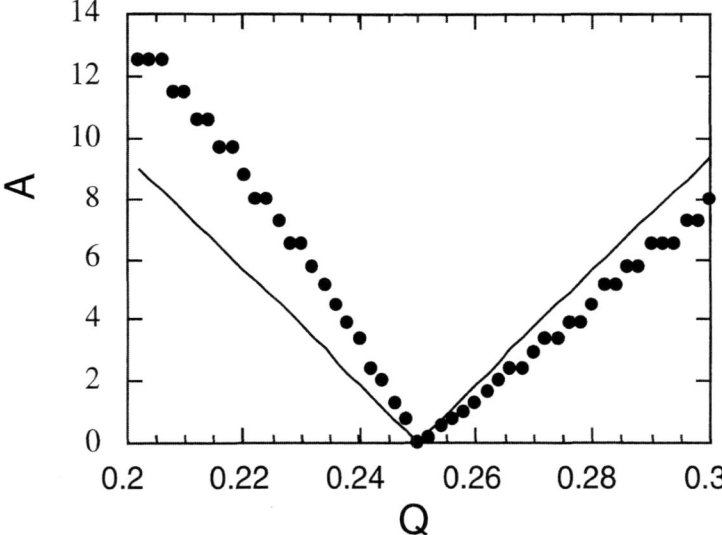

FIGURE 2. Dynamic aperture vs. the phase advance per cell $Q/2$: simulation result (dotted) and the crude analytical estimate of Eq. (10) (solid). For this example we have taken $k_1 = 1$.

Simulation results

In the following we present a series of simulation results for a realistic example, which approximates the situation over the first few kilometers in the NLC main linac. We assume normalized emittances of the order $\gamma \epsilon_x = 3 \times 10^{-6}$ m, and $\gamma \epsilon_y = 3 \times 10^{-8}$ m. The vertical emittance is slightly smaller than the NLC design emittance, and thus represents a worst case for the purpose of this study. The beam energy at the linac entrance is 10 GeV. The real linac accelerates the beam of course, and the spacing and strength of the quadrupoles increase along its length. We may, however, consider an "equivalent" linac with beam energy equal to that of the incoming beam in the real linac (10 GeV), without acceleration, and with strength and spacing of the quadrupoles equal to that of the magnets at the beginning of the real linac. The pole tip fields of the octupoles in the real linac would scale approximately as γ, if they are taken as constant in this "equivalent" linac.

The linac section considered consists of FODO cells with a length of 12.5 m, and a horizontal phase advance per cell close to 90 degree. Octupoles with an integrated strength of 4000 m^{-3} are placed at the focusing quadrupoles. At a beam energy of 10 GeV, this strength corresponds to a pole-tip field of 0.3 kG, for a 10 cm long octupole with 5 mm bore radius. The sign of the octupole field alternates from cell to cell, in accordance with our considerations above. For these parameters, collimation in the horizontal plane takes place at about 10–12 σ_x. This horizontal collimation section must be followed by an equivalent linac segment with octupoles at the defocusing quadrupoles, and with a vertical phase advance near 90 degree per cell. One advantage of separating the horizontal and vertical octupole sections is that the resonant collimation is most effective near a 90° phase advance. This phase advance can be established only in one plane at a time, because the NLC design foresees a 10° per cell phase-advance difference between the two planes, in order to reduce the sensitivity to skew quadrupole errors and to ions.

We present simulation results for collimation in the horizontal plane. Fig. 3 and Fig. 4 depict the emittance growth and the beam loss experienced by a beam that is injected into the linac with a betatron oscillation of varying amplitude. For injection oscillations less than about $10\sigma_x$ there is no significant emittance growth due to the octupoles. For larger oscillations, the beam size blows up quickly, while at the same time the transmission drops rapidly from 100% to roughly 0. For a $15\sigma_x$ incoming oscillation all particles are lost in the linac. The collimation amplitude is thus sharply defined.

Figure 5 illustrates that for larger oscillations the beam is lost completely in the linac. The positions of particle losses are spread out over a large region, as long as the betatron oscillation is less than about $40\sigma_x$. This is important in order to reduce the heat load for dedicated absorbers. At large oscillations the entire beam is lost within a few FODO cells after injection. Oscillations of this magnitude must be prevented by a machine protection system.

Alignment tolerances

We have also performed simulations to quantify the effects of octupole misalignments. The misalignments are assumed to be randomly distributed according to a Gaussian truncated at either $\pm 2\sigma_{misal}$ or $\pm 3\sigma_{misal}$. The value of σ_{misal} is assumed to be the same for both x and y. In Figure 6 we show the emittance blow-up factors in the two transverse planes as a function of σ_{misal}, for the cases of truncation at $2\sigma_{misal}$ or $3\sigma_{misal}$.

We see that to keep the emittance growth down to a few percent, the required tolerance is a standard deviation of about 200 μm. The octupoles must be tied to

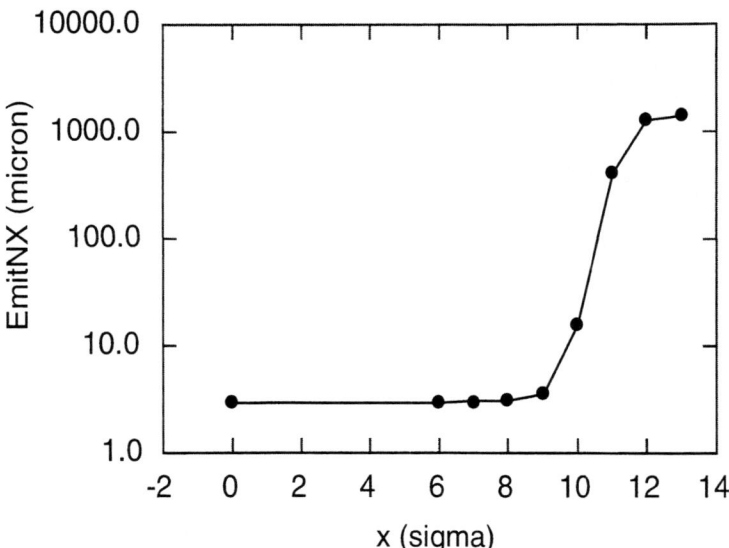

FIGURE 3. Emittance growth due to an incoming betatron oscillation: Shown is the horizontal emittance growth after about 160 FODO cells, as a function of the initial horizontal betatron amplitude in units of the rms beam size.

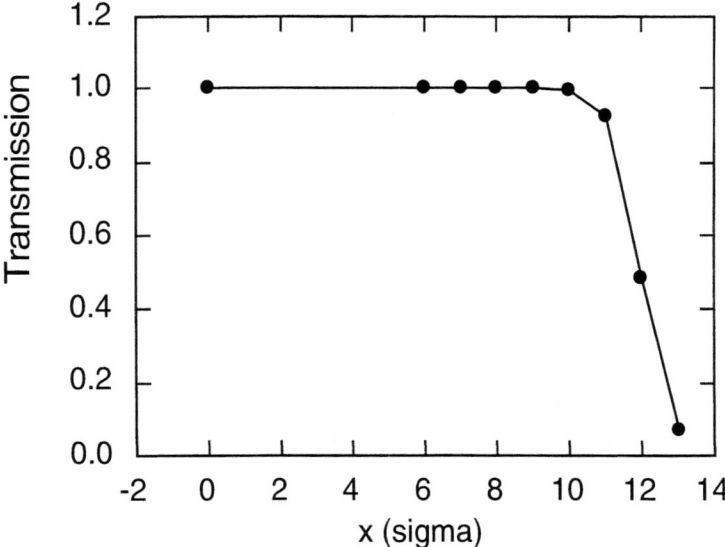

FIGURE 4. Beam loss due to an incoming betatron oscillation: Shown is the fraction of beam lost after about 160 FODO cells, as a function of the initial betatron amplitude in units of the rms beam size.

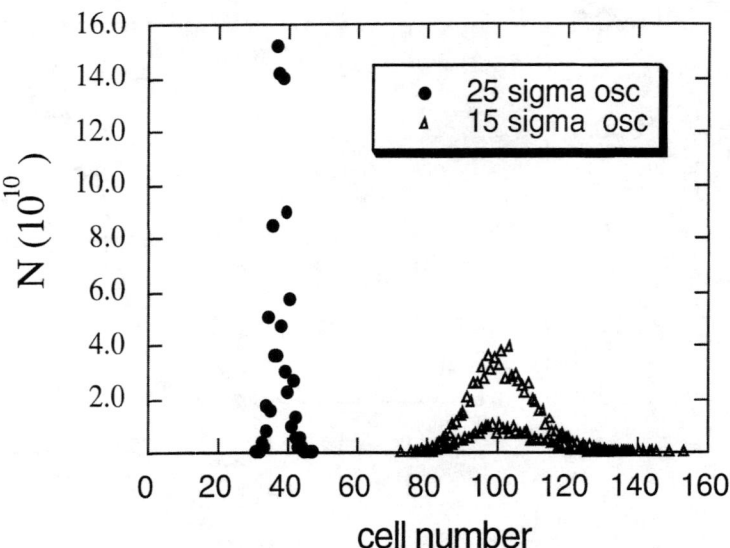

FIGURE 5. Distribution of lost particles along the NLC linac, considering incoming horizontal oscillations, with an amplitude of $15\sigma_x$ and $25\sigma_x$, respectively.

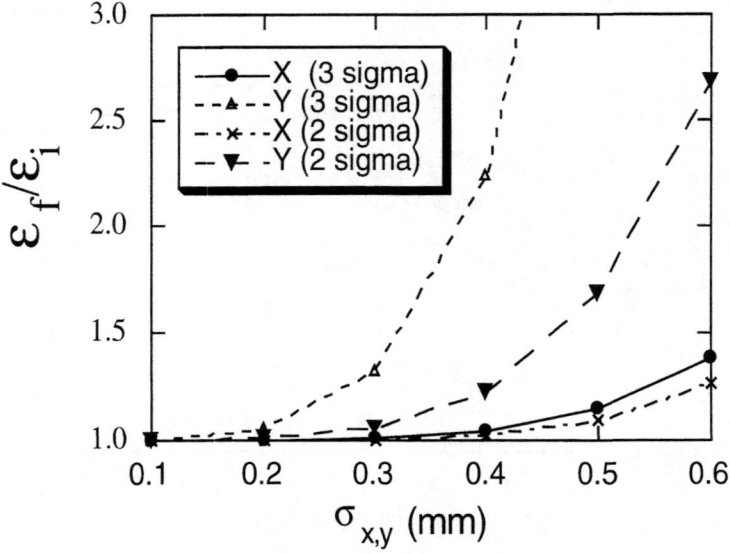

FIGURE 6. Emittance blow-up factor due to octupole misalignments

the quadrupoles at this level of precision. The quadrupoles themselves are aligned with respect to the beam to within 1–2 μm. At the SLC it was attempted to align the final-focus sextupoles with respect to the next quadrupoles with a precision of 50 μm, which proved to be too difficult a tolerance to maintain in the actual tunnel. However 150 μm is routinely achieved.

SIMULTANEOUS COLLIMATION IN BOTH TRANSVERSE PLANES

In this section, we present simulation results using an octupole configuration that allows simultaneous collimation in both x and y planes. The advantage of this scheme is that the halo is collimated in both transverse plane at once, which may save space or increase the collimation efficiency. There is an octupole at every quadrupole (defocusing and focusing), and there is a phase advance per FODO cell close to 90 degrees in both transverse planes. The basic linac cell for our proposed scheme is shown in Figure 7. The sign of k_{oct} alternates from FODO cell to FODO cell, where the kicks in x and y are related to k_{oct} according to:

$$\Delta x' = -\frac{1}{6}k_{oct}(x^3 - 3xy^2) ,$$

(11)

$$\Delta y' = -\frac{1}{6}k_{oct}(y^3 - 3x^2y) .$$

(12)

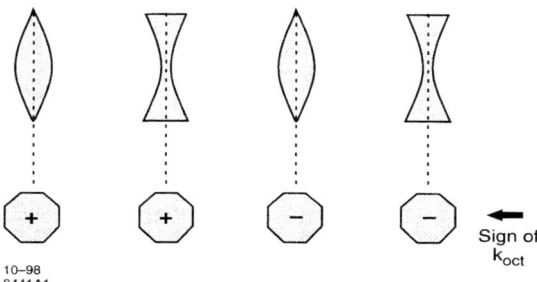

10–98
8441A1

FIGURE 7. Basic linac cell for octupole resonant collimation scheme. The significance of the sign of k_{oct} is discussed in the text.

As discussed previously, we consider an "equivalent" linac with beam energy 10 GeV. This constant FODO cell length is taken to be 12.4 m. We use stronger octupoles than in the preceding section, namely 40000 m^{-3} (which for a beam energy of 10 GeV, an inner bore radius of 5 mm and an octupole length of 10-cm, yields a 3-kG pole tip field). The beam sizes are $\sigma_x = 66\mu$m $(27\mu$m) and $\sigma_y = 3.3\mu$m $(8.0\mu$m) for the quadrupoles that are focusing (defocusing) in x.

To visualize the effects of collimation, we consider initial distributions consisting of rings in the $x - y$ plane; the distribution is taken to be uniform in such a ring and also in a rectangle in the $x' - y'$ plane that is cut off at $\pm n_x \sigma'_x$ and $\pm n_y \sigma'_y$, where $n_x(n_y)$ is the radius of the ring in units of $\sigma_x(\sigma_y)$. In Figure 8, we show the fate of the particles in such a halo ring after 350 FODO cells (the equivalent of the full length of the linac), as a function of halo ring radius. The solid curve (with o's) shows the percentage of particles that leave the halo ring and are collimated. The dashed curve (with x's) shows the percentage of particles still remaining in the halo ring. The dotted curve (with +'s) shows the percentage of particles that leave the halo ring, but do not reach the collimation radius of 4 mm. We see that all particles beyond 8-10σ are collimated. It may be possible to reduce the number of particles that are removed from the halo rings between approximately 2σ and 6σ but do not reach the collimation radius, by progressively weakening the octupoles along the linac.

In Figure 9, we show the pattern of loss of the particles in three different halo rings in the $x - y$ plane: (a) a ring from $4\sigma_x$ to $6\sigma_x$, (b) a ring from $6\sigma_x$ to $8\sigma_x$, (a) a ring from $8\sigma_x$ to $10\sigma_x$, as the particles travel down the linac. The losses are distributed over a substantial fraction of the linac length in all cases.

Blow-up of mis-steered beam

The beam density of a mis-steered beam that hits an absorber somewhere along the linac has to be sufficiently low to guarantee absorber survival; as noted earlier, a particle density of about $10^6 \mu m^{-2}$ suffices throughout the linac. Near the beginning of the linac, the nominal particle density in a full train of n_b bunches with N particles per bunch hitting perpendicular to an absorber surface would be about $n_b N / \sigma_x \sigma_y \sim 10^9 \mu m^{-2}$. In our scheme, only a fraction f of the train particles are lost on a given absorber (see Figure 10 below; at any given magnet f is at most a few percent). Thus, a nominal-emittance NLC bunch train needs to be blown up in transverse area by a factor of about $10^3 f$.

In Figure 10 we show the blow-up of the beam area for a beam that is mis-steered and lost somewhere along the linac, as a function of the location where it is lost. The phase advance per cell is 90° and k_{oct}=40000 m^{-3}. The beam is initially offset in x by $10\sigma_x$, and at the initial (focusing in x) quadrupole, the beam σ_x=66 μm and σ_y=3.3 μm. The vertical axis is the product of the standard deviations of the particle distributions in the two transverse directions. Note that the losses occur near the peaks of the beam area curve since these are also where the orbit excursions are largest. The beam area at the location of the losses blows up by a factor of over 100 after about 40-50 FODO cells (80-100 magnets on horizontal axis), and the fraction of beam particles lost at a time is not more than a few percent. Thus the above criterion for the blow-up factor is met, and so a mis-steered beam will not destroy the absorbers or accelerator structures.

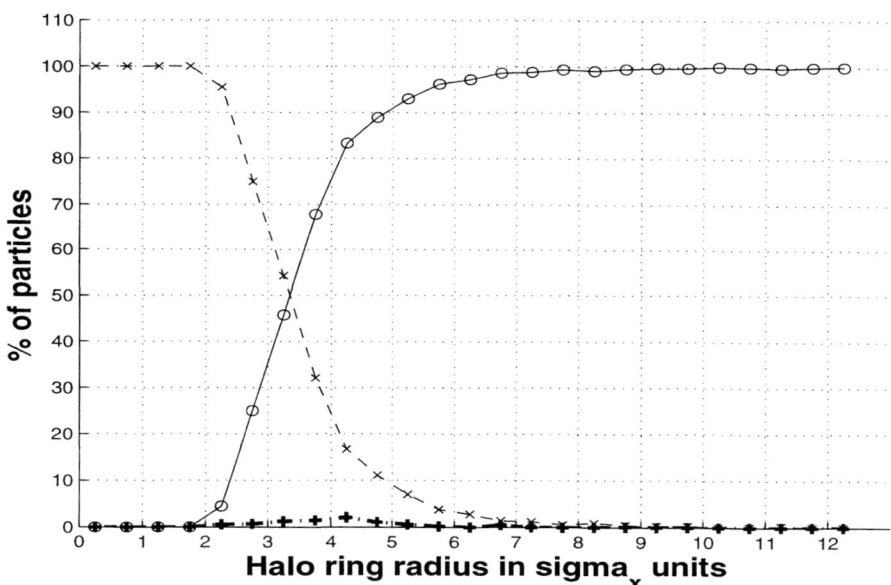

FIGURE 8. As a function of halo ring radius: Percent of particles remaining in halo ring after 350 FODO cells (dashed curve with x's), percent of particles that leave halo ring but do not reach collimation radius of 4 mm (dotted curve with +'s), percent of particles in halo ring that have been collimated (solid curve with o's)

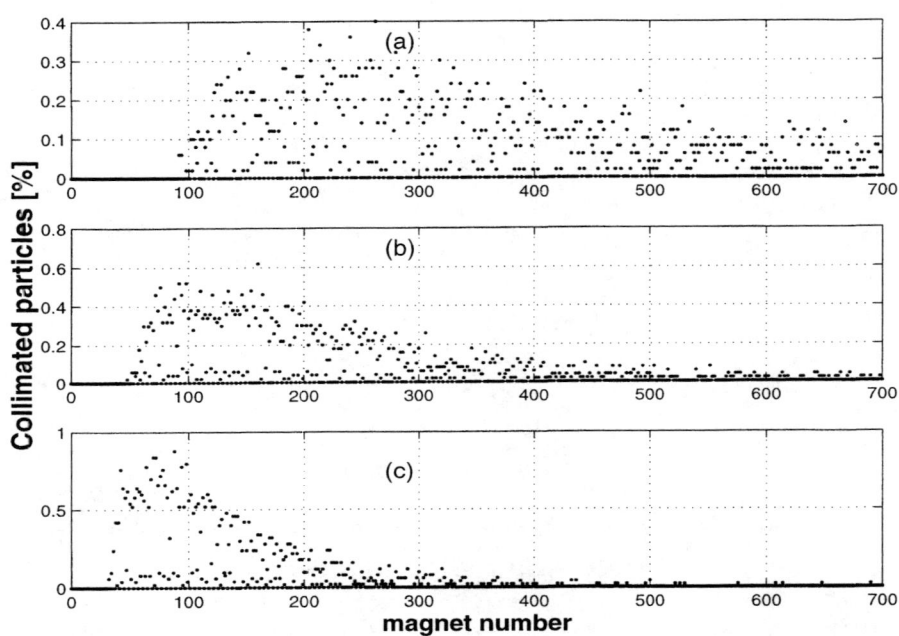

FIGURE 9. Pattern of loss of particles in three different halo rings, as a function of distance along the linac, for a collimation radius of 4 mm. (a) For halo ring from $4\sigma_x$ to $6\sigma_x$. (b) For halo ring from $6\sigma_x$ to $8\sigma_x$. (c) For halo ring from $8\sigma_x$ to $10\sigma_x$.

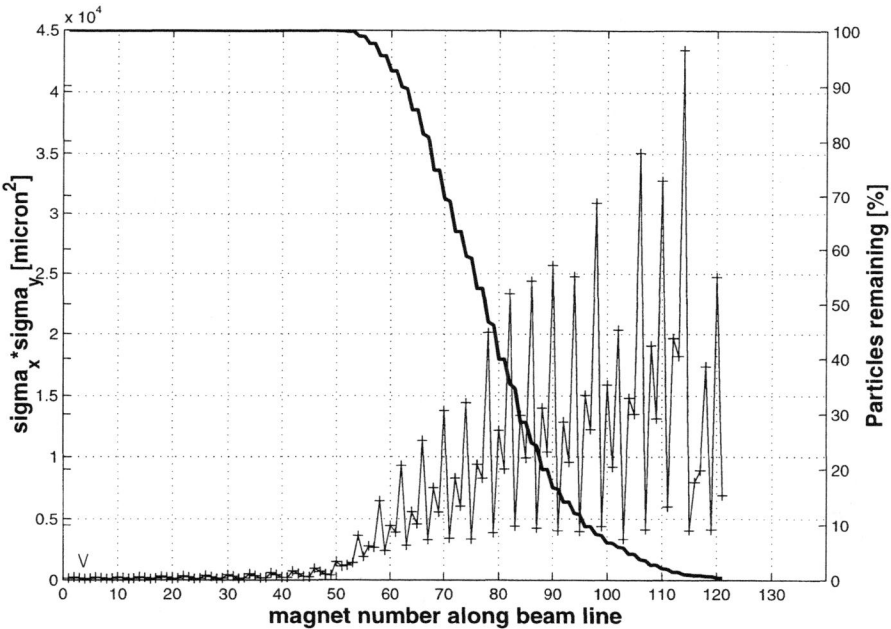

FIGURE 10. Beam blow-up (dotted curve) and loss (solid curve) as a function of quadrupole number along the linac, for a beam offset by $10\sigma_x$ ($= 660~\mu$m) in x.

CONCLUSIONS AND FUTURE WORK

The scheme presented here seems quite promising; it appears that it can both collimate the beam at the required collimation depth with reasonable octupole strengths, and it can blow up a mis-steered, nominal emittance beam by the required factor of 100 or more before hitting an absorber. This scheme can be integrated into the main linac and thus may allow removing or at least shortening dedicated collimation sections after the linac, which in the present NLC design are several kilometers in length. Further study is needed to determine the optimum distribution of x and y collimation in the linac. Another natural topic for further study is the use of dodecupoles instead of or in addition to octupoles, in which case one would expect a sharper collimation of the halo and a better preservation of the core.

A number of other issues remain to be looked into. This scheme inherently involves reducing the dynamic aperture of the linac; the implications of this for linac operations needs to be fully explored. There are concerns about spreading the lost particles over large regions, in particular the possibility of significant radioactivation. Another issue is the specification of linac absorbers, and whether these can be placed close to the beam without diluting the emittance. Finally, the issue of halo regeneration from particles in the core, including the effects of wakefields and beam-gas scattering, needs to be looked at more closely.

REFERENCES

1. Adolphsen, C., "Zeroth Order Design Report for the Next Linear Collider," SLAC-Report **474** (1966).
2. Zimmermann, F., *"New Final Focus Concepts at 5 TeV and Beyond,"* Eighth Advanced Accelerator Concepts Workshop, Baltimore, July 5-11 1998; SLAC-PUB-7883.
3. Tsoupas, N., Lankshear, R., Snead, C.L., Ward, T.E., Zucker, M., Enge, H.A., "Uniform Beam Distributions Using Octupoles," *Proceedings of IEEE Particle Accelerator Conference*, San Francisco (1991).

Overview of single-particle nonlinear dynamics

E. Todesco

CERN, LHC Division Geneva, CH 1211

Abstract. We give an overview of the single-particle non linear dynamics in circular accelerators. The main topics are: integration of equations of motion, fast symplectic tracking, dynamic aperture definition, long-term methods, quality factors and lattice optimization. Special emphasis is put on ideas and tools developed during the last decade.

INTRODUCTION

The single-particle nonlinear dynamics in accelerator physics has raised much interest during the last decades, both in superconducting hadron colliders whose magnets have strong nonlinearities, and in high performance light sources. In this paper we review recent work in this field, limiting ourselves to the analysis of the single-particle dynamics in a strict sense, i.e., excluding beam-beam interaction, space charge phenomena, or intrabeam scattering.

We will initially make some remarks about the modeling of a lattice and the approaches to analyse the dynamics, namely numerical integration and perturbative theory (section 1). Then, we briefly outline the ideas that have been developed to speed up the numerical integration through fast symplectic tracking (section 2). Having built the tools for integrating the equations of motion, one can carry out simulations to evaluate the so-called dynamic aperture (DA), i.e. the dimension of the domain in phase space where trajectories are stable. We point out some difficulties in the DA definition in section 3. For electron machines the damping time is of the order of $10^2 - 10^4$ turns and therefore only the short-term stability is determinant. For large hadron machines the requested stability time is of the order of $10^6 - 10^8$ turns: in section 4 we review some methods that have been developed to analyse this problem and some numerical tools for the phase space analysis.

The leading mechanisms that rule both short term and long term dynamic aperture are not yet understood. Moreover, the parameteric dependence on the lattice lay-out is unknown with the exception of few cases. We make some basic remarks about this problem in section 5. In section 6 we review some optimization techniques to increase the lattice dynamic aperture; we discuss the idea of replacing

CP468, *Nonlinear and Collective Phenomena in Beam Physics–1998 Workshop*, edited by S. Chattopadhyay, M. Cornacchia, and C. Pellegrini

the optimization of the DA with the optimization of a quality factor (QF) that is strongly correlated with the DA. These techniques have been widely used in beam dynamics both for protons and electrons machines. The knowledge of a good QF allows a theoretical understanding of the phenomena that rule the DA and either an analytical optimization or a speeding up of the numerical optimization. Finally, the basic practical techniques that can be used to improve lattice performance (changing the optics, inserting correcting elements, sorting the magnets) are outlined in section 7. Some open problems are given in section 8.

INTEGRATION OF EQUATIONS OF MOTION

The single-particle motion in a magnetic lattice is described by an Hamiltonian with three degrees of freedom: two transversal (x, y) and one longitudinal s. We restrict to the case of circular lattices, where the dependence on s is periodic. It is customary to use s as the independent time-like variable and therefore one ends with the Hamiltonian

$$H(x, p_x, y, p_y, t, \delta; s)$$

that describes the single-particle motion; here δ is the relative deviation of the particle energy. The nonlinearities in H can be summarized as follows

- In electron machines the main source of nonlinearities are the sextupoles that are used for chromatic correction, i.e., for reducing the change of the linear betatron frequencies for off-energy particles.

- In proton machines built with superconducting magnets the dipoles and the quadrupoles are affected by strong nonlinear errors.

Such a nonlinear Hamiltonian has no analytic solution and therefore one has to go for two complementary approaches.

- One can use numerical tools to provide an approximate solution to the equations of motion (tracking codes). Tracking is a precise tool to determine the particle trajectory in every regime (both weak and strong nonlinearities), but it has the drawback of being a 'black box' that provides little theoretical understanding of the nonlinear motion.

- One can use a perturbative approach to derive analytical information on some dynamical quantities such as detuning, phase space deformations, resonances, etc. Perturbative theory gives a worse approximation of motion compared to tracking, and fails when the nonlinearities become too strong; it has the advantage of providing an analytical understanding of the dynamics, even though much care has to be paid not to apply it beyond its validity limits.

It is general belief that the numerical integration should mantain the symplectic structure of the original Hamilton equations. This should be preserved also for

158

electron machines, where the non-Hamiltonian effects due to the radiation should be added in the framework of a symplectic scheme.

A very simple method to derive an explicit symplectic integrator is the kick approximation. The idea is to replace a nonlinearity which is diffused along an element with one or more delta 'functions', keeping constant the integral of the nonlinearity along the element. This approach is widely used to track large hadron machines. More refined symplectic schemes can be worked out for general cases [1–3]. For short machines the problem of constructing explicit symplectic integrators is more involved [4], since for instance fringe fields at the edge of the magnets become relevant.

In any case, the motion along the lattice is integrated through a series of explicit symplectic nonlinear maps. These maps can be written using the powerful formalism of Lie series, whose first applications to accelerator physics date back to the second half of the seventies [5–7]. One can produce Taylor expansion of these maps; the map coefficients are related to the nonlinear aberrations of the lattice. By composing and truncating these maps one obtains the truncated one-turn map; truncation is unavoidable as computer codes can deal only with a limited number of Taylor coefficients (order 20 for a four-dimensional mapping is already very challenging in terms of memory and processing time). The formalism of truncated power series and its implementation in computer codes has been known for decades in the field of celestial mechanics [8,9], and applied to accelerator physics in the eighties (see for instance [10]).

A lot of analytical work has been carried out to describe betatron motion. One can choose between the standard approach based on canonical perturbative theory for hamiltonian flows [11,12] and the discrete approach based on the truncated one-turn map and normal forms [13–15]. In the first case the perturbative parameters are usually the field gradients, whilst in the second one the series are in powers of the actions (i.e., amplitudes in phase space). This difference leads to a completely different ordering of the nonlinear contributions. We believe that the second approach in general is more suitable to describe the nonlinear motion. Moreover, effective codes have been written to compute the map perturbative expansion at arbitrary order [14,16]. This feature makes the map approach very attractive and powerful for the analysis of nonlinear betatron motion.

FAST SYMPLECTIC TRACKING

The numerical integration through symplectic tracking can be very onerous, especially in the case of large machines with several sources of nonlinearities. For hadron machines, this effect severely restricts the possibility of carrying out exhaustive studies of long-term stability. On the other hand, in the case of electron machines the dynamic aperture is determined by short-term phenomena, and the possibility of having fast tools to integrate the equation of motion allows a better optimization of the lattice parameters.

159

For hadron machines the symplectic property is crucial, and therefore tracking the truncated one-turn map is not adequate, since it is symplectic only up to the truncation order. In order to speed up the integration keeping the symplecticity, one can use two methods

- Build an explicit symplectic map whose lower orders agree with the truncated one-turn map.

- Fit an explicit symplectic map to an extensive set of tracking data obtained with the standard procedures. Once this onerous operation is carried out, the map can be used instead of tracking.

An explicit symplectic map can be built using two different tools

- As a composition of a sequence of explicit nonlinear maps [17–19].

- Using a mixed variable generating function one can implicitly define a symplectic map. If there is a good initial guess for the map (for instance the truncated map is usually a very good first guess), using a Newton method one can easily invert the equation and obtain an explicit symplectic map [20].

The dynamics at low amplitudes is usually very well reproduced through the symplectified map. At high amplitudes the agreement can be worse, but one should point out that the tracking code itself is already an approximate solution to the equations of motion.

A crucial point for applications is to have a method that is really faster compared to tracking. One has to choose a compromise between the speed and the accuracy; indeed, it is not trivial to quantify the minimum degree of accuracy required. These comparison are not easy also because different tracking codes easily have rather different speeds, depending on platforms and code optimizations. We observe that whilst these methods have been widely applied in the U.S., in Europe they seem much less popular.

DYNAMIC APERTURE (DA) EVALUATION

"[Dynamic aperture] is one of the most fundamental and important objects in beam dynamics. [...] Even, it seems that there is no good constructive definition of dynamic aperture." This quotation [21] from the home page of the tracking code SAD (Strategic Accelerator Design), developed at KEK, well underlines the difficulties of the dynamic aperture issue. From the point of view of numerical simulation, the dynamic aperture $D(N)$ is a measure of the set of initial conditions in the 4D transverse phase space (plus the parametric dependence on off-momentum) that are stable for at least N turns. It is rather surprising that not much effort has been developed to give a reasonable definition of DA and, in particular, to associate to the evaluated DA an error. The estimate of the error is crucial to check the validity of the DA improvements obtained by lattice optimization.

The major difficulty in the DA definition is that the first amplitude where particle loss occurs at N turns depends at least on three separate factors.

a) The first amplitude where particle loss occurs depends on the ratio of the linear invariants along which we start initial conditions. Usually one makes a radial scan

$$x = A \cos \kappa \qquad\qquad y = A \sin \kappa$$

along a fixed κ (usually $\kappa = \pi/4$, i.e., $x = y$), and the initial momenta (p_x, p_y) are set to zero. Indeed, according to κ one has different linear invariants and therefore the whole dynamics is different (for instance the detuning strongly depends on the κ). This is due to the four-dimensionality of the problem.

b) The first amplitude where particle loss occurs depends on the initial phases in the two planes (x, p_x) and (y, p_y). This is due to the deformation of the orbits, which close to the dynamic aperture are not direct product of two rotations.

c) Particle losses take place in chaotic regions where one has sensitivity to initial conditions. Therefore, to each initial condition one cannot simply associate an escape time, but rather a distribution of possible escape times. This distribution can be rather wide (some orders of magnitudes). For instance in Fig. 1 we plot the distribution of the escape times for 50 particle started around a very tiny neighbourhood of a chaotic initial conditions for the Hénon map.

Due to these effects, the amplitude where particle loss occurs can vary in a range of 10-30% in non pathological cases. Moreover, the scan in phase space is discrete and this adds another source of error in the DA estimation.

One can outline two different strategies to overcome this problem:

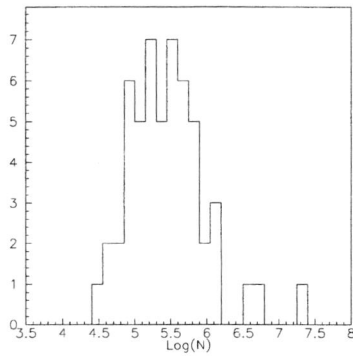

FIGURE 1. Escape times of 50 particles started around a chaotic initial condition at amplitude x=y=0.47 for the Hènon map, with linear frequencies 0.168 and 0.201.

- Fast but unprecise DA evaluation: one carries out tracking along one direction only, with one particle per initial condition (usually zero phases, and equal emittances). This is approach has been used for large machines where simulations are onerous. An estimate of the error associated to this DA evaluation is very hard to obtain, since the phase space dynamics has been analysed only in a very limited region.

- Averaging procedures: one carries out a richer and more onerous phase space sampling, and takes an average of the amplitude loss. In this way one can also give estimates of the error associated to this average. A procedure to take into account effects (a) and (b) has been given in [22], and error estimates have been given in [23]. One should associate to this dynamic aperture also a measure of the spread to provide a lower bound to the DA. In principle one could directly take the minimum amplitude over the phase space sampling, but this quantity is more unstable from a numerical point of view; moreover the error estimate is more troublesome.

TRANSVERSE PHASE SPACE, LONG-TERM STABILITY, DIFFUSION

Transverse phase space tools

During the last decade relevant advances have been carried out in the study of the phase space of 4D sympletic mappings that can model the transverse betatron motion. A major contribution has been given by frequency analysis tools [24] that through intensive tracking campaign allow to draw a complete picture of the global dynamics. Moreover, complementary analytical tools of resonant normal forms have been developed and arbitrary order codes which evaluate the resonance position and width are available [25].

In Ref. [24] a very precise numerical method has been presented to evaluate the nonlinear frequencies of time series. Moreover it has been shown that extensive tracking simulations over a two-dimensional grid of initial conditions in phase space (one typically sets the momenta to zero and scans along the physical coordinates) allow to give a numerical reconstruction of global dynamics in phase space. For each initial condition one evaluates the frequencies with high precision (at least $10^{-5} - 10^{-6}$); less than 10^4 are sufficient to get such a precision - see also [26] for a review of the methods and estimates of their errors. Then one can draw the results either in the frequency or in the action space.

- Tune footprint: the image of the uniform grid of initial conditions is plotted in the space of frequencies, and its deformation allows to see what are the strong resonances (see [24,27] for more details).

- Action print or resonance net: we plot in (x, y) only those initial conditions whose orbits are resonant; i.e., whose nonlinear frequencies satify a resonant condition up a given order. In this way one obtains in the physical space a picture of the resonant channels, and, contrary to the footprint, the width of the resonance is directly visible in the graph. More details on this method can be found in [28].

For instance in Fig. 2 we show the resonance net of a toy model (Hènon map plus an octupole). One can see that the main resonances that affect the motion are $(1, -1)$ and $(1, -4)$. The linear tune is set on 0.28 and 0.31. The short-term dynamic aperture is between 0.4 and 0.5 arbitrary units. One can see the relation of the resonances with the dynamic aperture. The plot also shows a wide net of high order resonances that cross each other also at very low amplitudes. The origin itself is resonant since the linear frequencies satisfy a 12th order resonant condition (8,4,1). Chaotic bands are also visible using this method (see [28]).

Unfortunately, most tracking codes are not yet equipped with automatic procedures to carry out simulations on a two-dimensional grid of initial conditions. It also would be a nice improvement to carry out the scan over the linear invariants and not over the coordinates, i.e. (x^2, y^2) instead of (x, y), as proposed in [24]; this

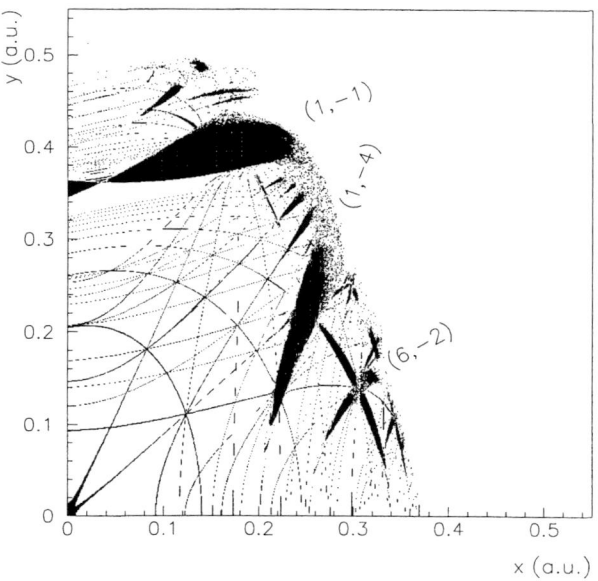

FIGURE 2. Initial conditions that give rise to resonant orbits (in black) for the 4D Hénon map plus an octupole, through tracking and frequency analysis. Initial momenta set to zero.

scan is more natural, and the grid is denser at high amplitudes, that are the more interesting for the dynamics.

The reconstruction of the net of resonances in the space of initial conditions can be realized also through the perturbative tools of normal forms. Running a resonant normal form for each resonance, one can evaluate the resonance position and width through semi-analytical tools, and obtain a perturbative reconstruction of the dynamics in phase space. As for all perturbative tools, results close to dynamic aperture can be not accurate, and one has to pay much attention to the truncation order. The advantage is that analytical minimization of resonances can be worked out, since all the parametric dependence can be included. More details can be found in [25].

Methods for long-term stability estimates

Sophisticated tools have been developed to evaluate the long-term stability. One can group them in three main areas.

Bounds on invariants

All stable orbits have two nonlinear invariants that are constant along the orbit, for instance the actions or the frequencies. On the other hand, invariants are not defined for the unstable orbits. If a method to evaluate the nonlinear invariants with infinite precision from tracking data would be available, one could easily distinguish stable from unstable orbits by evaluating the presence of a drift in the nonlinear invariants. For unstable particles, this drift can be extrapolated to longer times to get a stability estimate: fixing an outer domain where one assumes that a fast diffusion is present (short term dynamic aperture), one can compute the time necessary to the particle to go from the initial condition to this outer domain.

Therefore, in order to have realistic estimates, a method to evaluate the nonlinear invariants from tracking data with the highest precision must be worked out: in fact, any numerical error in the determination of the invariant will be erroneously considered as a diffusion in phase space.

The main source of error for most methods is the same: island structures associated to resonances. For instance, the invariant defined through nonresonant normal forms is very imprecise on resonances, and therefore it overestimates the diffusion in these regions: indeed, most resonances are stable at sufficiently low amplitudes. A similar phenomenon occurs if we consider the variation of the frequencies along the same orbit: on resonant orbits every algorithm that provides frequencies has a much larger error, and therefore using this method the diffusion is considerably overestimated along resonances.

A very elaborate and clever numerical method to construct a nonlinear invariant based on interpolating the action with splines has been proposed in [29,30], and has led to some reasonable estimates of stability times that have been verified through tracking [31].

Particle loss predictors

Also in this case one considers dynamical variables that can be extracted from short-term tracking data, as the variation of the actions or of the frequencies [27,28] along the orbit, or the rate of divergence of two nearby orbits [28,32,23]. The difference with respect to the previous approach is that one fixes a threshold, and assumes that all the particles whose orbits produce an indicator above the threshold will be lost and all the other ones are stable. No information is given on the time necessary to lose the stability.

The Lyapunov exponent has been very popular in celestial mechanics, and has also been used in accelerator physics. Results show that one can detect the border of instability using short term tracking data when no tune modulation is present; in the case of tune modulation the method does not seem to be effective [23].

Survival plots and DA extrapolation

A simple and effective way to plot the information contained in tracking data is given by survival plots, where the stability time is plotted versus the amplitude in phase space. Usually tracking is carried along one direction, i.e. equal emittances and zero phases, with very dense scan in the amplitude (see for instance [33]): these plots should show the trend of long-term dynamic aperture, but are usually rather irregular, and it seems rather difficult to work out a trend for times larger than the tracked ones.

An improvement of survival plots can be obtained if a dense scan along the amplitude in one direction is substituted by several, less dense scans along different ratio of emittances [28,23]: if we plot the averaged amplitude versus the stability time it turns out that it is well-interpolated by the three-parameter formula

$$D(N) = A + \frac{B}{\log^\kappa N}.$$

The formula has been shown to work for a wide variety of models which range from the 4D Hènon map to the LHC with tune modulation (see Fig. 3). The quantity A represents the dynamic aperture extrapolation for 'infinite times', and becomes negative when the tune modulation is increased. The exponent κ is arond 1.5 for the purely 4D case and decreases when the tune modulation becomes more relevant.

This formula gives on the one hand a clear model of the averaged phase space dynamics, providing a rate for the stability times which has the same logarithmic dependence on the time as the Nekhoroshev estimate. Moreover, it can be extrapolated to predict long-term dynamic aperture. For cases with tune modulation one can extrapolate of one-two order of magnitudes in the number of turns [23,34].

Diffusion

The possibility of fitting the dynamics of particle loss with a diffusion process, according to the ideas developed in [35], have been discussed and analysed for a long time. Both numerical simulations and experiments have been carried out, several times including multifrequency modulation of the linear frequencies that model tune ripple. Experimental data obtained at FNAL have been analysed in the framework of the diffusion equation [36]. In [37] the influence of two ripples on a simplified lattice model (Hènon map) has been analysed, using the concept of sidebands and overlapping. The effectiveness of the ripple correction to improve beam lifetime in the HERA has been shown in [38]. Different mechanisms that lead to transverse proton diffusion have been analysed in [39], where a diffusion equation and resonance parameters evaluated through perturbative theory were used to understand which mechanism could explain diffusion in the HERA. Experiments carried out at the SPS are reported in [40]. On the theoretical side, interesting results have been obtained by using the Neihstadt adiabatic theory [41]: for a two-dimensional phase space with a single resonance and a periodic modulation of the linear frequency, it has been proved that the process can be described in terms of

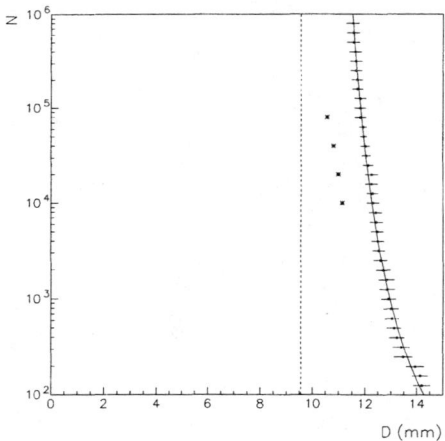

FIGURE 3. Averaged Dynamic Aperture versus stability time (survival plot) for an LHC model with imperfections, off momentum. Bars: tracking data. Solid line: interpolation. Circles: Lyapunov prediction.

166

a random walk in the action space [42].

In our opinion, notwithstanding the considerable amount of work that has been carried out during the last decade, we are far from a complete understanding of these phenomena. In particular, we point out that a good example of agreement between tracking simulations of dynamics at high amplitudes and a diffusion equation has not been reached, even in the simplest case of purely four-dimensional dynamics.

In fact, some authors did not believe in the possibility of using a diffusion equation to describe long-term dynamics [43]. This clearly depends also on the mechanisms that are involved in the long-term losses.

Chirikov theory of overlapping treats the case of a main resonance and of its sidebands created by tune modulation or other mechanisms. Indeed, in the general case of several resonances in a four-dimensional phase space, that cross each other also at low amplitudes (see Fig. 2), the Chirikov idea of overlapping becomes hard to apply. What is sure is that wide chaotic bands are a very relevant mechanism, but the origin of these bands is still unclear.

DA VS LATTICE PARAMETERS

The dynamic aperture depends on both linear (tunes, beta functions, phase advances, linear coupling, chromaticity ...) and nonlinear (multipolar errors, multipolar correctors, detuning, nonlinear chromaticity, ...) parameters. We limit ourselves to a few simple remarks.

- Lattice dominated by a multipole. If the dynamic aperture is dominated by the effect of a single multipole, one can derive a simple scaling that provides the dependence of the DA versus the multipole strength and versus the beta function where the multipole is located.

- DA versus fractional linear tune. Even for a simple model such as a linear lattice plus a sextupole, the dependence on the linear tune is extremely complex and is given by the intricate relation between resonances, detuning and nonlinearities (see Fig. 4). Only in the case of a linear tune close to an unstable resonance (for example, resonance (3,0) or (1,2) for a lattice with a normal sextupole) one can give an analytical estimate of the DA. In fact, in this case the stability is lost on the separatrix of the unstable resonance that can be evaluated through resonant perturbative theory [44]. In the generic case the situation is more complex and a general well known rule is to avoid to set the linear tune close to excited low order resonances (i.e., resonances from 3 to 5, see for instance [45]). Nevertheless high order resonances can be relevant to the dynamics (see for instance the case of resonance (7,0) in the LHC [46]). For electron machines, extensive studies of the DA as a function of the fractional tunes have been carried out using fast tracking (see for instance the so-called 'swamp plots' in [47].

167

- DA versus integer part of the tune. The integer part of the tune can be very relevant since it determines the phases between the nonlinear elements of the machine, leading to coherent sum or partial compensation. The integer part of the tune is not present in the linear part of the one-turn map (that contains only the fractional part), but its value determines the strength of the higher order map coefficients. Analytical approaches to the optimization of the integer part of the tune can be built through the perturbative theory based on hamiltonian flows [48].

QUALITY FACTORS AND CORRELATIONS

Since in the generic case the dependence of the DA on the lattice parameters is unknown, one has to use numerical integration (i.e., tracking) to evaluate the DA and to optimize it. Indeed, since many years the accelerator physicists community has widely used quality factors (QF) to understand and improve lattice performances.

- Definition and characteristics. A QF is a function of the dynamical variables (trajectory in phase space, nonlinear frequencies ...) and/or of the lattice parameters (multipoles, linear optics, ...). It can be either analytical, i.e. can be derived directly from the one-turn map coefficients, or numerical, i.e. evaluated through the postprocessing of a set of tracking data. The QF must have two fundamental characteristics: it must be fast to compute (compared to the DA) and for the set of considered different lattices it must have a good correlation with the DA.

- Examples of QFs. The nonlinear aberrations (i.e., some map coefficients; the norm of the map (i.e., a weigthed sum of map coefficients up to the truncation order); the tuneshift (either evaluated through perturbative series or computed

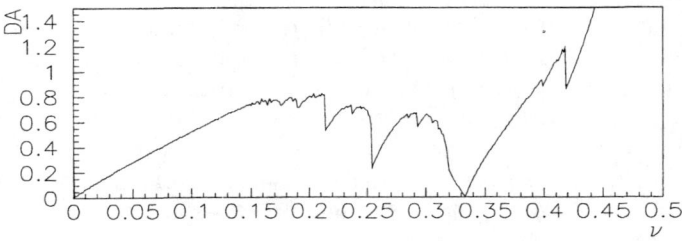

FIGURE 4. Dynamic aperture at 10000 turns for the 2D Hénon mapping versus the fractional part of the linear tune.

using tracking data); the 'resonances' (there are several ways to extract QF related to resonances, both numerical and perturbative); the smear, i.e. the deformation of the orbits with respect to the linear case [46]; the fixed points residuals [49].

- Correlation between QF and DA. Even though in several works the QF is established a priori, we believe that one should always check that the given QF is correlated with the DA, since the QF is model-dependent. For instance, in some lattices the detuning can be the driving mechanism of the DA and therefore is a good QF to choose, whilst for other cases it can be irrelevant. The correlation between QF ad DA cannot be established a priori with the present knowledge.

- What a QF can provide. Once a correlation between the QF and the DA has been shown through tracking a number of cases, one can use the QF instead of the DA for carrying out more onerous simulations, such as a wider statistical analysis of the lattice or of the lattice parameters. With this respect, the QF is a tool that allows to save CPU time. Another very relevant feature is that the QF gives an insight on the mechanism that rules DA and therefore can suggest analytical ways of carrying out optimizations [46,50]. Even though one must spend some CPU time to check the QF correlation with the DA, one obtains a better theoretical understanding of the lattice dynamics. In this respect, the QF approach is useful also for electron machines where the need of CPU time is not so stringent as for proton machines.

- Some applications. The minimization of the detuning has been used nearly one decade ago to correct the systematic errors in the SSC and in the LHC [51,52]. The correction of resonances has been used as criterion to sort the dipoles to compensate random errors [50,53]. QF related to the resonances have been also used at PEP-II to improve the lattice [47] and for the LHC to determine the lattice performances [46].

WAYS OF CORRECTING

The ways of optimizing a lattice can be summarized as follows:

- Changing the linear parameters of the lattice, that is the optics: tune, beta functions, position of the magnetic elements ...

- Inserting in the lattice some nonlinear elements that reduce the nonlinearity already present in the lattice. We distinguish between three main approaches:

  - Local correction: the corrector element is close to the error source. This is the case for instance of the sextupolar chromatic correctors in a regular lattice put close to the regular quadrupoles, which are the sources of chromaticity.

- Lumped correction: corrector elements not too far from the error source; this was used for the SSC and the LHC lattices when only two correctors per half cell were used to effectively reduce the nonlinearities originated by four dipoles [51,52,54].
- Global correction: only a few correctors are inserted, distant from the error sources.

Evidently, the last solution is the least expensive but it is rarely feasible. Also in the case of the local correction, there is a limit in the nonlinearity strength beyond which the correction becomes uneffective.

- Sorting the magnets: this method does not involve the construction of new hardware. If several magnets with the same function and hardware characteristics (such as for instance the cell dipoles) feature large random errors, one can install the magnets in a sequence that minimizes the nonlinear effects. The problem is very hard since one has to analyse a huge number of possible permutations. To apply this procedure one must measure all the magnets, and have some space to store a large batch of magnets. Several studies have been carried out for different machines [50,53,55–58].

SOME OPEN PROBLEMS

We conclude this overview by pointing out some topics that we believe are open problems in this field.

- About long-term stability: it is still not clear what the mechanism is that creates macroscopic chaotic bands and how resonances are related to it. Moreover, more work should be done in order to understand whether the loss of integrability in phase space is a sudden phenomenon or whether it is gradual.

- About dynamic aperture: is it necessary to go towards a probabilistic definition of the DA, i.e. asking that not all particles survive for a given time but that a very small percentage could escape ? This question is related to the solution of the previous one about the loss of integrability.

- About diffusion: does the diffusion equation really fit experimental or numerical data ?

- About quality factors: is it possible to know a priori (i.e., without tracking) whether a given QF is correlated with the dynamic aperture ?

- About correction: when is local correction necessary and when is global correction sufficient ?

- About experiments: what informations on nonlinear motion can we extract from experiments ? Good agreement on detuning has been reached (see for instance [59]). Some work has been carried out on measuring the map coefficients in experiments, and also to obtain experimentally a tune footprint (see for instance [60,61]). Both tasks are very challenging, the main aim being a more complete modelization of the machine and a better understranding of the dynamic aperture.

ACKNOWLEDGEMENTS

We wish to thank M. Cornacchia and C. Pellegrini for the kind invitation to the conference. We are very indebted to Prof. Turchetti and to W. Scandale for introducing us to this subject and for their constant help and support. We wish to thank M. Giovannozzi and for critical reading of the manuscript. This work is dedicated to A. della Monica.

REFERENCES

1. Forest, E., and Ruth, D., *Physica D* **43**, 105–17 (1990).
2. Yoshida, H., *Phys. Lett. A* **150**, 262–8 (1990).
3. Forest, E., Bengtsson, J., and Reusch, M., *Phys. Lett. A* **158**, 99–101 (1991).
4. Forest, E., Reusch, M., Bruhwiler, D., and Amiry, A., *Part. Accel.* **45**, 65–94 (1994).
5. Dragt, A., and Finn, J. M., *J. Math. Phys.* **17**, 2649–60 (1976).
6. Dragt, A., and Finn, J. M., *J. Math. Phys.* **20**, 2649–60 (1979).
7. Dragt, A., *Nucl. Instrum. and Methods Phys. Res., A* **258**, 339–354 (1987).
   Perturbative theory for betatron motion
8. Giorgilli, A., *Comp. Phys. Comm.* **16**, 331 (1979).
9. Servizi, G., and Turchetti, G., *Comput. Phys. Commun.* **32**, 201–7 (1984).
10. Berz, M., *Part. Accel.* **24**, 109 (1989).
11. Schoch, A., *CERN* **57–21** (1957).
12. Guignard, G., *CERN* **76–06** (1976).
13. Bazzani, A., Mazzanti, P., Servizi, G., and Turchetti, G., *Nuovo Cim. B* **102**, 51–80 (1988).
14. Berz, M., Irwin, J., and Forest, E., *Part. Accel.* **24**, 109 (1989).
15. Bazzani, A., Todesco, E., Turchetti, G., and Servizi, G., *CERN* **94–02** (1994).
16. Bazzani, A., Giovannozzi, M., and Todesco, E., *Comp. Phys. Commun.* **86**, 199–207 (1995).
    Fast symplectic tracking
17. Rangarajan, G., Dragt, A., and Neri, F., *Part. Accel.* **28**, 119–24 (1990).
18. Gjaja, I., *Part. Accel.* **43**, 133–44 (1994).
19. Shi, J., and Yan, Y., *Phys. Rev. E* **48**, 3943–51 (1993).
20. Berg, J. S., Warnock, R. L., Ruth, R. D., and Forest, E., *Phys. Rev. E* **49**, 722–739 (1994).
    Dynamic aperture definition
21. SAD web pages http://www-acc-theory.kek.jp/sad/tracking.html

22. Todesco, E., and Giovannozzi, M., *Phys. Rev. E* **53**, 4067 (1996).
23. Giovannozzi, M., Scandale, W., and Todesco, E., *Phys. Rev. E* **57**, 3432–43 (1998).

Transverse phase space, long-term stability, diffusion

24. Laskar, J., *Physica D* **67**, 257–81 (1992).
25. Todesco, E., Gemmi, M., and Giovannozzi, M., *Comp. Phys. Comm.* **106**, 169–80 (1997).
26. Bazzani, A., et al., *Part. Accel.* **52**, 147-77 (1996).
27. Laskar, J., Froeschlé, C., and Celletti, A., *Physica D* **56**, 253–69 (1992).
28. Giovannozzi, M., Scandale, W., and Todesco, E., *Part. Accel.* **56**, 195-225 (1997).
29. Warnock, R. L., *Phys. Lett.* **66**, 1803–6 (1991).
30. Warnock, R. L., and Ruth, R. D., *Physica D* **56**, 188–215 (1992).
31. Warnock, R. L., and Berg, J. S., *AIP Conf. Proc.* **395**, 423–45 (1996).
32. Schmidt, F., Willeke, F., and Zimmermann, F., *Part. Accel.* **35**, 249–256 (1991).
33. Yan, Y., *SSC* **500** (1991).
34. Böge, M., and Schmidt, F., *AIP Conf. Proc.* **405**, 201–210 (1997).
35. Chirikov, B. V., *Phys. Rep.* **52-5**, 263–379 (1979).
36. Chen, T., et al., *Phys. Rev. Lett.* **68**, 33-6 (1992).
37. Brüning, O., *Part. Accel.* **41**, 133–51 (1993).
38. Brüning, O., and Willeke, F., *Part. Accel.* **54**, 237–46 (1996).
39. Zimmermann, F., *Part. Accel.* **49**, 67–104 (1995).
40. Fischer, W., Giovannozzi, M., and Schmidt, F., *Phys. Rev. E* **55**, 3507–20 (1997).
41. Neishtadt, A., *Sov. J. Plasma Phys.* **12**, 568–73 (1986).
42. Bazzani, A., Brini, F., and Turchetti, G., *AIP Conf. Proc.* **395**, 129-38 (1997).
43. Gerasimov, A., *CERN-SL (AP)* **92–38** (1992).

DA vs. lattice parameters

44. Giovannozzi, M., *Phys. Rev. E* **53**, 6403 (1996).
45. Robin, D., and Laskar, J., *Part. Accel.* **54**, 193–202 (1995).
46. Papaphiloppou, Y., and Schmidt, F., these proceedings.
47. Yan, Y., Irwin, J., and Chen, T., *Part. Accel.* **55**, 17–26 (1996).
48. Talman, R., *LHC Project report* **197** (1998).

QF and lattice optimization

49. Wan, W., Cary, J. R., and Shasharina, S. G., *AIP Conf. Proc.* **395**, 407–422 (1996).
50. Giovannozzi, M., Grassi, R., Scandale, W., and Todesco, E., *Phys. Rev. E* **52**, 3093–101 (1995).
51. Neuffer, D., *Part. Accel.* **27**, 209–14 (1990).
52. Scandale, W., Schmidt, F., and Todesco, E., *Part. Accel.* **35**, 53–88 (1991).
53. Willeke, F., *DESY HERA* **87–12** (1987).
54. Neuffer, D., and Forest, E., *Phys. Lett. A* **135**, 197–201 (1989).
55. Gluckstern, R. L., and Ohnuma, S., *IEEE Trans. Nucl. Sci.* **NS-32**, 2314-16 (1985).
56. Ziemann, V., *Part. Accel.* **51**, 155-79 (1995).
57. Shi, J., and Ohnuma, S., *Part. Accel.* **56**, 227–47 (1997).
58. Bartolini, R., et al., *Nuovo Cim. B* **113**, 511 (1998).
59. Gareyte, J., Scandale, W., and Schmidt, F., *IOP Conf. Series* **131**, 235–48 (1993).
60. Irwin, J., *AIP Conf. Proc.* **326**, 662 (1995).
61. Terebilo, A., et al, these proceedings.

Dynamical Properties of a Model for Synchro-betatron Coupling

L. Bongini*, G. Franchetti**, G. Turchetti *

* Dip. di Fisica and INFN Bologna, ITALY
** GSI, Darmstadt, GERMANY

Abstract The long term stability of betatron motion in presence of multipolar nonlinearities has been extensively investigated for high energy accelerators and the Hénon maps has been proposed as a model. In high intensity accelerators the non linear effects due to space charge and synchro-betatron coupling are relevant. We propose a simple model to investigate the coupling between longitudinal and horizontal motion due to chromaticity. The model consists in a standard Chirikov map coupled to a Hénon map. Diffusion of orbits is observed and the structure of resonances is inspected by the frequency analysis.

INTRODUCTION

High intensity beams are facing new applications beyond basic research in nuclear physics [1,2]. In the absence of space charge the multipolar errors and the r.f. cavities exert nonlinear forces on the transverse and longitudinal motion. The space charge force has a defocusing effect on the transverse motion and couples it to the longitudinal one. A synchro-betatron coupling occurs also because any change of momentum displaces the closed orbit and in presence of sextupoles or higher multipoles the tune changes (chromatic effect).
To investigate the coupling we propose a simple 4D model consisting in a standard map which describes the effect of a thin RF cavity and a 2D Hénon map which describes the effect of a thin sextupole for a flat beam. We make the linear tune vary with the longitudinal momentum and require the map to be symplectic. This fixes the longitudinal displacement due to the coupling with the transversal motion. When a particle is almost synchronous and near to the corresponding closed orbit, the coupling effect is negligible. For large transverse displacements the coupling with the longitudinal motion becomes visible when a large island is encountered, because the frequency is slowly modulated and adiabatic invari-

CP468, *Nonlinear and Collective Phenomena in Beam Physics–1998 Workshop*,
edited by S. Chattopadhyay, M. Cornacchia, and C. Pellegrini

ance is lost close to the islands border, where a localized diffusion is observed. When the particle approaches the borderline of the bucket, namely the pendulum separatrix, the diffusion in the transverse phase plane is enhanced. A similar diffusion pattern occurs for a Hénon map whose linear frequency is stochastically perturbed. The frequency analysis and the tune-action map provide the global pattern of the resonances and of the chaotic regions.

THE SYNCHRO-BETATRON MAP

The design of high energy accelerators like LHC requires a stable beam for a high number turns ($\sim 10^8$ corresponding to 10^{11} crossings of FODO cells). As a consequence the dynamic aperture and the ripple induced diffusion in presence of multipolar errors have been actively investigated [3,4]. In the case of a high intensity LINAC [1] the number of FODO cells does not exceed 10^3 whereas the storage ring of HIDIF [2] has 10^2 cells visited during 10^2 turns. Since the number of FODO crossings does not exceed 10^4 we are not very concerned with the long term stability and only the short term dynamic aperture is relevant. It is well known that a particle with momentum p different from the design momentum p_s follows a different closed orbits and its linear tune is shifted by $\alpha\,(p-p_s)/p_s$. This tune shift may be due to the structure of the linear lattice and to the presence of nonlinear forces (sextupoles). We assume that the natural chromaticity, present in the linear lattice, is zero. In the case of a simple lattice with a single thin sextupole the tune shift due to the off momentum is given by [5]

$$\delta\omega_x = -\alpha\frac{p-p_s}{p_s} + O\Big(\,(\frac{p-p_s}{p_s})^2\,\Big), \qquad \alpha = k_2\beta\,D$$

where β and the dispersion D are evaluated at the sextupole position and k_2 is the sextupolar gradient. We propose the following model for the synchro-betatron coupling

$$\begin{pmatrix} j_{n+1} \\ \theta_{n+1} \end{pmatrix} = \begin{pmatrix} j_n - \lambda\sin(2\pi\,h\,\theta_n) \\ \theta_n + \eta\,j_{n+1} + \dfrac{\alpha}{2k_2^2\beta^3}\,f'(j_{n+1})\,(x_n^2 + (p_{x\,n} + x_n^2)^2) \end{pmatrix} \ \mathrm{mod}\ 1$$

$$\begin{pmatrix} x_{n+1} \\ p_{x\,n+1} \end{pmatrix} = R\big(\omega_x + \alpha\,f(j_{n+1})\big) \begin{pmatrix} x_n \\ p_{x\,n} + x_n^2 \end{pmatrix}$$

$$(1)$$

where h is the harmonic number equal to the ratio between the RF frequency and the revolution frequency. The function $f(j)$ is periodic with period 1 and such that $f(\jmath) = \jmath + O(\jmath^2)$. We choose $f(\jmath) = (2\pi)^{-1}\sin(2\pi\jmath)$ in all the numerical examples considered below; if $\lambda \ll 1$ the results are the same as for the linear function $f(\jmath) = \jmath$. We have assumed that in our lattice there is only one RF cavity and a sextupole nearby. The constant λ is given by

$$\lambda = \frac{eV\, c^2}{E_s\, v_s^2}$$

where V is the maximum potential difference of the cavity, E_s is the energy of the synchronous particle whose velocity and momentum are v_s and p_s. If there are N cavities then $\eta \to \eta/N$. The x, p_x coordinates we use are scaled with respect to the normalized Courant Snyder coordinates $\hat{x}, \hat{p}_x$, and $\jmath, \theta$ are the normalized longitudinal momentum and phase defined by

$$x = \hat{x}\, k_2\, \beta^{3/2}, \quad p_x = \hat{p}_x\, k_2\, \beta^{3/2}, \quad \jmath = \frac{p_s - p}{p_s} \equiv -\frac{\delta p}{p_s}, \quad \theta = \frac{\phi - \phi_s}{2\pi h} \equiv \frac{\delta \phi}{2\pi h}$$
(2)

where $\phi_s = 2\pi h\, n$ is the phase of the nearby synchronous particle after n turns, for which the potential jump $V\sin\phi$ across the cavity vanishes. For a realistic machine like SIS we have $\lambda \sim 10^{-3}$, $k_2 = 10^{-2}$ and $\eta \sim 1$, $h = 4$. Choosing $\beta = 10$, $D = 2$ we have $\alpha \sim 0.2$ and $\alpha/2k_2^2\beta^3 \sim 1$. In the mathematical model we consider the choice $2k_2^2\beta^3 = 1$ and $\eta = 1$ is made, leaving only λ and α as variable parameters.

Choosing $\alpha = 0$ this model reduces to the standard map and the Hénon map. Letting $\delta\phi_n = \phi_n - \phi_s$ be the phase shift with respect to a synchronous particle and $\delta E_n = E_n - E_s$ be the energy shift, one has [6]

$$\begin{cases} \delta\phi_{n+1} = \delta\phi_n - 2\pi h\, \eta \dfrac{\delta p_{n+1}}{p_s} \\ \delta E_{n+1} = \delta E_n + eV(\sin\phi_n - \sin\phi_s) = \delta E_n + eV\sin\delta\phi_n \end{cases}$$

Using the kinematic identity

$$\frac{\delta p_n}{p_s} = \frac{c^2}{v_s^2}\frac{\delta E_n}{E_s}$$

and the normalized variables (2) we can write the map as

$$\begin{cases} \jmath_{n+1} = \jmath_n - \dfrac{eV\, c^2}{E_s\, v_s^2}\sin(2\pi h\theta_n) \\ \theta_{n+1} = \theta_n + \eta\jmath_{n+1} \end{cases}$$
(3)

On the other hand the Hénon map written in Courant Snyder coordinates reads

$$\begin{pmatrix} \hat{x}_{n+1} \\ \hat{p}_{x\,n+1} \end{pmatrix} = R(\omega_x) \begin{pmatrix} \hat{x}_n \\ \hat{p}_{x\,n} + k_2\,\beta^{3/2}\,\hat{x}_n^2 \end{pmatrix}$$
(4)

175

The coupling has been chosen so that the map is symplectic. More generally we replace the linear tune shift $\alpha\jmath$ with the derivative of $f(\jmath)$ where $f(\jmath+1) = f(\jmath) = \jmath + O(\jmath^2)$ so that the map is continuous on the torus. This is not relevant for beam physics but the map, extended to any value of $\jmath$, remains continuous and bounded. To prove that this map is symplectic we show that it is the Poincaré map of the following time periodic Hamiltonian

$$H = \frac{\omega}{2}(\hat{x}^2 + \hat{p}_x^2) + \eta\frac{\jmath^2}{2} + \alpha f(\jmath)\frac{\hat{x}^2 + \hat{p}_x^2}{2} + \left(-\frac{\lambda}{2\pi h}\cos(2\pi h\,\theta) - k_2\beta^{3/2}\frac{\hat{x}^3}{3}\right)\delta_P(\sigma)$$

(5)

where $\delta_P(\sigma)$ is the periodic Dirac function of period 1 in $\sigma = s/L$, where L is the length of the FODO cell. Splitting H into $H_0 + H_1\delta_P(s)$ and introducing action and angles $\theta_x, \jmath_x$ for the transverse motion we have

$$H_0 = \eta\frac{1}{2}\jmath^2 + \omega\,\jmath_x + \alpha f(\jmath)\,\jmath_x, \qquad \jmath_x = \frac{\hat{x}^2 + \hat{p}_x^2}{2}$$

whose solution is given by

$$\theta(s) = \theta(0) + \left(\eta\jmath + \alpha f'(\jmath)\jmath_x\right)s, \qquad \theta_x(s) = \theta_x(0) + \left(\omega + \alpha f(\jmath)\right)s$$

The map from $s = n$ to $s = n+1$ of the unperturbed Hamiltonian immediately follows and composing it with the map corresponding to the impulsive contribution in Hamiltonian we obtain

$$\begin{pmatrix} \jmath_{n+1} \\ \theta_{n+1} \\ \hat{x}_{n+1} \\ \hat{p}_{x\,n+1} \end{pmatrix} = \begin{pmatrix} \jmath \\ \theta + \eta\jmath + \alpha f'(\jmath)\frac{\hat{x}^2 + \hat{p}_x^2}{2} \\ \\ R(\omega_x + \alpha f(\jmath)) \end{pmatrix}\begin{pmatrix} \jmath_n - \lambda\sin(2\pi h\,\theta_n) \\ \theta_n \\ \hat{x}_n \\ \hat{p}_{x\,n} + k_2\beta^{3/2}\hat{x}_n^2 \end{pmatrix}$$

(6)

Written in explicit form the map (6) reads

$$\begin{pmatrix} \jmath_{n+1} \\ \theta_{n+1} \end{pmatrix} = \begin{pmatrix} \jmath_n - \lambda\sin(2\pi h\,\theta_n) \\ \theta_n + \eta\,\jmath_{n+1} + f'(\jmath_{n+1})\frac{\alpha}{2}\left(\hat{x}_n^2 + (\hat{p}_{x\,n} + k_2\beta^{3/2}\,\hat{x}_n^2)^2\right) \end{pmatrix}\ \ \text{mod }1$$

$$\begin{pmatrix} \hat{x}_{n+1} \\ \hat{p}_{x\,n+1} \end{pmatrix} = R\left(\omega_x + \alpha f(\jmath_{n+1})\right)\begin{pmatrix} \hat{x}_n \\ \hat{p}_{x\,n} + k_2\beta^{3/2}\,\hat{x}_n^2 \end{pmatrix}$$

(7)

Replacing $\hat{x}, \hat{p}_x$ with x, p_x according to (2) the map (1) is recovered.

ACTION-FREQUENCY MAP ANALYSIS

We have introduced the map (1) by coupling the transverse motion to the longitudinal motion by the chromatic effect of the sextupole, introduced as a shift in the linear frequency due to the off momentum. The coupling of the longitudinal motion to the transverse one was automatically fixed by the requirement that the map is symplectic. When the coupling vanishes the usual description of the longitudinal and transverse motion is recovered. The model enables us to see how the increase of the transverse emittance influences the change of the longitudinal momentum. We first analyze the dynamical features of the map (1), with $\eta = 1$, $2k_2^2\beta^3 = 1$, $h = 1$, when the potential strength λ and the coupling α are varied in a physically reasonable interval. Dynamically the model is well defined even outside the bucket, namely the pendulum separatrix, and the restriction to the torus (by the mod 1 condition) makes the map bounded. If λ is large the map becomes chaotic and the model describes a random perturbation to the linear betatronic frequency if α is small. At $\lambda \simeq (2\pi)^{-1}$ the chaotic transition occurs since the last KAM curve is broken whereas for $\lambda \ll 0.1$ the map is a good integrator of the pendulum and the phase portrait is almost the same. When the coupling is switched on the phase plots show that the pendulum drive on the oscillator is significant and the reverse is even more important. In order to analyze the global aspects of the dynamics such as the location and strength of nonlinear resonances or the presence of chaotic regions it is convenient to have a two dimensional plot in frequency space or in action space, since the 2-dimensional or 3-dimensional projections do not allow to obtain a global view of the dynamics.

Analysis of 2D maps

We first describe the frequency analysis when the system is uncoupled. The frequency analysis for integrable systems is locally defined in every region delimited by a separatrix. For the Hénon map in the neighborhood of the origin, (almost all) the invariant curves are slightly distorted circles represented by

$$x = (2a)^{1/2}\cos\phi, \qquad p_x = -(2a)^{1/2}\sin\phi$$

$$\phi = \Theta + f(J,\Theta), \qquad a = J + g(J,\Theta),$$

(8)

where Θ, J are the angle-action coordinates and f, g are periodic functions. Expanding the invariant curves in a Fourier series we write

$$x - i\,p_x = \sum_k c_k(J)e^{ik\Theta}$$

(9)

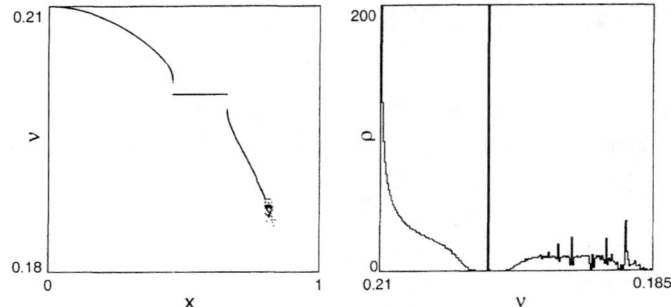

FIGURE 1. Frequency plots for the Hénon map on the line $p_x=0$ for linear frequencies $\omega/2\pi=0.21$. On the left is shown $\nu(x)=\Omega(x)/2\pi$, on the right the histogram of the frequency density $\rho(\nu)$

The orbits of the map with a non-resonant frequency are dense on a close curve and their Fourier representation is given by

$$x - i\,p_x = \sum_k c_k e^{ik(n\Omega(J)+\Theta(0))} \tag{10}$$

We choose the points uniformly distributed on a half line issued from the origin, for instance the positive x axis. The frequency of each orbit issued at the point $(x,0)$ is $\Omega = f(x)$ and the density of points in frequency space is

$$\rho(\Omega) = \rho_0 \sum_i \left| \frac{df}{dx} \right|_{x_i}^{-1}, \qquad f(x_i) = \Omega$$

If Ω is locked to a resonance Ω_* for $x_{min} \le x \le x_{max}$ then $f(x) \sim \Omega_* - |x - x_{min}|^{1/\alpha}$ with $\alpha > 1$ for $x < x_{min}$. The density $\rho(\Omega)$ vanishes as $(\Omega - \Omega_*)^{\alpha-1}$ by approaching Ω_*. There is an empty region around Ω_* and ρ diverges at Ω_*, see figure 1, where the tune $\nu(x) = \Omega/2\pi$ plot and the histogram $\rho(\nu)$ are shown. The action is given by

$$J = \frac{1}{2\pi} \oint p\,dq = \frac{1}{2\pi} \int_0^{2\pi} p(\Theta) \frac{\partial q}{\partial \Theta}\,d\Theta = \frac{1}{2} \sum_k k|c_k|^2 \tag{11}$$

and is defined at every point where Ω is non resonant. The map $\Omega(J)$ is known for every trajectory with non resonant Ω since the tracking points are dense on the closed orbit. For a resonance, the frequency Ω is locked and corresponding action interval is $[J_{min}, J_{max}]$ where the ends correspond to the actions of the inner and outer separatrix.

Defining J by the sum (11) over the resonant Fourier components we find that its value falls in the interval $[J_{min}, J_{max}]$, see figure 2. This is not surprising by

178

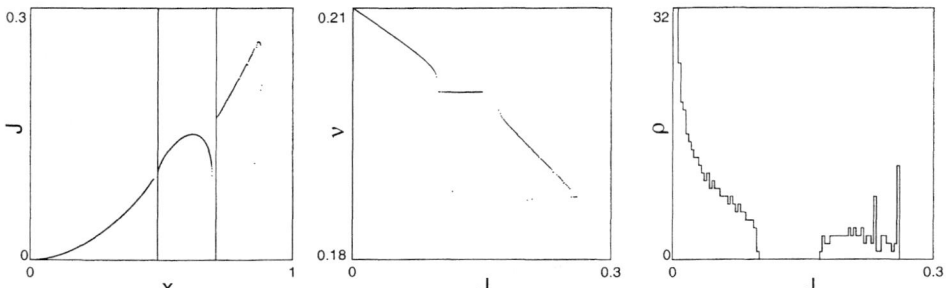

FIGURE 2. Action $J(x)$ of the Hénon map with linear frequency $\omega/2\pi=0.21$ for the points of the line $p_x=0$ (left), tune-action map $\nu(J)$ (center), action density $\rho(J)$ (right)

considering the interpolated orbit, which cannot reproduce the islands having to be connected but falls between the inner and the outer separatrix.

As a consequence the plot $\Omega(J)$ has holes where the frequency is locked and the width of the hole is the resonance width, see figure 2. The analysis extends to any 4D map.

DYNAMICAL FEATURES OF THE MODEL

We describe the dynamical behavior of the map (2.2) where $2k^2\beta^3 = 1$ and $h = 1$. The linear tune of the Hénon map is chosen to be $\omega_x/2\pi = 0.21$ so that the nonlinear tune decreases and locks on the resonance 5 as we move out from the origin. The chain of islands intersects the x axis on the interval $[0.45, 0.65]$ and the dynamic aperture is at $x = 0.8$. We have analyzed the 2D projections of the orbits for $\lambda = 0.001$ by varying the coupling from a very small $\alpha = 10^{-3}$ up to the physical value $\alpha = 10^{-1}$.

The normalized coordinates (x, p_x) were chosen in the interval $[-1, 1]$; since $\theta, \jmath$ vary in the interval $[-1/2, 1/2]$, in order to have the same range in both phase planes we plot $(2\theta, 2\jmath)$. The linear frequency of the standard map is $\omega_\jmath/2\pi = (\lambda/2\pi)^{1/2} = 0.0126$ and the amplitude of the island corresponding to the pendulum oscillations is $\Delta\jmath = 2(\lambda/2\pi)^{1/2} = 0.252$.

The figures 3 shows for $\alpha = 0.01$ the projection of the orbits in the x, p_x plane for two initial conditions in the other plane $\jmath_0 = 0, 2\theta_0 = 0.1$, close to the elliptic fixed point, and $\jmath_0 = 0, 2\theta_0 = 0.99$ close to the hyperbolic fixed point. The projections in the $\theta, \jmath$ phase plane are also shown for two distinct initial conditions $x_0 = 0.2, p_{x\,0} = 0$ close to the origin and $x_0 = 0.7, p_{x\,0} = 0$ close to the dynamic aperture.

The frequencies for all the orbits for initial conditions in the region $0 \le x_0 \le$

179

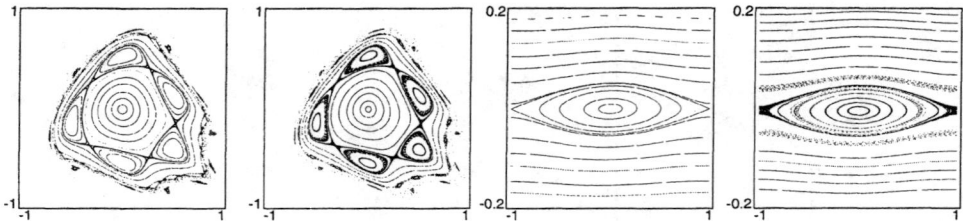

FIGURE 3. Phase plots of the Hénon map with $\omega_x/2\pi=0.21$, $\alpha=0.01$, $\lambda=0.001$. Phase plane x, p_x with initial conditions: $\jmath_0=0$, $2\theta_0=0.1$ (left), $\jmath_0=0$, $2\theta_0=0.99$ (center left) Phase plane 2θ, $2\jmath$ with initial conditions $x_0=0.1$, $p_{x\ 0}=0$ (center right), $x_0=0.7$, $p_{x\ 0}=0$ (right)

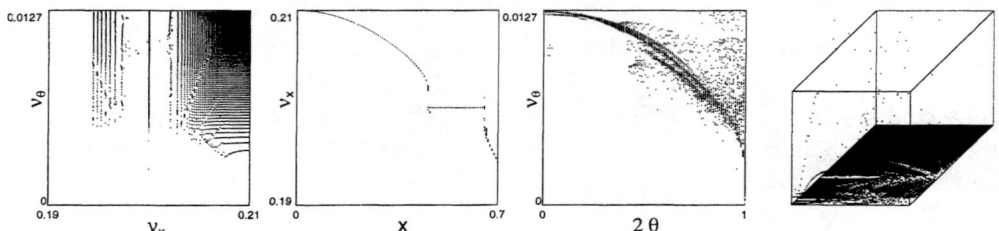

FIGURE 4. Frequency analysis for $\omega_x/2\pi=0.21$, $\alpha=0.01$, $\lambda=0.001$. Plot in the ν_x, ν_θ plane (left), plot ν_x, x (center left), plot ν_θ versus 2θ (center right), error plot in the plane x, 2θ with a scale $[0,0.01]$ (right)

0.7, $0 \le 2\theta_0 \le 0.99$ and $p_{x\ 0} = \jmath_0 = 0$ have also been computed chosing a uniform grid of 100×100 and 2000 iterations for each initial condition (the 600 MH CPU time is 1 minute since only one harmonic is evaluated). The results are shown in the figures 4. For a very weak coupling $\alpha = 0.001$ the projection of the orbits in the x, p_x plane is almost unaffected, whereas in the θ, j plane a deformation is visible when the initial condition in the (x, p_x) plane is chosen inside the islands. For the intermediate coupling $\alpha = 0.01$, see figure 3, the chaotic layers near the separatrices in both planes are well visible. The width of the layer in the x, p_x plane is the largest for initial conditions close to the separatrix in the $(\theta, \jmath)$ plane. The width of the chaotic layer in the $(\theta, \jmath)$ plane becomes significant when we cross the island and approach the dynamic aperture in the (x, p_x) plane. We show also an error plot of the Fourier reconstruction of the orbit: points with large errors are scattered and correspond to chaotic orbits.

We have also considered another example with a linear frequency $\omega_x/2\pi = 0.175$ and $\lambda = 0.18$, above the critical value $\lambda_c \simeq (2\pi)^{-1}$. In this case the linear frequencies are comparable since $\omega_j/2\pi = 0.178$; the coupled $1:1$ resonance and many others appear. Since λ is large, the map is a bad integrator of the pendu-

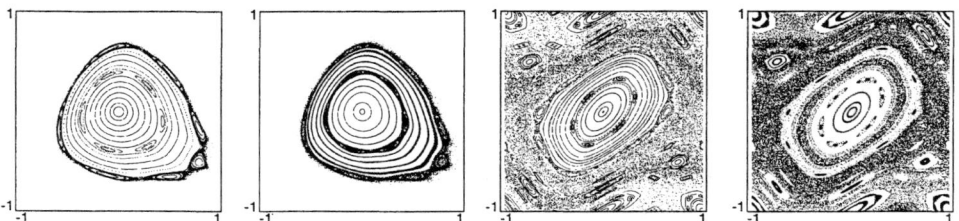

FIGURE 5. Phase plots for =0.175, α=0.01, λ=0.18. Phase plane x,p_x with initial conditions: $\jmath_0$=0, $2\theta_0$=0.1 (left), $\jmath_0$=0,$2\theta_0$=0.99 (center left) Phase plane 2θ, $2\jmath$ with initial conditions x_0=0.1, p_x $_0$=0 (center right), x_0=0.7, p_x $_0$=0 (right)

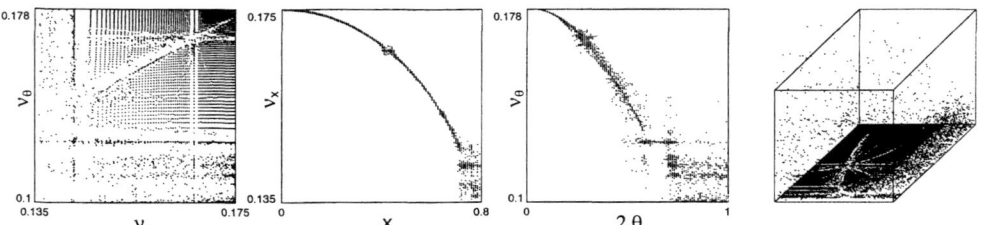

FIGURE 6. Frequency analysis for $\omega_x/2\pi$=0.175, α=0.01, λ=0.18. Plot in the ν_x, ν_θ plane (left), plot ν_x,x (center left), plot ν_θ,2θ (center left), error plot in the plane $x,2\theta$ with a scale [0,0.1] (right)

lum, whose linear tune $(\lambda/2\pi)^{1/2} = 0.169$ differs from $\omega_j/2\pi$. The separatrix is replaced by large stochastic layer and initial conditions there produce a random perturbation of the linear frequency of Hénon map, see figures 5. Many resonant structures are visible as confirmed by the frequency analysis, see figure 6.

A Physical Example

We have considered the mapping (1) for a set of the parameters compatible with SIS namely $k_2 = 0.01$, $\beta = 10, D = 1$ so that $\alpha = 0.1$ and the coefficient $(2k_2^2\beta^2)^{-1}$ which multiplies the coupling term in the longitudinal plane is equal to 5 rather than to 1, the value chosen in the previous examples. For comparison we have condered first a case where the dispersion is very small $D = 0.1$ so that $\alpha = 0.01$. The dynamic aperture in the normalized variables is $A = 0.6$ along the x axis.

In the weakly coupled case $\alpha = 0.01$ as long as we remain up to 1/6 the dynamic aperture, the coupling effect is negligible in the longitunal cordinates; the effect becomes apreciable at 1/2 of the dynamic aperture and has a strong randomizing effect close to the dynamic aperture. Due to the absence of resonances of appre-

181

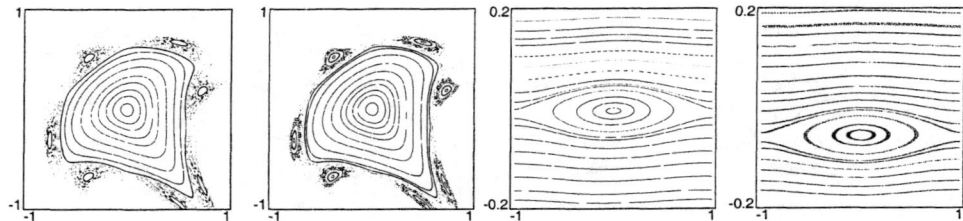

FIGURE 7. Phase plots for $\omega_x/2\pi=0.295$, $\alpha=0.1$, $\lambda=0.001$. Phase plane x,p_x with initial conditions: $j_0=0$, $2\theta_0=0.1$ (left), $j=0$, $2\theta=0.99$ (center left) Phase plane θ, j with initial conditions $x_0=0.1$, $p_{x\ 0}=0$ (center right), $x_0=0.3$, $p_{x\ 0}=0$ (right)

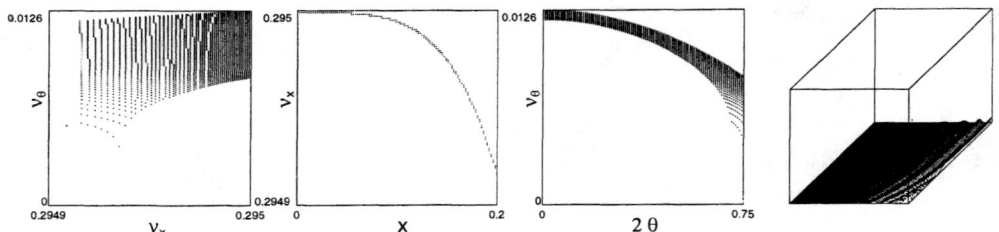

FIGURE 8. Frequency analysis for $\omega_x/2\pi=0.295$, $\alpha=0.1$, $\lambda=0.001$. Plot in the ν_x, ν_θ plane with initial conditions restricted to $0<x_0<0.2$, $|2\theta_0|<0.75$ for (left), plot ν_x,x (center left), plot $\nu_\theta,2\theta$ (center left), error plot in the plane $x,2\theta$ (right) with a scale $[0,0.0001]$

ciable width the effect of coupling in the transverse plane is small, also when we are close the separatrix in the longitudinal phase plane (this is partly due to the factor $(2k_2^2\beta^2)^{-1}$ which is larger than 1 and enhances the coupling effect in the longitudinal phase plane).

For strong coupling $\alpha = 0.1$ below $1/6$ of the dynamic aperture, namely for $p_{x\ 0} = 0, 0 < x_0 < 0.1$, the orbits in the longitudinal phase plane are weakly affected at least for initial conditions $0 < 2\theta_0 < 0.9$, $j_0 = 0$, see figure 7. Up to $1/3$ of the dynamic aperture namely $p_{x\ 0} = 0, 0 < x_0 < 0.2$ the longitudinal orbits are still regular at least for $0 < 2\theta_0 < .75$, $j_0 = 0$. Approaching the dynamic aperture the longitudinal motion becomes very chaotic and the fixed point is displaced. In Figure 8 we show the frequency plots for initial conditions in the range $0 < x_0 < 0.2$, $p_{x\ 0} = 0$ and $0 < 2\theta_0 < .75$, $j_0 = 0$. Approaching the dynamic aperture the longitudinal motion becomes very chaotic.

Action Frequency Map and Resonances

In order to have a better understanding of the dynamics when it is more intricate we have carried a more refined frequency analysis choosing a 600×600 grid

182

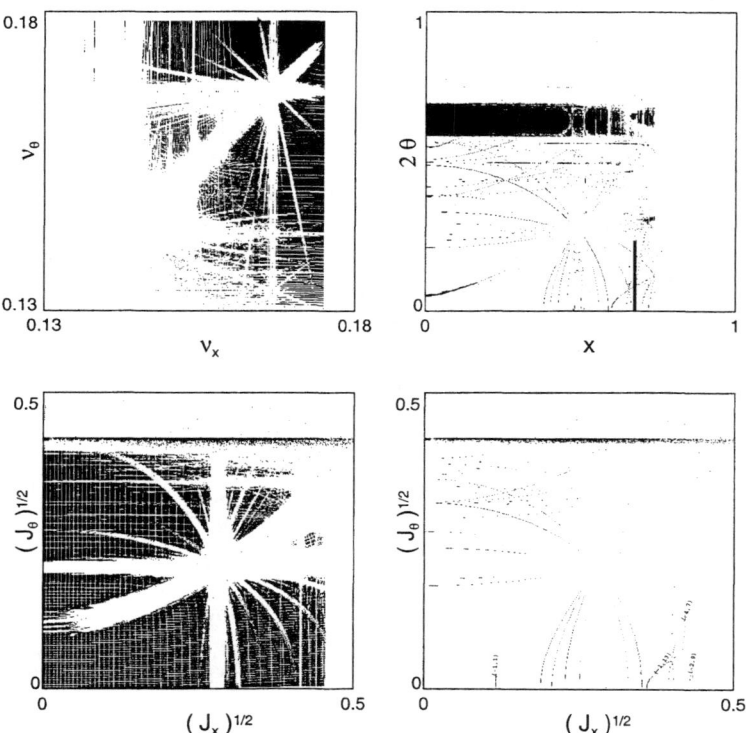

FIGURE 9. Frequency analysis for $\omega_x/2\pi=0.175\ \alpha=0.01,\ \lambda=0.18$. Plot in the $\nu_x,\ \nu_\theta$ plane of the points with a discrepancy less than 0.01 (upper left). Resonance lines in the $x,2\theta$ plane (upper right). Plot of nonresonat tori in the actions plane ($\sqrt{J_x},\ \sqrt{J_\theta}$ (lower left). Resonance lines in the actions plane ($\sqrt{J_x},\ \sqrt{J_\theta}$ (lower right)

and evaluating the harmonics up to order 10 (600 MH CPU time $\sim$ 7 hours). We have chosen $\omega_x/2\pi = 0.175$ and $\lambda = 0.18$ and $\alpha = 0.01$. The frequency plot ν_x, ν_θ, see figure 9 is obtained by dropping the points where the mean discrepancy between the signal and its reconstruction is larger than 0.01. A resonance plot consists of all the points in the $x, 2\theta$ plane whose orbits are resonant.

The actions for all the nonresonant orbits have been evaluated. In the actions space $J_x^{1/2}$, $J_\theta^{1/2}$ we plot all the points which correspond to the non resonant orbits. The resonances in this plot are empty channels. Finally we have also determined the values of the action given by the algorithm (11) within the resonances (excluding the 1:1). The lines so obtained fall within the empty channels of the previous plot and are close to the resonance lines in the $x, 2\theta$ plane, as one should expect close to the stable elliptic point from normal form theory [7].

CONCLUSIONS

To summarize, the proposed model seems adequate to describe the synchro-betatron coupling. As we should expect far enough from the dynamic aperture the coupling effect is small and the longitudinal motion is scarcely affected. By increasing the transverse amplitude, the longitudinal orbits are distorted and then become chaotic. The frequency analysis proves to be useful to determine the safety regions. The longitudinal motion affects the transverse one when resonances are present. In this case the islands become chaotic especially when the initial condition are chosen close to the separatrix in the longitudinal phase plane. The diffusion process can be analyzed in this case following well established techniques. From the dynamical point of view the model is rich and allows to explore different regimes. For instance when λ approaches the critical value $\lambda = 1/2\pi$ the linear frequencies become comparable and many coupled resonances appear as for coupled Hénon maps. A large stochastic layer appears in this case. For intial conditions chosen in that region the transverse motion is almost the same as for a random perturbation of the linear betatronic frequency. This model can be used also to explore how a random perturbation in the longitudinal coordinates determines a diffusion of the emittance.

ACKNOWLEDGMENTS

We are indebted to Dr. Bazzani for discussions on the frequency analysis.

REFERENCES

1. Prome M. "Proceedings of the 1996 International Linac Conference", Geneva, CERN 96-07, p.9 (1996).

2. Hofmann H. I. "Proceedings of EPAC", Stiges 10-14 June, *Institute of Physics Publishing*, Bristol, p. 225. (1996).

3. Giovannozzi M., Scandale W., Todesco E. *Part. Accel.* **56**, 195 (1997).

4. Fisher M., Giovannozzi M., Schmidt F. *Phys. Rev.* **E 55**, 3507 (1997).

5. Franchetti G., Turchetti G. "The micromaps description of a beam with space charge", Proceedings of the Workshop "Nonlinear and Collective Phenomena in Beam Physics" Arcidosso 1-4 September 1996, this volume.

6. Dome G. "Theory of RF acceleration" from "CAS Cern Accelerator School" 1985, Turner E. S., CERN 87-03 (1987).

7. Bazzani A., Todesco E., Turchetti G., Servizi G. "A normal form approach to the theory of nonlinear betatronic motion" CERN Yellow Report 94-02 (1994).

A Hénon Map Approach to the Transverse Dynamics of Off Momentum Particles

G. Franchetti*, G. Turchetti**

Dipartimento di Fisica, Università di Bologna

** INFN Sezione di Bologna, ITALY
* GSI, Darmstadt, GERMANY

Abstract. We present a method to investigate the off momentum effects on the single particle transverse dynamics in a non linear lattice. We show that the one turn map approach is suitable to describe the coupling between the nonlinearity and the off momentum. We prove that for a linear lattice with a single sextupole in the thin lens approximation, the Hénon map description can be proposed. The computation of the tune shift and non linear dispersion is presented. It is shown that the off momentum dynamic aperture can be related to the dynamic aperture of the Hénon map with the shifted tune. Extensions to the four dimensional case are briefly outlined.

INTRODUCTION

The transverse dynamics of particles in a non linear lattice has been successfully analyzed by using the symplectic one turn map approach. This model is suitable to describe the non linear tune shifts, the structure of resonances, and the dynamic aperture. For a flat beam moving in a ring of N identical FODO cells with thin sextupoles, the one turn map is the N-th iterate of a map which is conjugated, up to a scaling, with the standard two dimensional Hénon map, if the off momentum effects are neglected. Indeed the transfer map of each cell is quadratic and a linear change of coordinates (a area preserving Courant-Snyder transformation followed by a scaling dependent on the sextupole strength) allows to write it as the two dimensional Hénon map [1,2], which depends only on the linear phase advance ω_x. When the off momentum effects are relevant, as for high intensity operations, an extension of the previous model is required. The design particle with longitudinal momentum p_0 follows a closed orbit, which

CP468, *Nonlinear and Collective Phenomena in Beam Physics–1998 Workshop*,
edited by S. Chattopadhyay, M. Cornacchia, and C. Pellegrini
© 1999 The American Institute of Physics 1-56396-862-2/99/$15.00

corresponds to the fixed point of the one turn map. In a linear lattice a particle with off momentum $\delta p = (p - p_0)/p_0$ follows a different closed orbit and the fixed point in the one turn map is displaced. The presence of a nonlinear force determines a further shift of the fixed point and changes the linear tune [3]. As a consequence the non linear tune shift and the dynamic aperture vary with δp.

We discuss in detail the transformation leading for any δp to the standard Hénon map. As a consequence the change with δp of the dynamical variables of the map is analytically determined. For instance it is simple to obtain the dependence of the new linear tune $\overline{\omega}_x$ as a function of the old one ω_x and δp, the sextupole contribution to the dispersion $\delta x / \delta p$, and to relate the dynamic aperture of the off momentum map with the standard Hénon map with linear frequency $\overline{\omega}_x$.

The present scheme extends to the transverse motion in the x, y plane. The same correspondence with a standard four dimensional Hénon map is established. The major difference arises from the structure of the 4D map, which depends on 3 parameters (linear frequencies ω_x, ω_y and ratio β_y/β_x) and has four fixed points. In some cases an interchange of stability of the fixed points may occur implying a linear coupling and a vertical dispersion due to the non-linearity.

THE LINEAR OFF-MOMENTUM MAP

If the particle longitudinal momentum differs from the design one, $\delta p = (p - p_0)/p_0 \neq 0$, the closed orbit changes. Denoting by a dot the derivative with respect to the arc length s the equation of motion is

$$\ddot{x} + k_x(s)x = \frac{\delta p}{\rho(s)}$$

where $k_x(s) = \rho^{-2}(s) - k(s)$, having denoted with ρ the radius of curvature. We denote by $D(s)\delta p$ the particular solution of the equation where

$$\ddot{D} + k_x(s)D = \frac{1}{\rho(s)}, \qquad D(s) = D(s + \ell)$$

where ℓ is the length of the reference orbit. The coordinates of the new closed orbit are $x_c(s) = D\delta p$ and $p_{x,c}(s) = \dot{D}\delta p$.

The general solution $x(s) - D(s)\delta p$ of the homogeneous equation for the horizontal plane is the usual solution of the Hill equation and the change after one turn at a section $s = s_0$ is given by the one turn map L_x, defined by the product of the linear maps for the individual elements. The quantity $x(s) - D(s)\delta p \equiv x(s) - x_c$ represents the particle coordinate with respect the closed orbit. Assuming the

186

reference orbit is stable $\mathrm{Tr}\, \mathsf{L}_x = 2\cos\omega_x$ we can write $\mathsf{L}_x = \mathsf{W}_x R(\omega_x)\mathsf{W}_x^{-1}$ so that

$$\begin{pmatrix} x - D\delta p \\ p_x - \dot{D}\delta p \end{pmatrix}' = \mathsf{W}_x R(\omega_x)\mathsf{W}_x^{-1} \begin{pmatrix} x - D\delta p \\ p_x - \dot{D}\delta p \end{pmatrix}$$

where D is the dispersion function, $\dot{D}$ is its derivative (both evaluated at $s = s_0$) and the prime denote the coordinates evaluated after one turn. The matrix W_x has the form

$$\mathsf{W}_x = \begin{pmatrix} \beta_x^{1/2} & 0 \\ -\alpha_x\beta_x^{-1/2} & \beta_x^{-1/2} \end{pmatrix}$$

The 2D one turn map consequently reads

$$\begin{pmatrix} x \\ p_x \end{pmatrix}' = \mathsf{W}_x R(\omega_x)\mathsf{W}_x^{-1} \begin{pmatrix} x - D\delta p \\ p_x - \dot{D}\delta p \end{pmatrix} + \delta p \begin{pmatrix} D \\ \dot{D} \end{pmatrix} \tag{1}$$

The fixed point is now $\mathbf{x}_f = (D\delta p, \dot{D}\delta p)$ and $x_f = x_c(s_0)$. Introducing the Courant-Snyder coordinates by the transformation

$$\mathbf{X} = \mathsf{W}_x^{-1}\mathbf{x}, \qquad \mathbf{x} = \begin{pmatrix} x \\ p_x \end{pmatrix}, \qquad \mathbf{X} = \begin{pmatrix} X \\ P_x \end{pmatrix} \tag{2}$$

The linear map (1) takes the form

$$(\mathbf{X} - \mathbf{X}_f)' = R(\omega_x)(\mathbf{X} - \mathbf{X}_f) \tag{3}$$

where $\mathbf{X}_f = (X_f, P_{x\,f})$ denotes the fixed point $\mathbf{x}_f$ changed according to equation (2). Translating the origin at the fixed point $\mathbf{X}_f$ the map becomes a pure rotation.

THE NON-LINEAR OFF MOMENTUM MAP

The effect of sextupole of length ℓ_S in the thin length approximation for the case of a flat beam can be easily evaluated using the following expression for the force

$$F_{\text{sext}} = \ell_S K_2 \frac{x^2}{2}\delta(s - s_0)$$

Evaluating the one turn map at the left hand of the sextupole and defining $k_2 = \ell_S K_2/2$ the one turn map at $s = s_0 - 0$ is given by the composition of the one turn map (1) with a nonlinear kick $K(x, p_x) = (x, p_x + k_2 x^2)$.

$$\begin{pmatrix} x \\ p_x \end{pmatrix}' = \mathsf{W}_x R(\omega_x)\mathsf{W}_x^{-1} \begin{pmatrix} x - D\delta p \\ p_x + k_2\, x^2 - \dot{D}\delta p \end{pmatrix} + \delta p \begin{pmatrix} D \\ \dot{D} \end{pmatrix}$$

187

This map is suitable to describe any linear lattice with a linear part L_x which is conjugated to a rotation by the transformation W_x (easy to implement in a computer code). In the Courant-Snyder coordinates the map takes the form

$$
\begin{pmatrix} X - X_f \\ P_x - P_{xf} \end{pmatrix}' = \mathsf{R}(\omega_x) \begin{pmatrix} X - X_f \\ P_x - P_{xf} + k_2\beta_x^{3/2}\,X^2 \end{pmatrix}
\tag{4}
$$

Following the same procedure as in the previous section we translate the coordinate system to the linear fixed point and scale them according to

$$
\hat{\mathbf{X}} = k_2\beta_x^{3/2}(\mathbf{X} - \mathbf{X}_f), \qquad \mathbf{X} = \begin{pmatrix} \hat{X} \\ \hat{P}_x \end{pmatrix}
$$

Letting

$$
\hat{\mathbf{X}}_f = k_2\beta_x^{3/2}\mathbf{X}_f = \begin{pmatrix} k_2\beta_x\delta p\, D \\ k_2\beta_x\delta p(\alpha D + \beta_x \dot{D}) \end{pmatrix}
$$

be the fixed point in the new scaled coordinates, the off momentum one turn map reads

$$
\begin{pmatrix} \hat{X} \\ \hat{P}_x \end{pmatrix}' = \mathsf{R}(\omega_x) \begin{pmatrix} \hat{X} \\ \hat{P}_x + (\hat{X} + \hat{X}_f)^2 \end{pmatrix}
\tag{5}
$$

NEW FIXED POINT AND DISPERSION

The stable fixed point of the map (5) $\hat{X}_f^*, \hat{P}_{xf}^*$ is given by

$$
\hat{X}_f^* = -\hat{X}_f + \tan\frac{\omega_x}{2}\left(1 - \sqrt{1 - \frac{2\hat{X}_f}{\tan\dfrac{\omega_x}{2}}}\right), \qquad \hat{P}_{xf}^* = -\tan\frac{\omega_x}{2}\,\hat{X}_f^*
$$

where the condition for its existence is given by

$$
\hat{X}_f < \frac{1}{2}\tan\left(\frac{\omega_x}{2}\right)
$$

The stable fixed point does not exist in a neighborhood of the integer tunes $\omega_x = 2\pi n$ with n integer and the length of this interval can be estimated if δp is small enough, by using a first order expansion which gives

$$
\omega_x \in [2\pi n - \Delta\omega_x,\ 2\pi n + \Delta\omega_x], \qquad \Delta\omega_x = 4k_2\beta_x D\,|\Delta p|
$$

where Δp is the momentum spread of the beam.

In the range of ω_x where the fixed point exists, if the ratio $|\hat{X}_f/\tan(\omega_x/2)|$ is much smaller than 1, we can expand $\hat{X}_f^*$ according to

$$X_f^* = \frac{\hat{X}_f^2}{2\tan\frac{\omega}{2}} + \dots$$

In the original Courant-Snyder coordinates the fixed point is given by

$$X_f^* = X_f + k_2\beta_x^{3/2}\frac{X_f^2}{2\tan\frac{\omega}{2}} + \dots$$

In the initial coordinates the position of the fixed point reads

$$x_f^* = x_f + k_2\beta_x\frac{x_f^2}{2\tan\frac{\omega}{2}} + \dots, \qquad x_f = D\delta p$$

As a consequence we introduce a nonlinear dispersion defined as

$$\mathcal{D} = \frac{x_f}{\delta p} = D + k_2\beta_x\frac{D^2}{\tan\frac{\omega}{2}}\delta p + \dots$$

RECOVERING THE STANDARD HÉNON MAP

In order to bring the map (5) to a standard Hénon map we translate the origin to the nonlinear fixed point $\hat{\mathbf{X}}_f^*$

$$\overline{\mathbf{X}} = \hat{\mathbf{X}} - \hat{\mathbf{X}}_f^*, \qquad \overline{\mathbf{X}} = \begin{pmatrix} \overline{X} \\ \overline{P}_x \end{pmatrix}$$

Separating the linear and quadratic part of the map we have

$$\begin{pmatrix} \overline{X} \\ \overline{P}_x \end{pmatrix}' = R(\omega_x)\left[\begin{pmatrix} \overline{X} \\ \overline{P}_x + 2\overline{X}(\hat{X}_f + \hat{X}_f^*) \end{pmatrix} + \begin{pmatrix} 0 \\ \overline{X}^2 \end{pmatrix} \right]$$

The linear part $\overline{\mathsf{L}}_x$ of the map can be conjugated to a rotation if $|\mathrm{Tr}\,(\overline{\mathsf{L}}_x)| < 2$. Denoting by $\overline{\mathsf{W}}_x$ the corresponding similarity transformation and by $\overline{\omega}_x$ the new rotation angle we write

$$\overline{\mathsf{L}}_x \equiv R(\omega_x)\begin{pmatrix} 1 & 0 \\ 2(\hat{X}_f + \hat{X}_f^*) & 1 \end{pmatrix} = \overline{\mathsf{W}}_x R(\overline{\omega}_x)\overline{\mathsf{W}}_x^{-1}$$

where

$$\overline{W}_x = \begin{pmatrix} \overline{\beta}_x^{\,1/2} & 0 \\ -\overline{\alpha}_x \overline{\beta}_x^{\,-1/2} & \overline{\beta}_x^{\,-1/2} \end{pmatrix}$$

and the parameter $\overline{\beta}_x$ is given by

$$\overline{\beta}_x = \frac{(\overline{L}_x)_{12}}{\sin \overline{\omega}_x} = \frac{\sin \omega_x}{\sin \overline{\omega}_x}$$

Equating the trace $\operatorname{Tr} \overline{L}_x = 2 \cos \overline{\omega}_x$ we can write

$$\cos \overline{\omega}_x = \cos \omega_x + \sin \omega_x \, (\hat{X}_f + \hat{X}_f^*) \qquad (6)$$

Performing the last linear transformation defined by the matrix $\overline{W}_x^{\,-1}$ followed by the scaling of $\overline{\beta}_x^{\,3/2}$ we define the new coordinates

$$\begin{pmatrix} \mathcal{X} \\ \mathcal{P}_x \end{pmatrix} = \overline{\beta}_x^{\,3/2} \overline{W}_x^{\,-1} \begin{pmatrix} \overline{X} \\ \overline{P}_x \end{pmatrix}$$

and taking into account that $\overline{W}_x^{\,-1} R(\omega_x) = R(\overline{\omega}_x) \overline{W}_x^{\,-1}$ the map takes the final form of the standard 2D Hénon map, which reads

$$\begin{pmatrix} \mathcal{X} \\ \mathcal{P}_x \end{pmatrix}' = R(\overline{\omega}_x) \begin{pmatrix} \mathcal{X} \\ \mathcal{P}_x + \mathcal{X}^2 \end{pmatrix}$$

DISCUSSION OF THE RESULTS

We first investigate the tune shift $\overline{\omega}_x - \omega_x$ as a function of ω_x and δp. To this end it is convenient to recall that the tune is defined only when the following conditions are satisfied

$$i) \ \tan \frac{\omega_x}{2} > 2 k_2 \beta_x D \, \delta p, \quad ii) \ \left| \cos \omega_x + \cos^2 \frac{\omega_x}{2} \left(1 - \sqrt{1 - \frac{2 k_2 \beta_x D \, \delta p}{\tan \frac{\omega_x}{2}}} \right) \right| < 1$$

the first one corresponding to the existence of the closed orbit, the second to its stability. To the first order in δp the tune $\overline{\omega}_x$ is given by

$$\overline{\omega}_x = \omega_x - k_2 \beta_x D \delta p - \frac{1}{2} \left(\frac{1}{\tan \omega_x} + \frac{1}{\tan(\omega_x/2)} \right) (k_2 \beta_x D)^2 \delta p^2$$

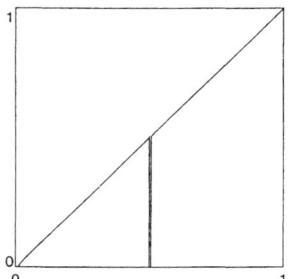 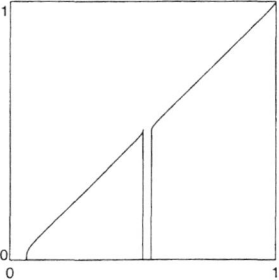 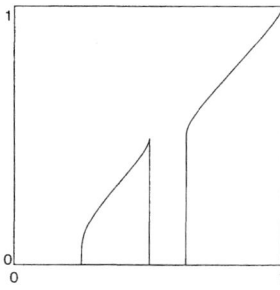

FIGURE 1. Behavior of the tune $\overline{\omega}_x/2\pi$ with respect to the linear tune $\omega_x/2\pi$ for three different values of the off momentum $\delta p=0.01$ (left), $\delta p=0.05$ (center) $\delta p=0.25$ (right)

In Figure 1 we show the tune $\overline{\omega}_x$ as a function of ω_x at three different values of the off momentum for a FODO cell with a thin sextupole. The following parameters were chosen $k_2 = 0.1$, $\beta_x = 10$, $D = 2$, and the off momentum values are $\delta p = 0.01$, 0.05, 0.25. Only the first one is realistic in a high intensity machine for beam compression operations. The higher values where chosen so as to enhance the gaps where the closed orbit does not exist or is unstable and the deviation of $\overline{\omega}_x$ from ω_x.

In order to have an insight of the dynamical changes introduced by the off momentum we compare the orbits (Figure 2) of the on momentum map (in scaled Courant-Snyder coordinates) with the off momentum one turn map given by equation (5). The values of k_2, β_x, D are the same quoted above and the chosen off momentum value is $\delta p = 0.05$. The phase portraits correspond to linear tunes $\omega_x/(2\pi) = 0.1$, 0.4, 0.51. 0.6. We recall that in the interval $[0, 0.0625]$ the fixed point does not exist since condition i) is not fulfilled, whereas in the interval $[0.5, 0.532]$ the fixed point is hyperbolic, since condition ii) is violated. For the lowest frequency it appears that the stability region is considerably reduced by the off momentum, since we are close to the value where the fixed point disappears. At the mirror tunes 0.4, 0.6 with respect to the central tune 0.5 the symmetry of the Hénon map is broken, and the tune shift effect is visible. The phase portrait for the tune 0.51 shows that the fixed point stability is changed by the off momentum. In all these cases the displacement of the fixed point is very small and can be perceived only for the lowest value of the tune.

One of the main advantages of relating the off momentum map to the standard Hénon map is that the corresponding dynamic apertures are related, which is particularly useful in the four dimensional case. We define the dynamic aperture as the radius of the disc with the same area as the domain of stable points. Denoting with $\mathcal{A}(\omega_x)$ the dynamic aperture of the on Hénon map and with $A(\omega_x, \delta p)$ the

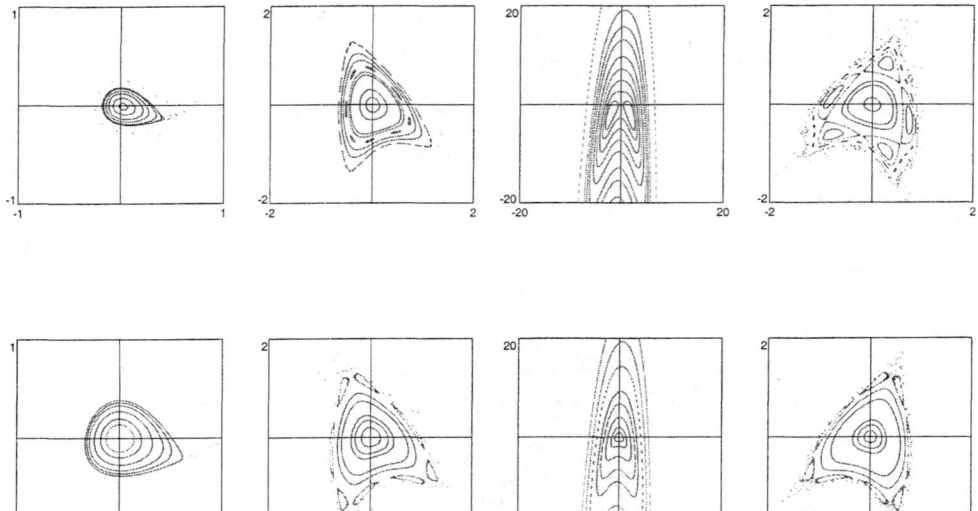

FIGURE 2. Comparison of the phase portraits of the off momentum map $\delta p=0.05$ for the following values of the tunes $\omega/(2\pi)=(0.1,0.4,0.51,0.6)$ (top left to right) with the corresponding on momentum portraits (bottom)

dynamic aperture of the off momentum map (4) we have the following relation

$$A(\omega_x, \delta p) = \left(\frac{\sin \overline{\omega}_x}{\sin \omega_x} \right)^{3/2} \frac{1}{\beta_x^{3/2} k_2} \, \mathcal{A}(\overline{\omega}_x)$$

where the dynamic aperture of the Hénon map is now computed for the the shifted frequency $\overline{\omega}_x$ of the off momentum map, given by equation (6). Since A is defined as the square root of an area, the scaling from the $(\mathcal{X}, \mathcal{P}_x)$ variables to (X, P_x) is simply $k_2 \, \beta^{3/2} \, \overline{\beta}_x^{\,3/2}$ since $\det \overline{W}_x = 1$. We have computed separately $A(\omega_x, \delta p)$ and $\mathcal{A}(\overline{\omega}_x)$ verifying that their ratio agrees with the above expression within the numerical accuracy of our computation. In order to visualize the effect of the off momentum on the dynamic aperture in Figure 3 we compare the results for $\delta p = 0, \ 0.05, 0.25$. It can be noticed that the dynamic aperture is zero for low frequencies where the closed orbit is lost and that the symmetry with respect to the central tune 0.5 is progressively lost as δp increases. If ω_x is fixed, $\overline{\omega}_x$ is function of the off momentum. As a consequence the tune of the standard Hénon map may cross a dangerous resonance for a certain δp. The crossing the 1/3 unstable resonance, where the dynamic aperture vanishes, is shown by figure 4.

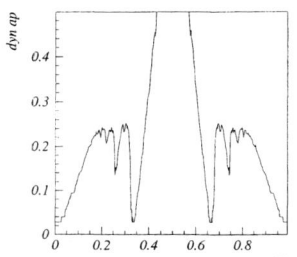 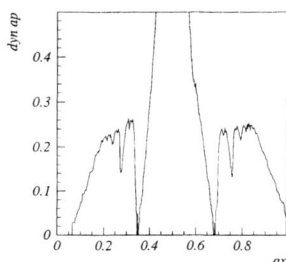 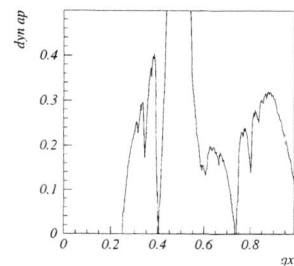

FIGURE 3. Comparison of the the dynamic aperture for the off momentum values $\delta p=0$ (left), $\delta p=0.05$ (center), $\delta p=0.25$ (right)

CONCLUSIONS

We have presented a method to include the off momentum effect in any one turn map. This method allows to evaluate the tune shift produced by the off momentum and the quadratic non linearities as well as the nonlinear contribution to the dispersion. For a single sextupole contribution the dynamic aperture of the off momentum map is related to the dynamic aperture of the standard Hénon map with the shifted tune.

The analytical procedure was described for the case of a flat beam but the methods applies to the four dimensional case as well, even though the discussion of the stability of the fixed points is more involved. In that case a new effect arises since, for some values of the linear tunes, a stability exchange of the fixed points can produce a vertical dispersion.

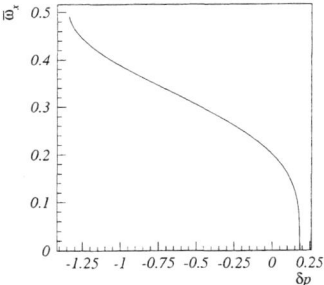 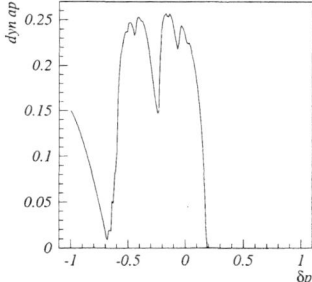

FIGURE 4. Effect of the off momentum on the frequency $\overline{\omega}_x$ (left) for $\omega_x/2\pi=0.2$ and it's consequences on the dynamic aperture (right). The dynamic aperture is zero exactly for an off momentum corresponding to a tune of $1/3$

RFERENCES

1. Hénon M. "Numerical study of quadratic area preserving mappings" *Q. Appl. Math* **27**, 291 (1969).

2.Bazzani A., Todesco E., G. Turchetti G., Servizi G. "A normal form approach to the theory of nonlinear betatronic motion" *CERN Yellow Report* 94-02 (1994).

3. Guiducci S. "CAS: Fifth general accelerator course" E. Turner CERN 94-01 (1994).

WORKING GROUP ON
CREATION AND MANIPULATION
OF HIGH PHASE DENSITY BEAMS

Quantum-like Description of Modulational and Instability and Landau Damping in the Longitudinal Dynamics of High-Energy Charged-Particle Beams

D. Anderson[1], R. Fedele[2], V.G. Vaccaro[2],
M. Lisak[1], A. Berntson[1], S. Johansson[1]

[1] Department of Electromagnetics, Chalmers University of Technology,
S-41296 Göteborg, Sweden

[2] Dipartimento di Scienze Fisiche, Università "Federico II" and INFN
Complesso Universitario di M. S. Angelo, Via Cintia, I-80126 Napoli, Italy

Abstract. Within the framework of the thermal wave model (TWM), a quantum-like description of longitudinal coherent instabilities of high-energy charged-particle beams in the presence of non-negligible resistive part of the coupling impedance is presented. It is shown that, similarly to previous quantum-like investigations in which only a purely reactive impedance was considered, the longitudinal coherent instability can be described in terms of a modulational instability associated with the nonlinear Schrödinger equation (NLSE) also in the present case. In addition, by using the Wigner transform to carry out the analysis in phase-space, the role of Landau damping is considered in connection with the above instability, showing that TWM is capable of reproducing all the results of the conventional theory of the coherent instability as well as of predicting new results (in particular, the possible existence of a *quantum-like Landau damping*), connected with the crucial role of thermal noise introduced by the emittance in the resonance condition between waves and particles in the beam. This new result generalizes the present conventional theory of the longitudinal coherent instability and may be related to the very recent new phenomena observed in the context of nonlinear collective particle beam dynamics.

I INTRODUCTION

Nonlinear collective effects that take place in particle beam dynamics constitute, at the present time, a very large body of the accelerator phenomenology [1,2]. In particular, coherent instabilities have already been recognized to be

CP468, *Nonlinear and Collective Phenomena in Beam Physics–1998 Workshop*,
edited by S. Chattopadhyay, M. Cornacchia, and C. Pellegrini

very important for the practical design of accelerating machines in their pioneering investigations [3]. Today, accelerator physics provides for a very well established theory and very powerful methodologies for describing coherent instabilities [4] as well as designing the machines (f.i., see [5]). In this context, very important subjects of both plasma theory (such as Landau damping [6] and plasma wave instabilities [7,8]) and control system theory (such as Nyquist diagrams [2,9,10]) are regularly applied.

Furthermore, recently discovered new phenomena, deeply connected with nonlinear collective phenomenology of particle beams [11,12], represent new important insights to be included in the above theoretical descriptions.

On the other hand, recently, new approaches for describing the nonlinear collective particle beam dynamics in this more extended way have been proposed [13], [14], [15], [16] and it seems that, in principle, they may include the above new phenomena. In particular, the Thermal Wave Model (TWM) [17], which basically is a *quantum-like* description of the *classical* charged-particle beam dynamics, seems to be suitable for correctly describing coherent instabilities [16,18] but which also yields new insights and predictions that go beyond the present conventional theory.

TWM has been formulated in a way which is fully similar to the one that Gloge and Marcuse [19] used to *transit* from electromagnetic optics to wave optics. In fact, using the classical correspondence between electromagnetic optics, electron optics, and quantum mechanics, the TWM formulation has been developed to *transit* from *geometrical electron optics* [20] to a *wave-like* (or *quantum-like*) electron optics, and has been applied to a number of problems of beam transport and dynamics in both conventional and plasma-based accelerators [16,18], [21], [22], [23]. In the TWM description, the beam properties are described by a complex valued beam wave function (BWF) which satisfies a Schrödinger-like evolution equation where the beam emittance plays the role of Planck's constant. The square modulus of the BWF represents the beam density profile.

The beam instabilities occurring (for certain combinations of parameters) when the coupling impedance between the beam and its surroundings is purely reactive, has been recovered within the TWM approach as the classical modulational instability [16,18] of the corresponding nonlinear Schrödinger equation (NLSE).

However, the more general case, when the resistive part of the coupling impedance becomes important, has not yet been analyzed within the framework of the TWM.

In this paper, we use the TWM to give a quantum-like description of the longitudinal coherent instabilities, taking into account the resistive part of the coupling impedance. According to the TWM, this is done assuming an evolution equation for the BWF which is a sort of generalized nonlinear Schrödinger equation including a nonlinear integral term in addition to the classical nonlinear cubic one. In fact, the reactive part of the coupling impedance corresponds to a nonlinear potential term in the above Schrödinger-like equation which is proportional to the squared modulus of BWF (i.e. the cubic nonlinearity of the nonlinear Schrödinger equation) [16], whilst the resistive part corresponds to a potential term which is proportional to the integral of the squared modulus of BWF. We show that coherent instabilities can be described in terms of the modulational instability also in the case of non negligible resistive part of the coupling impedance. The results obtained in the present work are in full agreement with previous results for the coherent instability based on conventional approaches and provide for further proof of the usefulness and consistency of the TWM approach but also give new insight which may be connected with recent new phenomena in particle beam dynamics [11,12].

In the next Section, in order to emphasize the role played by the nonlinear Schrödinger equation in the nonlinear collective dynamics of charged-particle beams in both plasmas and accelerators, we briefly review the main results recently obtained within the context of TWM on this subject. In Section 3, the formulation of the problem that we want to solve in this paper is presented. In particular, we present the appropriate nonlinear Schrödinger equation to be used for the instability analysis. In fact, in Section 4, the analysis of this equation is carried out in the conventional way i.e. first the appropriate stationary solution is found. A linear equation for small perturbations is then derived and it is shown that in the presence of a resistive part of the impedance, the perturbations are always unstable, in contrast to the purely reactive case where the perturbations are unstable only in certain parameter regimes. The instability results in the general case are summarized in a conventional form given as the contour plots in the (Z_r, Z_i) plane for constant instability growth rate, where Z_r and Z_i are the inductive and resistive parts, respectively, of the coupling impedance. In Section 5, the above analysis is also generalized to include the effects of a finite longitudinal width of the beam (bunched beams). This effect, which has not previously been consistently analyzed within the TWM formalism, is demonstrated to give rise to a stabilizing effect on the instability. The role of Landau damping is also considered, pointing out the difficulty to derive this effect, in the configuration space, by using the present mathematical methods suitable to describe the modulational instability of the nonlinear Schrödinger equation. However, this difficulty is overcome in Section 6, where a transition to the phase space is performed by means of the Wigner transform. The phase-space analysis, fully equivalent to the one given by nonlinear Schrödinger equation clearly show, in agreement with the conventional description, the existence of Landau damping and its stabilizing effect against the coherent (i.e. modulational) instability. Additionally, it is shown that there exists a *quantum-like effect* in the interaction between particles and collective modes in the beam which does not occur in the conventional description. This effect is due to the crucial role played by the thermal noise (through the emittance) in affecting the resonance condition in the wave-particle interaction leading to the Landau damping. Finally, the conclusions are summarized in Section 7.

II BRIEF REVIEW OF THE MAIN RESULTS OF THE TWM DESCRIPTION OF NONLINEAR COLLECTIVE EFFECTS IN PARTICLE BEAM DYNAMICS

The main features and results of TWM applied to the problems in which nonlinear collective effects are involved can be summarized as follows.

- TWM assumes that the transverse (longitudinal) dynamics of a (relativistic) charged particle beam travelling in a medium with velocity βc is governed by a Schrödinger-like equation of the form (for simplicity, in 1-D case) [17]

$$ i\epsilon \frac{\partial \Psi}{\partial s} = - \frac{\epsilon^2}{2} \frac{\partial^2}{\partial x^2} \Psi + U(x,s)\Psi \quad , \tag{1} $$

where $U(x,s)$ is a dimensionless potential energy normalized with respect to $m_0 \gamma \beta^2 c^2$ (m_0 and γ being the particle rest mass and the relativistic factor $(1 - \beta^2)^{-1/2}$), and x is the transverse (longitudinal) coordinate. Note that ϵ, which is the transverse (longitudinal) emittance, replaces Planck's constant $\hbar$, while $s \equiv ct$ plays the role of time. U is a dimensionless effective potential energy which accounts for the interaction between the beam and the surroundings. In general, $U(x,s)$ is the sum of an external potential energy, $U_{ext}(x,s)$, and a self–interaction potential energy (wake–potential energy), $U_s(x,s)$, namely

$$ U(x,s) = U_{ext}(x,s) + U_s(x,s) \quad . \tag{2} $$

$\Psi(x,s)$ is the BWF and its squared modulus, i.e. $|\Psi(x,s)|^2$, is interpreted as the transverse (longitudinal) density profile. Thus, provided that the following condition

$$ \int_{-\infty}^{\infty} |\Psi(x,s)|^2 \, dx < \infty \quad , \tag{3} $$

is satisfied, the transverse (longitudinal) number density $n_b(x,s)$ can be written as

$$ n_b(x,s) \equiv n_{b0} \, |\Psi(x,s)|^2 \quad , \tag{4} $$

where n_{b0} is a positive constant. In principle, $U_s(x,s)$ depends on the beam density and should be expressed in terms of $|\Psi(x,s)|^2$ through the equations for the wake field (namely, for the wake potential). Once this dependence is taken into account, (1) becomes a nonlinear Schrödinger equation where the potential term corresponding to the self-interaction is a nonlinear functional of Ψ. Given the above dependence, this potential term accounts for the collective effects.

• When the medium is specified to be a cold unmagnetized plasma, and the beam has a density much smaller than the unperturbed plasma density (propagation in an overdense plasma), the local charge-neutrality condition holds [24], which implies the following equation for the wake-potential energy U_s (1-D case) [21,22]:

$$
\left(\frac{\partial^2}{\partial x^2} - k_p^2 \right) U_s = \frac{4\pi q^2 n_{b0}}{m_0 \gamma \beta^2 c^2} |\Psi(x,s)|^2 \approx \frac{4\pi q^2 n_{b0}}{m_0 \gamma c^2} |\Psi(x,s)|^2 \quad , \quad (5)
$$

where $k_p \equiv \frac{2\pi}{\lambda_p} = \frac{4\pi q^2 n_p}{m_0 c^2}$ is the plasma wave number (q being the charge of the particles).

• When purely transverse dynamics is considered, the 2-D version of (1) with (2) and (5) constitute a system of coupled equations describing the transverse self-consistent interaction of a relativistic charged particle beam with a plasma.
(i). *Self-focusing.* If the beam spot size is larger than the plasma wavelength (viz. $k_p R >> 1$) and no external field is considered, the following cubic nonlinear Schrödinger equation is obtained [21]:

$$
i\epsilon \frac{\partial}{\partial s} \Psi = -\frac{\epsilon^2}{2} \nabla_\perp^2 \Psi - \frac{n_b}{n_p \gamma} |\Psi|^2 \Psi \quad . \quad (6)
$$

Equation (6) is similar in structure to the equation which describes the self-focusing of an electromagnetic beam in a nonlinear medium [25]. Thus, a simple criterion for finding the threshold of self-focusing can be established: $\epsilon^2 / R_0^2 \approx n_b / (2 n_p \gamma)$, which becomes $\beta_\perp \equiv v_{th}/c \approx 0.7 (n_b / n_p \gamma)^{1/2}$ where v_{th} is the electron (positron) thermal velocity. This is the threshold for the Weibel (or the filamentation) instability [26]. It is interesting to observe that a non self-consistent aberrationless solution of (6), of Gaussian form $\Psi \propto exp(-r^2/R^2)$, in this case corresponds to the following envelope equation :

$$
\frac{d^2 R}{ds^2} + \frac{2 n_b / (n_p \gamma)}{R} - \frac{\epsilon^2}{R^3} = 0 \quad , \quad (7)
$$

which takes into account the self-force and is similar to the well known envelope equation which is valid for a fully neutralized relativistic beam [2].

(ii). *Self-pinching.* If the beam spot size is smaller than the plasma wavelength (viz $k_p R << 1$), the following integro-differential nonlinear Schrödinger equation is obtained [21]:

$$i\epsilon \frac{\partial}{\partial s}\Psi = -\frac{\epsilon^2}{2}\nabla_\perp^2 \Psi + 2K\,\Psi \int_0^r \frac{dr'}{r'} \int_0^{r'} r''|\Psi(r''s)|^2\,dr'' \quad . \tag{8}$$

The corresponding envelope equation is [21]:

$$\frac{d^2 R}{ds^2} = \frac{\epsilon^2}{R^3} - \frac{2K}{R} < \int_0^r |\Psi(r's)|^2\,r'\,dr' > \tag{9}$$

where $K \equiv 2\pi q^2 n_b/(m_0\gamma\beta^2 c^2)$ is the focusing strength and $\langle ... \rangle$ denotes the quantum-like average with respect to Ψ [21]. For an initial Gaussian profile, Eq.n (9) gives the following equilibrium condition: $\epsilon^2/R_0^2 = (1/2)KR_0^2$, which is the well known Bennett self-pinch equilibrium condition [27]. It is worth to observe that an aberrationless approximation of (9) (we have again $\Psi \propto e^{-r^2/R^2}$) leads to the well known envelope equation:

$$\frac{d^2 R}{ds^2} + KR - \frac{\epsilon^2}{R^3} = 0 \quad , \tag{10}$$

which has been used to describe the optics of an overdense plasma lens in the linear regime for the final focus in linear colliders [24,28,29].

- It has been shown that to describe the longitudinal beam self-interaction in the presence of a large amplitude plasma wave in cold plasmas the following Schrödinger-like equation can be assumed [23] ($q \neq e$):

$$i\tilde{\epsilon}\frac{\partial\Psi}{\partial s} = -\frac{\tilde{\epsilon}^2}{2}\frac{\partial^2\Psi}{\partial x^2} + \frac{1}{2}K_p x^2\Psi - \frac{q^2}{e^2}\frac{n_b}{n_p\gamma^3}|\Psi|^2\,\Psi \quad , \tag{11}$$

where $\tilde{\epsilon} \equiv \epsilon/\gamma^2$, and K_p is the plasma wave strength [23]. The corresponding envelope equation in the aberrationless approximation is thus

$$\frac{d^2\sigma}{ds^2} + K_p\sigma - \frac{\xi}{\sigma^2} - \frac{\epsilon^2}{4\gamma^4\sigma^3} = 0 \tag{12}$$

where $\sigma(s)$ is the bunch length and $\xi \equiv (q^2/e^2)n_b/(\sqrt{2\pi}\gamma^3 n_p) > 0$. Eq.n (12) gives the following equilibrium equation:

$$\alpha\left(\frac{\sigma}{\sigma_0}\right)^4 - \delta\left(\frac{\sigma}{\sigma_0}\right) - 1 = 0 \quad , \tag{13}$$

where σ_0 is the bunch length when the self interaction is absent (but, of course, K_p must be positive), $\alpha \equiv K_p/|K_p|$ and $\delta \equiv \xi/(|K_p|\sigma_0^3)$. Since $\sigma(s)$ is a positive function and ξ is a positive constant, it can easily be seen that the equilibrium solutions σ_0' associated with (12) exist only for $K_p > 0$, and satisfy the condition $\sigma_0' > \sigma_0$. The competition between the effect on the particles due to the plasma wave potential well and the self interaction produces a new equilibrium condition which results in a *bunch lengthening* according to (13) [23]. For example, for $\delta = 1$, $\sigma_0' \approx 1.22\,\sigma_0$, for $\delta = 10$, $\sigma_0' \approx 2.19\,\sigma_0$, and for $\delta = 100$, $\sigma_0' \approx 4.64\,\sigma_0$ [23].

• The nonlinear longitudinal dynamics of a relativistic particle bunch in circular accelerating machines, assuming a purely reactive coupling impedance Z_i and neglecting the radiation damping effects, has been described by the following NLSE [16]:

$$i\epsilon\eta\frac{\partial\Psi}{\partial s} = -\frac{\epsilon^2\eta^2}{2}\frac{\partial^2\Psi}{\partial x^2} + \frac{1}{2}K'x^2\Psi - \eta\frac{qI}{2\pi E_0}\left(\frac{Z_i}{n}\right)|\Psi|^2\Psi \quad , \qquad (14)$$

where K is the radio frequency cavity strength, I is the beam current, and E_0 is the synchronous particle energy, and η is the slip factor. In the aberrationless approximation, the corresponding envelope equation is:

$$\frac{d^2\sigma}{ds^2} + K'\sigma - \frac{\xi'}{\sigma^2} - \frac{\epsilon^2\eta^2}{4\sigma^3} = 0 \quad , \qquad (15)$$

where $\xi' \equiv \eta qI(Z_i/n)/[(2\pi)^{3/2}E_0]$. Eq.n (15) coincides with Sacherer's envelope equation, which leads to the potential well bunch lengthening [16]. When the RF is switched off, the following important results have been recovered.

(i). *Coherent instability condition for monochromatic beams.* In the quantum-like context, this condition corresponds to the *Lighthill criterion* for modulational instability [16,18]: $\eta Z_i > 0$.

(ii). *Stability condition for bunched beams.* The following well known inequality, which determines to find the stability region, has been recovered for a purely reactive impedance [16]:

$$\left|\frac{Z_i}{n}\right| \leq \mathcal{F}\frac{|\eta|E_0}{q}\frac{\sigma_{p0}^2}{I} \quad , \qquad (16)$$

where σ_{p0} is the r.m.s. momentum spread at the equilibrium, and $\mathcal{F} \approx 2\pi$. These values are in agreement with the predictions of the conventional theory [2,4].

• For a bunched beam, in both plasmas and accelerating machines, a solitary solution has been found [16,23]. According to the Lighthill criterion, it has been pointed out that a soliton formation would be the natural evolution of the initial beam density modulation toward a self bunching which asymptotically gives a soliton-like envelope wave.

III THE NONLINEAR SCHRÖDINGER EQUATION FOR AN ARBITRARY COUPLING IMPEDANCE

Let us consider a charged-particle beam travelling in a circular accelerating machine. Neglecting the transverse dynamics as well as the radiation damping effects, the longitudinal evolution of the BWF, $\Psi(x, s)$, according to the TWM [17], can in general be written as [16]:

$$i\epsilon\frac{\partial\Psi}{\partial s} = \frac{1}{2}\frac{\eta}{\beta^2}\epsilon^2\frac{\partial^2\Psi}{\partial x^2} + \frac{\Psi}{(E_0/q)\beta cT_0}\int_0^x \mathcal{U}(x's)\,dx' \quad , \tag{17}$$

where $\mathcal{U}(x, s)$ denotes the self consistent voltage describing in general the interaction of the beam with the surroundings (other notation is standard, see [2,4,30]).

The voltage, $\mathcal{U}(x, s)$, is related to the charge line density, $\lambda(x, s)$, according to

$$\mathcal{U}(x, s) = e\beta cZ_r\lambda(x, s) + e\beta cE_0\frac{Z_i}{n}\frac{\partial\lambda}{\partial x} \quad , \tag{18}$$

and the system is self-consistently closed by the relation

$$\lambda(x, s) = \frac{N}{2\pi R_0}|\Psi(x, s)|^2 \quad , \tag{19}$$

(R_0 being the radius of the synchronous particle orbit). Thus, using Eq.s (17)-(19), we can write the equation for the BWF in the form:

$$i\frac{\partial\Psi}{\partial s} = \alpha\frac{\partial^2\Psi}{\partial x^2} + \kappa|\Psi|^2\Psi + \mu\,\Psi\int_0^x |\Psi(x', s)|^2\,dx' \quad , \tag{20}$$

where the coefficients in Eq.n (20) are given by

$$\alpha = \frac{\eta\epsilon}{\beta^2} \quad , \tag{21}$$

$$\kappa = \frac{q^2N}{2\pi\epsilon E_0T_0}\frac{Z_i}{n} \quad , \tag{22}$$

and

$$\mu = \frac{q^2N}{2\pi\epsilon E_0T_0R_0}Z_r \quad . \tag{23}$$

Our problem consists in making a stability (instability) analysis, in order to show that the modulational instability associated with the solutions of (20) coincides with the coherent instability, extending in this way what has been done in previous papers [16,18].

IV INSTABILITY ANALYSIS AND ITS COMPARISON WITH CONVENTIONAL APPROACHES

According to the previous section, in the case of $\mu = 0$, Eq.n (20) has already been investigated for the stability of small perturbations on a stationary background solution. Thus, we directly consider here the case $\mu \neq 0$. In this case (20) includes the resistive integral nonlinearity.

(i). It is easy to show that a stationary CW-solution of Eq.n (20) exists of the form

$$\Psi(x, s) \ = \ \Psi_s(x, s) \ = \ \Psi_0 \, e^{i\phi_0(x,s)} \quad , \tag{24}$$

where the phase $\phi_0(x, s)$ varies as

$$\Phi_0(x, s) \ = \ \lambda_1 s \ + \ \lambda_3 s^3 \ + \ \nu x s \quad , \tag{25}$$

with

$$\lambda_1 = -\kappa \Psi_0^2, \quad \lambda_3 = -\mu \Psi_0^2, \quad \text{and} \quad \nu = \frac{1}{3}\alpha\mu^2 = \frac{1}{3}\alpha\mu^2\Psi_0^4 \quad . \tag{26}$$

(ii). The dynamics of small perturbations on the stationary solution is investigated by writing

$$\Psi(x, s) \ = \ (\Psi_0 \ + \ \delta\Psi(x, s)) \ e^{i\phi_0(x,s)} \quad . \tag{27}$$

Inserting this ansatz into Eq.n (20), the following coupled system for the perturbations $\delta\Psi$ and $\delta\Psi^*$ is obtained (the partial derivatives with respect to s and to x are still denoted with the subscript notation):

$$i\delta\Psi_s \ = \ \alpha\,\delta\Psi_{xx} \ + \ \kappa\Psi_0^2(\delta\Psi \ + \ \delta\Psi^*) \ - \ 2i\alpha\mu s\Psi_0^2\delta\Psi_x \ + \ \mu\Psi_0^2\int_0^x (\delta\Psi \ + \ \delta\Psi^*)\,dx' \quad , \tag{28}$$

$$-i\delta\Psi_s \ = \ \alpha\,\delta\Psi_{xx} \ + \ \kappa\Psi_0^2(\delta\Psi \ + \ \delta\Psi^*) \ + \ 2i\alpha\mu s\Psi_0^2\delta\Psi_x^* \ + \ \mu\Psi_0^2\int_0^x (\delta\Psi \ + \ \delta\Psi^*)\,dx' \quad . \tag{29}$$

Introducing the new functions

$$u \ = \ \delta\Psi \ - \ \delta\Psi^* \quad \text{and} \quad v \ = \ \delta\Psi \ + \ \delta\Psi^* \quad , \tag{30}$$

the following simplified system can be obtained from Eq.s (28) and (29):

$$iv_s \ = \ \alpha u_{xx} \ + \ 2\kappa\Psi_0^2 u \ - \ 2i\alpha\mu s\Psi_0^2 v_x \ + \ 2\mu\Psi_0^2\int_0^x u\,dx' \quad , \tag{31}$$

205

$$iu_s = \alpha v_{xx} - 2i\alpha\mu s \Psi_0^2 u_x \quad, \tag{32}$$

(*iii*). Since the lower limit of the integration in Eq.s (31) and (32) is arbitrary (or equivalently if we take the derivative of Eq.n (31) with respect to x), we realize that we can look for solutions of the form

$$u, v \sim e^{-i\Omega x} \quad. \tag{33}$$

This implies that Eq.s (31) and (32) can be written as

$$iv_s + 2\alpha\mu s\Omega\Psi_0^2 v = -\alpha\Omega^2 u + 2\kappa\Psi_0^2 u + i\,\frac{2\mu\Psi_0^2}{\Omega}\, u \quad, \tag{34}$$

and

$$iu_s + 2\alpha\mu s\Omega\Psi_0^2\, u = -\alpha\Omega^2 v \quad. \tag{35}$$

It is clear from Eq.s (34) and (35) that u and v must include a quadratic s-dependence in the phase which can be split off the solutions by the transformation

$$v = V(s)\, e^{i\alpha\mu\Omega s^2\Psi_0^2} \quad, \text{ and } \quad u = U(s)\, e^{i\alpha\mu\Omega s^2\Psi_0^2} \quad. \tag{36}$$

The coupled equations for V and U then become

$$V_s = i\left(\alpha\Omega^2 - 2\kappa\Psi_0^2 - i\frac{2\mu}{\Omega}\Psi_0^2\right)U \quad, \text{ and } \quad U_s = i\alpha\Omega^2 V \quad. \tag{37}$$

Since this new system has constant coefficients we can look for solutions of the form

$$V, U \sim e^{iks} \quad, \tag{38}$$

and the corresponding conventional determinant condition for nontrivial solutions finally yields the dispersion relation, $k = k(\Omega)$, for the perturbation as:

$$k^2 = \left(\alpha\Omega^2\right)^2\left(1 - \frac{2\kappa\Psi_0^2}{\alpha\Omega^2} - i\frac{2\mu}{\alpha\Omega^3}\Psi_0^2\right) \quad. \tag{39}$$

Separating k into real and imaginary parts, according to $k = k_r + ik_i$, we obtain from Eq.n (39)

$$k_r^2 - k_i^2 = \left(\alpha\Omega^2\right)^2\left(1 - \frac{2\kappa\Psi_0^2}{\alpha\Omega^2}\right) \quad, \tag{40}$$

and

$$k_r \, k_i \; = \; -\left(\alpha\Omega^2\right)^2 \frac{2\mu}{\alpha\Omega^3} \Psi_0^2 \quad, \tag{41}$$

from which we can solve for k_i to obtain

$$k_i^2 \; = \; \frac{\left(\alpha\Omega^2\right)^2}{2} \left[-\left(1 - \frac{2\kappa\Psi_0^2}{\alpha\Omega^2}\right) + \sqrt{\left(1 - \frac{2\kappa\Psi_0^2}{\alpha\Omega^2}\right)^2 + \left(\frac{4b\mu\Psi_0^2}{\alpha\Omega^3}\right)^2} \, \right] \quad. \tag{42}$$

We note that when $\mu = 0$, Eq.n (42) reduces to

$$k_i^2 \; = \; \frac{\left(\alpha\Omega^2\right)^2}{2} \left(\frac{2\kappa\Psi_0^2}{\alpha\Omega^2} - 1 \right) \quad \text{when} \quad \frac{2\kappa\Psi_0^2}{\alpha\Omega^2} > 1 \quad, \tag{43}$$

and

$$k_i^2 \; = \; 0 \quad \text{when} \quad \frac{2\kappa\Psi_0^2}{\alpha\Omega^2} < 1 \quad, \tag{44}$$

as it should, cf [16]. It follows directly from Eq.n (42) that $k_i^2 > 0$, *whenever* $\mu \neq 0$, *i.e.* *the CW-solution is always unstable when resistive effects are included.* This is in full agreement with the result obtained by the conventional approach for a coasting beam, see e.g. [16].

Furthermore, if we consider the instability growth rate, k_i, as given, we can view Eq.n (42) as a relation between μ and κ, or equivalently, between Z_i and Z_r. Since $\kappa \sim Z_i$ and $\mu \sim Z_r$, it is convenient to introduce in Eq.n (42) the normalizations

$$\overline{Z_i} \; = \; \frac{2\kappa\Psi_0^2}{\alpha\Omega^2} \; \sim \; \frac{Z_i}{\eta n} \quad, \tag{45}$$

$$\overline{Z_r} \; = \; \frac{\mu\Psi_0^2}{\alpha\Omega^3} \; \sim \; \frac{Z_r}{\eta n} \quad, \tag{46}$$

and

$$\overline{k_i} \; = \; \frac{k_i}{\alpha\Omega^2} \quad, \tag{47}$$

Eq.n (42) can then be rewritten in the lucid form:

$$\overline{Z_i} \; = \; 1 + \overline{k_i}^2 - \frac{1}{4\overline{k_i}^2}\overline{Z_r}^2 \quad. \tag{48}$$

This implies that the curves for constant instability growth rate are parabolas in the $(\overline{Z_r}, \overline{Z_i})$ (or equivalently (Z_r, Z_i)) space.

For a comparison with results found for the coherent instability obtained by conventional techniques, it is convenient to rewrite the dispersion relation obtained by the TWM approach, Eq.n (39), as

$$k^2 = k_0^2 + k_1^2 \quad , \tag{49}$$

where

$$k_0^2 = \left(\alpha\Omega^2\right)^2 \quad , \tag{50}$$

and

$$k_1^2 = -2\alpha\Omega^2\left(\kappa + i\frac{\mu}{\Omega}\right)\Psi_0^2 \quad . \tag{51}$$

Using the fact that

$$\kappa + i\frac{\mu}{\Omega} = \frac{Nq^2}{2\pi\epsilon E_0 T_0 n}\left(Z_i + i\frac{n}{\Omega R_0}Z_r\right) \quad , \tag{52}$$

and noting that [2,4] $\Omega = -n/R_0$, $\omega_0 = 2\pi/T_0$, and $R_0 = \beta c/\omega_0$, we can rewrite k_1^2 as

$$k_1^2 = \left(\frac{n\omega_0}{\beta c}\right)^2 \frac{qI_0}{2\pi\beta^2 E_0} i\frac{\eta Z}{n} \quad . \tag{53}$$

In the high current limit, when $k_1^2 \gg k_0^2$ (as studied in [16,18]), the dispersion relation, Eq.n (49), reduces to

$$k^2 \approx k_1^2 \quad , \tag{54}$$

which is identical to the result obtained by conventional techniques in Ref. [2,4].

The present analysis includes the stabilizing influence of the linear dispersion on the modulational instability. However, in the appropriate limit of a high current beam, the predictions of the TWM approach and conventional approaches agree completely, not only for the case of a purely reactive impedance as shown previously [16] , but also in the general case when the impedance contains a resistive part as well as a reactive part. This result provides further proof that the TWM approach is a convenient and alternative description of the dynamics of high energy charged-particle beams in accelerators.

V   EFFECTS OF FINITE BEAM ENERGY SPREAD ON THE MODULATIONAL INSTABILITY

When analyzing the effect of finite beam energy spread on the coherent modulational instability, it is more convenient to use a slightly different approach. Instead of the approach used in Section 4, we will start by separating $\Psi(x, s)$ into real amplitude and phase according to:

$$\Psi(x,s) = A(x,s) e^{i\Theta(x,s)} \quad . \tag{55}$$

Inserting this ansatz into Eq.n (20), and separating real and imaginary parts, one obtains the following system for A and Θ:

$$A_s = \alpha \left(2A_x\Theta_x + A\Theta_{xx} \right) \quad , \tag{56}$$

$$-\Theta_s = \alpha \left(\frac{A_{xx}}{A} - \Theta_x^2 \right) + \kappa A^2 + \mu \int_0^x A^2 \, dx' \quad . \tag{57}$$

In this case, the zero-order solutions for A_0 and Θ_0 are determined by

$$A_{0s} = \alpha \left(2A_{0x}\Theta_{0x} + A_0\Theta_{0xx} \right) \quad , \tag{58}$$

$$-\Theta_{0s} = \alpha \left(\frac{A_{0xx}}{A_0} - \Theta_{0x}^2 \right) + \kappa A_0^2 + \mu \int_0^x A_0^2 \, dx' \quad . \tag{59}$$

In the CW-case, $A_0 = constant \Rightarrow A_{0x} = A_{0xx} = A_{0s} = 0$ and $\Theta_0(x,s)$ must satisfy simultaneously two equations, viz

$$\Theta_{0xx} = 0 \quad , \tag{60}$$

$$-\Theta_{0s} = -\alpha\Theta_{0x}^2 + \kappa A_0^2 + \mu \, x \, A_0^2 \, dx' \quad . \tag{61}$$

Integrating Eq.n (60) we obtain

$$\Theta_0(x,s) = \lambda_1 s + \lambda_3 s^3 + \nu x s \quad , \tag{62}$$

where λ_1, λ_3, and ν are given by Eq.s (26). The equation for the perturbations read

$$\delta A_s = \alpha \left(2\delta A_x\Theta_{0x} + A_0\delta\Theta_{xx} \right) \quad , \tag{63}$$

$$-\delta\Theta_s = \alpha \left(\frac{\delta A_{xx}}{A_0} - 2\Theta_{0x}\delta\Theta_x \right) + 2\kappa A_0\delta A + 2\mu A_0 \int_0^x \delta A \, dx' \quad . \tag{64}$$

Since $\Theta_{0x} = \nu s$, we can assume

$$\delta A \ , \ \delta\Theta \sim e^{-i\Omega x} \quad , \tag{65}$$

which implies that Eq.s (63) and (64) can be written as

$$\delta A_s = -\alpha \left(2is\nu\Omega\delta A + A_0\Omega^2\delta\Theta \right) \quad , \tag{66}$$

$$-\delta\Theta_s \;=\; \alpha\left(-\frac{\Omega^2}{A_0}\delta A \;-\; 2i\nu s\Omega\delta\Theta\right) \;+\; 2\kappa A_0\delta A \;+\; 2i\,\frac{\mu A_0}{\Omega}\,\delta A \quad. \tag{67}$$

The terms proportional to $2is\nu\Omega\alpha$ can be transformed away, cf Eq.s (36), and the remaining system becomes

$$\delta A_s \;=\; -\alpha A_0\Omega^2\delta\Theta \quad, \tag{68}$$

$$-\delta\Theta_s \;=\; \left(-\frac{\alpha\Omega^2}{A_0} \;+\; 2\kappa A_0 \;+\; 2i\,\frac{\mu A_0}{\Omega}\right)\delta A \quad, \tag{69}$$

and finally assuming s-variation according to $\exp(iks)$ we obtain the dispersion relation:

$$k^2 \;=\; \alpha\Omega^2\left(\frac{\alpha\Omega^2}{A_0} \;-\; 2\kappa A_0^2 \;-\; 2i\,\frac{\mu A_0^2}{\Omega}\right) \quad, \tag{70}$$

i.e. the same result as in Eq.n (39). In the case of finite beam energy spread, new terms appear in the linearized equations for δA and $\delta\Theta$, Eq.ns (66) and (67). One stabilizing effect is proportional to $A_{0xx}/A_0 = -F/a^2$, where a is the characteristic longitudinal width of the beam and F is a form factor of order unity) which depends on the actual longitudinal density profile. Including this term, the dispersion relation becomes:

$$k^2 \;=\; \alpha^2\Omega^2\left(\Omega^2 \;+\; \frac{F}{a^2} \;-\; 2\frac{\kappa}{\alpha}A_0^2 \;-\; 2i\,\frac{\mu A_0^2}{\Omega}\right) \quad, \tag{71}$$

Eq.n (71) can be written in the form as Eq.n (49), i.e. as

$$k^2 \;=\; k_0^2 \;+\; k_1^2 \quad, \tag{72}$$

where k_1^2 is defined as before, Eq.n (51), but where k_0^2 now includes a contribution from the beam energy spread and is given by

$$k^2 \;=\; \left(\alpha^2\Omega^2\right)^2\left(1 \;+\; \frac{F}{a^2\Omega^2}\right) \quad. \tag{73}$$

This implies that the curves in the (Z_r, Z_i) plane corresponding to constant growth rate, k_i, still are parabolas, viz

$$\overline{Z}_i \;=\; \Gamma^2 \;+\; \overline{k}_i^{\,2} \;-\; \frac{1}{4\overline{k}_i^{\,2}}\overline{Z}_r^{\,2} \quad, \tag{74}$$

but where however now

$$\Gamma^2 \;=\; 1 \;+\; \frac{F}{a^2\Omega^2} \quad. \tag{75}$$

The result expressed by Eq.s (74) and (75) which implies that the finite longitudinal length of the beam provides a stabilizing effect on the instability which manifests itself by an upward shift of the parabolic level curves, thus extending the stability region upward along the $\overline{Z_i}$ axis.

However, contrary to the results reviewed in [16,18], the level curves remain parabolic and no two-dimensional stability region appears around the origin in the $(\overline{Z_r}, \overline{Z_i})$ plane. The reason for this discrepancy is that the present analysis of NLSE, which properly recovers the coherent instability in terms of the modulational instability in the CW case, does not reproduce the Landau damping for the case of a finite beam length. In fact, for moderate instability growth rates, Landau damping will be strong enough to deform the level curves away from parabolic form and even create a region of stability around the origin of the (Z_r, Z_i) plane. However, for stronger instabilities, Landau damping will be negligible and the level curves regain their parabolic shape and conform with the present predictions. This explanation is amply confirmed by a comparison with the contour plots presented in [2,4]. Nevertheless, in the next section, we show that the present quantum-like description is indeed capable of reproducing also Landau damping, provided that the instability analysis is carried out in the phase space. This is done using the Wigner transform [31]. It should be emphasized that what is done in the phase space in terms of this transformation should be fully equivalent to the analysis in the configuration space in terms of the NLSE. A more rigorous analysis of the modulational instability for the generalized NLSE will be given in a forthcoming work.

VI THE QUANTUM-LIKE LANDAU DAMPING AND ITS ROLE IN THE INSTABILITY ANALYSIS

We want to transit from the longitudinal instability description in configuration space, with the equation

$$i\epsilon\eta\frac{\partial\Psi}{\partial s} = -\frac{\epsilon^2\eta^2}{2}\frac{\partial^2\Psi}{\partial x^2} + U(x,s)\Psi \quad , \tag{76}$$

where

$$U(x,s) = \frac{q}{E_0\beta cT_0}\int_0^x \mathcal{U}(x',s)\,dx' \tag{77}$$

(see also Eq.n (17)), to the one that may be carried out in the phase space. To this end, let us introduce the following Wigner-like function [32]:

$$\rho_w(x,p,s) = \frac{1}{2\pi\epsilon|\eta|}\int_{-\infty}^{\infty}\Psi^*\left(x+\frac{y}{2},s\right)\Psi\left(x-\frac{y}{2},s\right)\exp\left(i\frac{py}{\epsilon\eta}\right)dy \quad , \tag{78}$$

where $p \equiv dx/ds = -\eta q \mathcal{U}(x, s)/(2\pi E_0 R_0)$ is the momentum conjugated to x. The following normalization condition is also assumed:

$$\int \rho_w(x, p, s) \, dx \, dp = 1 \quad . \tag{79}$$

We observe that, if Ψ satisfies the (76), thus ρ_w satisfies the following von Neumann-like equation [32]:

$$\left\{ \frac{\partial}{\partial s} + p \frac{\partial}{\partial x} + \frac{i}{\epsilon\eta} \left[U \left(x + i \frac{\epsilon\eta}{2} \frac{\partial}{\partial p} \right) - U \left(x - i \frac{\epsilon\eta}{2} \frac{\partial}{\partial p} \right) \right] \right\} \rho_w = 0 \quad , \tag{80}$$

which can be cast in the form:

$$\frac{\partial \rho_w}{\partial s} + p \frac{\partial \rho_w}{\partial x} = \sum_{\alpha=0}^{\infty} \frac{(-1)^\alpha}{(2\alpha+1)!} \left(\frac{\epsilon\eta}{2} \right)^{2\alpha} \frac{\partial^{2\alpha+1} U}{\partial x^{2\alpha+1}} \frac{\partial^{2\alpha+1} \rho_w}{\partial p^{2\alpha+1}} \quad . \tag{81}$$

By using (77), (81) becomes:

$$\frac{\partial \rho_w}{\partial s} + p \frac{\partial \rho_w}{\partial x} = \left(\frac{q\eta}{2\pi E_0 R_0} \right) \sum_{\alpha=0}^{\infty} \frac{(-1)^\alpha}{(2\alpha+1)!} \left(\frac{\epsilon\eta}{2} \right)^{2\alpha} \frac{\partial^{2\alpha+1} \mathcal{U}}{\partial x^{2\alpha+1}} \frac{\partial^{2\alpha+1} \rho_w}{\partial p^{2\alpha+1}} \quad . \tag{82}$$

The beam current is now introduced as follows [2]:

$$I(x, s) \equiv q\beta c \lambda_0 \int_{-\infty}^{\infty} \rho_w(x, p, s) \, dp \quad . \tag{83}$$

Note that the Fourier transform of $\mathcal{U}$ and $I(x, s)$ are connected by the coupling impedance Z.

Linearizing around the equilibrium state (i.e., $\rho_w = \rho_0(p)$, $\mathcal{U} = \mathcal{U}_0 = 0$, and $I = I_0 = 0$):

$$\rho_w(x, p, s) = \rho_0(p) + \rho_1(x, p, s) \quad , \tag{84}$$

$$\mathcal{U}(x, s) = \mathcal{U}_1(x, s) \quad , \tag{85}$$

$$I(x, s) = I_1(x, s) \quad , \tag{86}$$

we have:

$$\frac{\partial \rho_1}{\partial s} + p \frac{\partial \rho_1}{\partial x} = \left(\frac{q\eta}{2\pi E_0 R_0} \right) \sum_{\alpha=0}^{\infty} \frac{(-1)^\alpha}{(2\alpha+1)!} \left(\frac{\epsilon\eta}{2} \right)^{2\alpha} \frac{\partial^{2\alpha+1} \mathcal{U}_1}{\partial x^{2\alpha+1}} \rho_0^{(2\alpha+1)} \quad , \tag{87}$$

$$I_1(x, s) \equiv e\beta c \lambda_0 \int_{-\infty}^{\infty} \rho_1(x, p, s) \, dp \quad , \tag{88}$$

212

where $\rho_0^{(2\alpha+1)} \equiv d^{2\alpha+1}\rho_0/dp^{2\alpha+1}$. By assuming for ρ_1, U_1, and I_1 solutions of the form:

$$\rho_1(x,p,s) = \widetilde{\rho_1}(\chi,p,\omega) \, \exp\left(i\chi x - i\omega s\right) \quad , \tag{89}$$

$$U_1(x,s) = \widetilde{U_1}(\chi,\omega) \, \exp\left(i\chi x - i\omega s\right) \quad , \tag{90}$$

$$I_1(x,s) = \widetilde{I_1}(\chi,\omega) \, \exp\left(i\chi x - i\omega s\right) \quad , \tag{91}$$

respectively, and introducing the impedance definition, we finally get the following dispersion relation:

$$1 \; = \; i\alpha_0 Z(\chi,\omega) \int_{-\infty}^{\infty} \frac{\rho_0\left(p + \epsilon\eta\chi/2\right) - \rho_0\left(p - \epsilon\eta\chi/2\right)}{\epsilon\eta\chi} \, \frac{dp}{\chi p - \omega} \quad , \tag{92}$$

where $\alpha_0 \equiv q^2 \beta c \eta \lambda_0/(2\pi E_0 R_0)$.

A preliminary analysis of this dispersion relation can be carried out as follows.

- We can take the limit of small χ, but keeping ϵ and ω finite (f.i., $v_{ph}/c \equiv \omega/\chi >> 1$). Since in this case

$$\frac{\rho_0\left(p + \epsilon\eta\chi/2\right) - \rho_0\left(p - \epsilon\eta\chi/2\right)}{\epsilon\eta\chi} \approx d\rho_0/dp \quad , \tag{93}$$

Eq.n (92) becomes:

$$1 \; = \; i\alpha_0 Z(\chi,\omega) \int_{-\infty}^{\infty} \frac{d\rho_0/dp}{\chi p - \omega} \, dp \quad , \tag{94}$$

which coincides with the dispersion relation of the conventional theory [2] and, thus, reproduces all the coherent instability results for coasting beams of the conventional theory for small χ but including the Landau damping.

- Eq.n (92) shows the existence of new effects which should be significant for large χ that are not included in the conventional theory. In this preliminary analysis we only point out that Landau damping is intrinsically included in the TWM description of coherent instability. Generalizing the conventional theory, TWM seems to show the existence of a quantum-like Landau damping. However, this very novel subject will be investigated more carefully in a forthcoming work.

- In order to recover, as an example, some of the results given in the previous sections in terms of NLSE in the configuration space, let us consider the case of monochromatic beam, which means:

$$\rho_0(p) \propto \delta(p) \quad . \tag{95}$$

In this case, although the instability is present, Landau damping is not working due to the absence of the momentum spread. In fact, the dispersion relation (92) becomes now:

$$1 = -\frac{i\alpha_0 Z(\chi,\omega)}{\epsilon\eta\chi} \left[\frac{1}{\epsilon\eta\chi^2/2 + \omega} + \frac{1}{\epsilon\eta\chi^2/2 + \omega} \right] \quad , \tag{96}$$

which can be cast in the form:

$$\omega^2 = \frac{\epsilon^2\eta^2\chi^4}{4} + i\alpha_0\chi Z \quad . \tag{97}$$

Expressing both ω and Z in their complex representations, viz

$$\omega = \omega_R + i\omega_I \quad and \quad Z = Z_R + iZ_I \quad , \tag{98}$$

we get the following relation:

$$Z_I = -\frac{\delta_0}{4\omega_I^2} Z_R^2 + \frac{\omega_I^2}{\delta_0} + \frac{\epsilon^2\eta^2\chi^2}{4} \quad , \tag{99}$$

where $\delta_0 \equiv \alpha_0\chi = q^2\beta c\eta\chi\lambda_0/(2\pi E_0 R_0)$, which is formally identical to (48) given in Section 4. In particular, for small χ (99) becomes, still in accordance with Section 4, the following relation:

$$Z_I \approx -\frac{\delta_0}{4\omega_I^2} Z_R^2 + \frac{\omega_I^2}{\delta_0} \quad , \tag{100}$$

which is in full agreement with the corresponding instability equation for a monochromatic coasting beams given by conventional description.

VII   CONCLUSIONS, REMARKS, AND PERSPECTIVES

In this paper, an investigation of longitudinal coherent instability has been carried out within the context of TWM. The interaction of the beam with its surroundings (and with itself) has been expressed in terms of a (nonlinear) potential in a Schrödinger-like equation.

The corresponding equation for the beam wave function constitutes a new generalized Schrödinger equation, which, as far as we know, has not been analyzed before. In this paper, this equation has been analyzed for the stability of small perturbations of a constant amplitude backgroud beam. The above investigation is inserted in a field already explored of several nonlinear phenomena described by different kinds of NLSE. In order to give an idea to the reader about this scenario, we have reviewed the main results obtained with TWM concerning the nonlinear collective particle beam dynamics in both plasmas and conventional machines (see Section 2) [16,18], [21]- [23]. In particular, we have reviewed the main results concerning the longitudinal coherent instability for a coasting beam in the conventional machines when the interaction between the beam and the surroundings is modelled in terms of a purely reactive impedance [16,18]. In this case the instability reduces to the classical modulational instability of the conventional Schrödinger equation for cubic nonlinearity.

However, in this paper we have extended this problem to the more general case of non-negligible resistive part of the coupling impedance for coasting beam as well as for beams with finite size. In the first case, we have found that the perturbation are always unstable and the instability growth rate is found in terms of the real and the imaginary parts of the coupling impedance (Z_R and Z_I, respectively).

The results are summarized in terms of curves in the (Z_r, Z_i) plane, corresponding to constant instability growth rate. These curves are found to be parabolas, in full qualitative as well as quantitative agreement with results of previous conventional techniques for analyzing coherent instabilities of high-energy charged-particle beams.

In the second case, the analysis has been extended to include the effects of a longitudinal extent of the background beam. This is shown to give rise to a stabilizing effect on the modulational instability, but does not, within the present analysis, change qualitatively the form of the curves for constant instability growth rate, which remains parabolic.

On the other hand, in the conventional approaches, it has been found that for a finite energy spread of the beam, Landau damping will become an important effect and will deform the level curves for small instability growth rates, even to the point of creating a two-dimensional region of stability around the origin in the (Z_r, Z_i) plane. However, for stronger instabilities, Landau damping becomes negligible and the level curves regain their parabolic form.

However, if the analysis within configuration space, as expresses by the NLSE, is generalized into a phase space description by means of a Wigner-like formalism, the TWM approach predicts new important results concerning coherent instabilities for the beam dynamics. In fact, we have shown that the results of instability given by the conventional theory can be recovered also for finite energy spread, provided to transit to the phase space. The resulting phase-space description is fully equivalent to the one given by the

NLSE in the configuration space, although it seems to be simpler in the phase space. In this way, we have given a preliminary phase-space description of the collective interaction of the beam with the surroundings, modelled in terms of an arbitrary coupling impedance. With this analysis we have obtained a linear dispersion relation which shows the existence of a more general Landau damping (we have called it *quantum-like Landau damping*) which for the case of small χ reproduces all the results of the conventional theory.

Remarkably, in the limit of $\epsilon \to 0$, the dispersion relation (92) coincides exactly, for arbitrary χ, with the one given by the conventional theory. However, ϵ cannot in principle be reduced to zero, because it accounts for the thermal noise which is very important and competes with the resonance in the wave-particle interaction (i.e. Landau damping). Consequently, TWM takes realistically into account, due to the thermal noise, eventual displacements from the exact resonance condition. Since the inhomogeneity wave parameter is χ, the variation of $\rho_0(p)$ cannot be estimated in regions of momentum space with size smaller than $\epsilon|\eta|\chi$. This limitation transforms the usual derivative appearing in the conventional theory, as given by (94), into the finite difference ratio as give in (92).

In future work the explicit dispersion relation for the case of arbitrary χ should be found for several distribution functions $\rho_0(p)$.

We conclude that the presently obtained results further validate the TWM approach as a consistent alternative description of the dynamics of high-energy charged-particle beams in accelerators.

REFERENCES

1. Month, M., and Turner, S. (Ed.s), *Frontiers of Particle Beams; Observation, Diagnosis and Correction*. Proc. Joint US-CERN School on Particle Accelerators. Capri, Italy, October 20-26, 1988 (Springer-Verlag, Berlin, 1989).

2. Lawson, J., *The physics of charged particle beams*, (Clarendon Press, Oxford, 1988), 2nd edition, and references therein.

3. Nielsen, C. E., Sessler,A. M., and Symon, K. R., *High Energy Accelerators and Instrumentation* (CERN, Geneva, 1959), p. 239; for a historical review on this subject see Ref. [2] and references therein.

4. Ruth, R. D., *An overview of collective effects in circular and linear accelerators*, in Ref. [1] ; Hofman, A. *Single-beam coherent phenomena - longitudinal*, CERN 77-13, November 1976, p. 139.

5. Chattopadhyay, S., Cornacchia, M. and Pellegrini, C. (Ed.s), Proc. of Workshop on *Nonlinear Dynamics in Particle Accelerators: Theory and Experiments*. Arcidosso, Italy, September 4-9,1994 (AIP Press, Woodbury, New York, 1995).

6. Landau, L. D., *J. Phys. USSR*, **10**, 25 (1946).

7. Sturrock, P. A. *Plasma Physics*, (Cambridge University Press, Cambridge, 1994).

8. Schmidt, G. *Physics of High Temperature Plasmas*, (Academic Press, New York, 1979).

9. Wang, J. M., *Beam Transfer and Landau Damping*, in Ref. [1] .

10. Chanel, M. *Beam Instabilities and Their Cures*, in E.D. Maletić E. D., and Ruggiero , A. G. (Ed.s), *Crystalline Beams and Related Issues* (World Scientific, Singapore, 1996); see also Kohaupt, R. D. *Cures for Instabilities*, in Ref. [1] .

11. Ruggiero, F., *A Review of New Manifestations of Collective Effects*, Proc. of the EPAC98, Stockholm, 22-26 June, 1998 (Institute of Physics Publishing, Bristol and Phyladelphia, 1998), p. 18, and references therein.

12. Spentzouris, L. K., Ostiguy, J. F., and Colestock, P. L., *Phys. Rev. Lett.* **76**, 620 (1996); Colestock, P. L., and Spentzouris, L. K., in The Tamura Symposium Proc., Austin, Texas, 1994 AIP Conf. Proc. 356 (AIP, Woodbury, New York, 1996).

13. Schamel, H. *Phys. Rev. Lett.* **79**, 2811 (1997); Schamel, H., Physica Scripta **T75**, 23 (1998).

14. Colestock, P. L., Spentzouris, L. K., and Tzenov, S. (1998), these Proceedings and references therein.

15. Fedele R., and Shukla, P. K., (Ed.s), *Quantum-like Models and Coherent Effects*, (World Scientific Publ., Singapore, 1995).

16. Fedele, R., Palumbo, L., and Vaccaro, V. G., Proc. of EPAC92, Berlin, 24-28 March, 1992, Henke, H., Homeyer, H., and Ch. Petit-Jean-Genaz (Ed.s) (Editions Frontieres, Singapore, 1992) p. 762; Fedele, R., Miele, G., Palumbo, L., and Vaccaro, V. G., *Phys. Lett.* **A179**, 407 (1993).

17. Fedele R., and Miele, G., *Il Nuovo Cimento* **D13**, 1527 (1991).

18. Migliorati M., and Palumbo, L., *Coherent instabilities in particle accelerators: conventional and novel approaches*, in Ref. [15] ; Fedele, R., *Physica Scripta*, **T63** 162 (1996).

19. Gloge, D., and Marcuse, D., *J. Opt. Soc. Am.* **59**, 1629 (1969).

20. Sturrock, P. A., *Static and Dynamic Electron Optics*, (Cambridge University Press, Cambridge, 1955).

21. Fedele, R., and Shukla, P. K., *Phys. Rev.* **A45**, 4045 (1992).

22. Fedele, R., Shukla, P. K., and Vaccaro, V. G., *Journal de Physique IV, C6-II*, **5** 119 (1995).

23. Fedele, R., and Vaccaro, V. G., *Physica Scripta*, **T52** 36 (1994).

24. Chen. P., *Particle Accelerators.*, **20**, 171 (1987); P. Chen *Phys. Rev.*, **A45** R3398 (1991).

25. Anderson D., and Lisak, M., *Pulse propagation determined by the nonlinear Schrödinger equation: a variational approach*, in Ref. [15] and reference therein; Anderson, D., and Lisak, M., *Physica Scripta*, **T63** 69 (1996) and references therein.

26. Su, J. J., Katsouleas, T., Dawson, J. M., Jones, M., and Keinigs, *IEEE Trans. Plasma Sci.* **PS-15**, 192 (1987); Weibel, E. S., *Phys. Rev. Lett.* **2**, 83 (1959).

27. Bennett, W. H., *Phys. Rev.*, **45** 89 (1934).

28. Su, J. J., Katsouleas, T., Dawson, J. M., and Fedele R., *Phys. Rev.*, **A41** 3321 (1990).

29. Barov, N., and Rosenzweig, J. B., *Phys. Rev.* **E49**, 4407 (1994).

30. Palumbo, L., and Vaccaro, V. G., *Wake Field and Mesurements*, in Ref. [1] .

31. Wigner, E., *Phys. Rev.*, **40** 749 (1932).

32. Fedele, R., Galluccio, F., Man'ko, V. I., and Miele, G., *Phys. Lett.* **A209**, 263 (1995); Fedele, R., and Man'ko, V. I., *Phys. Rev.* **E58**, 992 (1998).

High Intensity Beams
for Heavy Ion Driven Inertial Fusion
by Multiturn Stacking [1]

R.W. Hasse

GSI Darmstadt, D-64291 Darmstadt, Germany

Abstract. A European study group presently is investigating the possibility and the layout of a heavy ion driven fusion ignition facility (HIDIF). We present this scenario and perform PIC simulations of multiturn stacking in the accumulator rings. By stacking twenty 400mA beams of $^{209}\mathrm{Bi}^{1+}$ ions the current is increased to about 8A with particle losses of about 5% and an increase of emittance by a factor of 10.

I HIDIF SCENARIO

The HIDIF study group includes members of CERN Geneva, DENIM Madrid, ENEA Frascati, FZK Karlsruhe, GSI Darmstadt, KFA Jülich, MPQ Garching, RAL Chilton and University of Frankfurt and is coordinated by G. Plass (CERN). The group is convinced that heavy ion driven inertial fusion with the advantage of a high repetition rate (50...100Hz) over laser induced inertial fusion will be one of the future energy sources.

Therefore the aim of the studies is to present a working scenario of a driver capable to deliver a pulse of total energy 3MJ, of pulse power 500TW and of length 6ns onto an indirectly driven target. The results will be published in due course as an interim report [1] and a list of parameters can be found in the web [2].

The schematic of the scenario is shown in Fig. 1. The high current needed is achieved by telescoping three species of ions, $^{209}_{83}\mathrm{Bi}^{1+}$ and one heavier and one lighter in mass by 10%. These ions are offset in velocity by 5% and, thus, do not interact.

The ions are generated in 16×3 ion sources, funneled and accelerated in four injector linac stages and finally accelerated to 10GeV in the main injector linac with a beam current of 400mA. Then the beam is split and two symmetric systems follow. Each part consists of 2×3 accumulator rings where 20 turns are stacked, thus increasing the current to 8A containing 12 bunches each. Then there follow 2×12×3 synchronization stages for the delay of the bunches, then 3 induction linac

---

[1] Proc. 16*th* ICFA Beam Dynamics Workshop on Nonlinear and Collective Phenomena in Beam Physics, Arcidosso, Italy, Sept.1-5,1998, AIP Conf. Series, 1999

bunchers for bunch rotation each having 24 beam lines. At the end the species are merged and concentrated on the target station by final focusing systems.

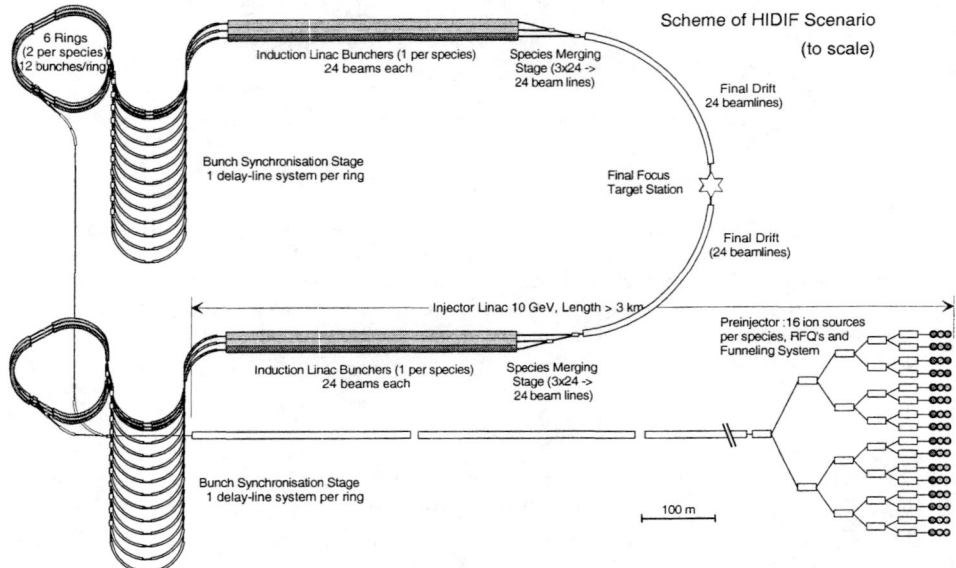

FIGURE 1. Proposed layout of the HIDIF scheme (From [1])

II MULTITURN INJECTION

In the HIDIF system 20 turns from the linac are stacked in the accumulator storage rings. The specifications demand that the emittances should increase from $4...5\pi$ mm mrad to not more than about 50π mm mrad with at most 2% total loss of particles. It has been shown in [3] that 3% loss can be reached with closed orbit bumps changing from turn to turn and with an inclined plane septum. It is the aim of this work to perform similar particle simulations with a rectangular corner septum. After careful optimization of the lattice we have achieved similar results with losses of about 5% presumably due to the corner of the septum [4].

HIDIF has a number of 'triangular' accumulator rings. In the simulations they are simplified as having three sections with 13 cells which simply consist of a single FODO cell. Due to the envisaged linac current of 400 mA the shift in phase advance of the injected beam and the Laslett tune shifts are predicted by solutions of the K-V envelope equation to be $\Delta\sigma_{x,y} \approx -0.3^0$ per cell or $\Delta Q_{x,y} \approx -0.032$. The beam and cell properties of the optimized and fine tuned lattice studied are listed in the following table:

Table 1:

- charge number $Z = 1$
- mass number $A = 209$
- energy per nucleon $E/A = 47.84$ MeV
- velocity $\beta = 0.3193$
- current $I = 400$ mA
- circumference $L = 429$ m
- stiffness $B\rho = 209.77$ Tm
- field gradients B'_x, $B'_y = 9.542, -9.657$ kG/cm
- beam diameter 1 cm
- emittances $\epsilon_x = \epsilon_y = 4\,\pi$ mm mrad
- β-functions $\beta_x, \beta_y = 7.6458,\ 7.5009$ m/rad
- α-functions $\alpha_x,\ \alpha_y = -0.59069, 0.58826$
- tunes $Q_x,\ Q_y\ = 8.730,\ 8.858$

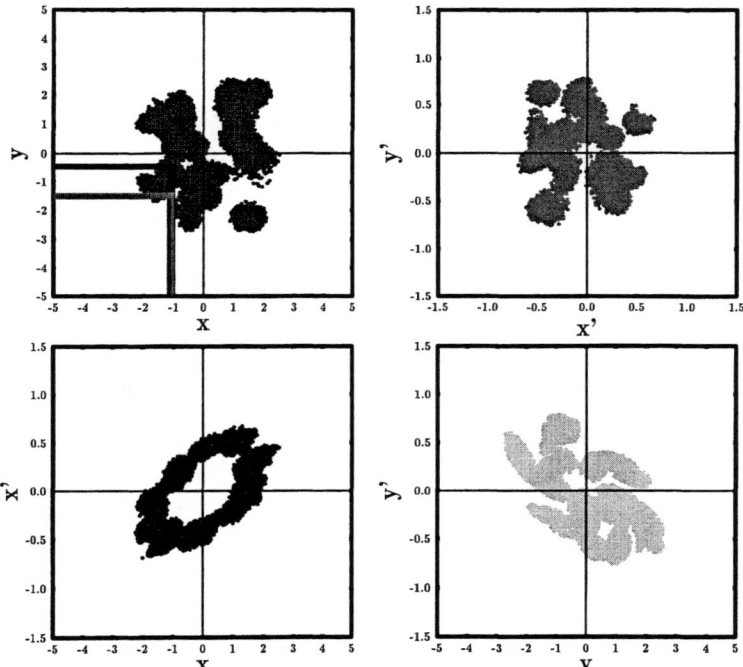

FIGURE 2. Phase space scatter plots of 10000 particles after 20 turns. The upper and lower boxes in the upper left plot indicate the position of the corner septum at the beginning and after the 20 turns.

The following scheme has emerged because this seems to be the closest stacking process with least particle losses: After one turn the first beamlet is bumped to the locus (-15mm, -10mm) measured from the center. The corner septum admits a

221

beam diameter of 10mm so that the corner of the septum is located at (-10mm, -5mm). The following 19 turns are then bumped with vertical offsets of -0.5mm each from the respective previous location so that the 20*th* turn is bumped to (-15mm, -19.5mm). In the simulations, however, not the beam is moved but the septum.

In [4] we have performed simulations in a lattice with tunes similar to the ones used in ref. [3], i.e. $Q_x = 8.65$, $Q_y = 8.78$ and by using initial K-V particle distributions. These studies resulted in particle losses after 20 turns of about 15%. In switching to a waterbag distribution and fine tuning the lattice slightly away from the above given tunes we now obtain a minimum loss of 5%.

The 2D particle simulations are performed with a PIC code in two space and two velocity coordinates including a Poisson solver on a rectangular grid. 500 macro particles in a waterbag initial distribution are used per beamlet and a grid of 51×51 points. Time steps correspond to 1 cm azimuthal steps.

Fig. 2 shows the phase space distribution of the macro particles after 20 injections with full space charge. Here the box indicates the position of the septum at the beginning (upper line) and at the end (lower line). The beamlets although overlapping in coordinate space (upper parts) are still well separated in phase space. Just 5% of the particles are lost at the septum, see Fig. 3. As can be seen from

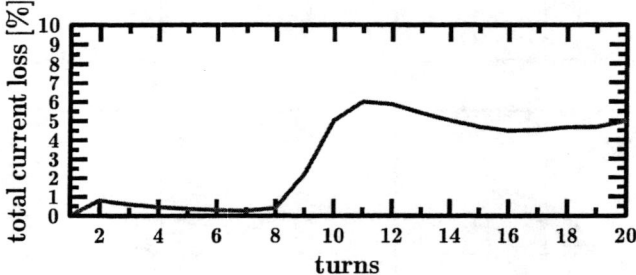

FIGURE 3. Total current loss (number of scraped off particles over total accumulated number) at successive turns for the lattice of Table 1.

Fig. 4, after injection of the 20 turns 95% of the stacked particles have emittances of $\epsilon_x, \epsilon_y \approx 60\pi$ mm mrad roughly in agreement with the specifications. These emittances are obtained by fitting tilted ellipses to the final phase space distributions.

After even more turns (without further injection) the particles then are pushed towards the center of the phase space. The slower particles at the outside form spiral like halos which will be scraped off at the corner septum. They have lost memory of their original position and fill the hollow space at the center. The final emittances do not increase very much.

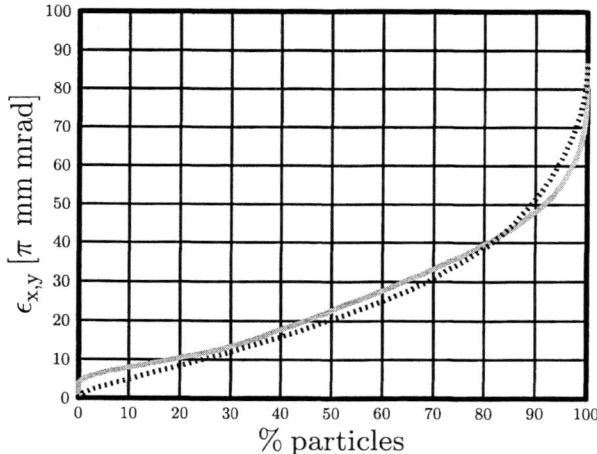

FIGURE 4. Final horizontal (full line) and vertical (dotted line) emittance vs. number of particles left after the 20*th* turn.

III SUMMARY

We have studied a case of stacking of 20 beamlets in a simplified version of the present HIDIF lattice. Closest stacking resulted in about 5% loss and in an increase of emittance of a factor of 15. This has to be compared with a factor of $\sqrt{20} = 4.5$ by Liouvilles theorem which yields horizontal and vertical phase space dilution factors of 5 and 3.3. These values are only slightly higher than the specifications of 2% loss and emittance increase of a factor of 12.5. Wider stacking would result in less losses but also in a larger increase of emittance.

In summary, we believe that due to space charge limits not more than 20 turn should be stacked in the HIDIF ring. A smaller number of turns, about 15, together with a careful optimization of the lattice could result in no losses at the septum.

REFERENCES

1. HIDIF Interim Report, *GSI report GSI-98-06, GSI Darmstadt 1998*
2. http://www.gsi.de/~hidif/HIDIF/parameters.html
3. Prior, C.R. and Rees, G.H., Proc. 12*th* Int. Symp. on Heavy Ion Inertial Fusion, Heidelberg 1997, *Nuclear Instruments and Methods* **415** (1998) 357
4. Hasse, R.W., and Hofmann, I., *in Proceedings [3]* p. 478

Amplification of Beam Acceleration in a Plasma by Plasma Instability

V. A. Lebedev

Jefferson Lab, 12000 Jefferson Ave., Newport News, VA 23606, USA.

Abstract. Although achieving of high accelerating field in a plasma has been demonstrated experimentally, a practical use of such a scheme for building a large accelerator is questionable. A novel scheme of beam acceleration by a plasma wave is considered in this article. The scheme is based on an initial excitation of a plasma wave by a probe beam with comparatively modest intensity. This seed excitation is then amplified by plasma instability, so that the test beam which follows the probe beam with a small delay will be accelerated by the plasma wave with an amplitude significantly exceeding the initial amplitude of the wave. Because of small interaction between the synchronization beam and the plasma, such a scheme allows one to excite a plasma over large length and, consequently, to build a large accelerator.

INTRODUCTION

Beam acceleration by plasma wave first suggested in reference (1) has created new horizons in achieving high accelerating gradients in linear accelerators. In comparison with a general linear accelerator based on an electromagnetic wave propagated in a waveguide, it allows one to reach an order of magnitude higher accelerating gradient of about 1 GeV/m. Achieving such high gradients has been recently demonstrated experimentally by a few groups (2, 3), but many technical and scientific problems have to be resolved before such an accelerator can be built.

Plasma acceleration has a serious advantage in comparison with classical accelerators: it does not involve high electromagnetic fields on vacuum chamber walls and therefore does not have a problem of high voltage breakup. In general, the plasma accelerator is based on plasma excitation by an intense laser (1) or electron (4) pulse. We will call this the probe bunch. Then, after a short delay, when the amplitude of the plasma wave reaches the maximum, the accelerated bunch is injected. We will call that the test bunch. While creating the initial plasma does not represent great difficulties, both electron and plasma excitations have the common problem of creating a sufficiently intense probe bunch. To resolve this problem one needs to reduce the amount of energy pumped into the plasma, which requires a smaller electromagnetic field volume and, consequently, smaller wavelength.

There is another basic reason pushing us to smaller wavelength. It is determined by properties of the plasma oscillations. To make an estimate we consider the flat plasma wave propagating with phase velocity equal to the light velocity, c, in a boundless plasma. For simplicity we will use non-relativistic formulas. In this case from the

CP468, *Nonlinear and Collective Phenomena in Beam Physics–1998 Workshop*,
edited by S. Chattopadhyay, M. Cornacchia, and C. Pellegrini

equations $\operatorname{div}\mathbf{E} = 4\pi n$ and $m\dot{\mathbf{v}} = e\mathbf{E}$ one can deduce that the electric field amplitude is $E_{max} = 4\pi e n_e \xi / k$, while the amplitude of the electron velocity oscillations is $\tilde{v} = eE_{max}/m\omega$. Here n_e is the electron density, e is the electron electric charge, _ is the relative density perturbation, $\xi = \Delta n_e / n_e$, and k is the wave vector[1]. Taking into account that $k = \omega/c$, the oscillation frequency is equal to the plasma frequency, $\omega_p = \sqrt{4\pi n_e e^2 / m}$, and expressing values through _ and the wavelength, $\lambda = 2\pi / k$, one obtains

$$\tilde{v} = c\xi \ ,$$

$$E_{max} = \frac{2\pi mc^2 \xi}{e\lambda} \ . \tag{1}$$

The first equation shows that the large relative density perturbation, _, yields relativistic motion of plasma electrons, and thus additionally increases the motion non-linearity, which limits _ to about $\xi \leq 0.1$. The second equation can be rewritten in practical units as follows, $\lambda_{[mm]} = 3.21\xi / E_{max[GeV/m]}$, that for $E_{max} = 1$ GeV/m and $\xi = 0.1$ yields _=0.32 mm.

Practical use of so small wavelength in a high-energy accelerator creates two fundamental problems. First, how one can phase the accelerating voltage of different accelerator sections. Second, how one can suppress harmful focusing effects due to transverse components of the accelerating field. The second issue is additionally complicated by the fact that the focusing effects are different along the accelerating bunch. The accelerating scheme considered in this article addresses these two issues as well as how to excite the plasma wave without creating a very intense probe bunch.

The main problems of the considered before schemes arise from the fact that the test bunch performs two functions. It carries the energy for plasma excitation and it excites and synchronizes the plasma wave. It can work well for a small accelerator, but with increased accelerator size it will require thousands of intense high-energy probe bunches for plasma excitation. This makes an accelerator too expensive and therefore unrealistic. To resolve the question of section synchronization one needs to separate these two functions. The basic idea is in the following. One creates an unstable plasma but before the instability has developed, a low-intensity probe bunch is injected, making a seed excitation. The probe bunch excites many modes, but the plasma only amplifies the mode with the correct field configuration, thus creating a high-quality accelerating wave. After the amplitude of the plasma wave reaches the maximum, the test bunch is injected for acceleration. To create the required plasma properties it is immersed into a longitudinal magnetic field.

To formulate the main requirements for such plasma acceleration we will start our consideration from basic low frequency analysis of plasma properties when the vertex part of the electromagnetic field is omitted (Section 1). This model is comparatively

---

[1] To simplify formulas I will use the SGS system through the article.

simple and well describes waves with phase velocities less than light velocity. Then, we will discuss a two-beam instability as a candidate to make the plasma unstable (Section 2). Finally, we will consider possible parameters of the suggested accelerator (Section 3).

1. LONGITUDINAL WAVES IN PLASMA COLUMN

Let's consider a plasma column with radius a inside a vacuum chamber with radius b as shown in figure 1. Neutral plasma has a uniform density distribution across the column and is immersed into longitudinal magnetic field B_0. The electron density n_e and the ion density n_- satisfy to the neutrality condition, $Z_i n_i - n_e \ldots \sum_\alpha Z_\alpha n_\alpha = 0$. The motion of the electrons and ions in such a system is described by the following system of equations:

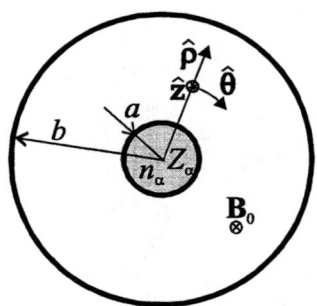

FIGURE 1. Coordinate frame and plasma column layout.

$$\frac{\partial \mathbf{v}_\alpha}{\partial t} + \left(\mathbf{v}_\alpha, \frac{\partial}{\partial \mathbf{x}}\right)\mathbf{v}_\alpha = \frac{Z_\alpha e}{m}\left(-\frac{\partial}{\partial}\varphi + \frac{\mathbf{v}_\alpha \times \mathbf{B}_0}{c}\right) ,$$

$$\frac{\partial n_\alpha}{\partial t} + div(n_\alpha \mathbf{v}_\alpha) = 0 , \qquad (2)$$

$$\Delta\varphi = -4\pi e \sum_\alpha Z_\alpha n_\alpha .$$

Linearizing these equations and looking for an axial symmetric solution,

$$\begin{vmatrix} \mathbf{v}(r,z,\theta,t) \\ n_\alpha(r,z,\theta,t) \\ \varphi(r,z,\theta,t) \end{vmatrix} = \begin{vmatrix} \mathbf{v}(r) \\ n_\alpha(r) \\ \varphi(r) \end{vmatrix} e^{i(\omega t - kz)} , \qquad (3)$$

one obtains (see Ref. (5)) the dispersion equation

$$f(k,\omega) \ldots \sqrt{-\varepsilon_\perp(\omega)\varepsilon_\parallel(\omega)} \frac{J_1\left(ka\sqrt{-\frac{\varepsilon_\parallel(\omega)}{\varepsilon_\perp(\omega)}}\right)}{J_0\left(ka\sqrt{-\frac{\varepsilon_\parallel(\omega)}{\varepsilon_\perp(\omega)}}\right)} - \frac{I_1(ka)K_0(kb) + I_0(kb)K_1(ka)}{I_0(kb)K_0(ka) - I_1(ka)K_0(kb)} = 0 \quad .(4)$$

226

Here $J_0(x)$ and $J_1(x)$ are the Bessel functions, $I_0(x)$, $I_1(x)$, $K_0(x)$ and $K_1(x)$ are the modified Bessel functions, $\varepsilon_{\parallel}(\ )$ and $\varepsilon_{\perp}(\ )$ are the longitudinal and transverse plasma dielectric permittivities,

$$\varepsilon_{\parallel}(\omega) = 1 - \frac{1}{\omega^2} \sum_{\alpha} \omega_{p\alpha}^2 \ ,$$

$$\varepsilon_{\perp}(\omega) = 1 - \sum_{\alpha} \frac{\omega_{p\alpha}^2}{\omega^2 - \Omega_{\alpha}^2} \ ,$$

(5)

and $\omega_{p\alpha}$ and Ω_{α} are the plasma and Larmor frequencies,

$$\omega_{p\alpha} = \sqrt{\frac{4\pi n_{\alpha} Z_{\alpha}^2 e^2}{m_{\alpha}}} \ ,$$

$$\Omega_{\alpha} = \frac{e Z_{\alpha} B_0}{m_{\alpha} c} \ .$$

(6)

Only fast plasma oscillations, related to electrons, are important for the instability considered below, and therefore we will eliminate below the ion contribution into the dielectric permittivities, leaving only the electron contribution with the electron plasma frequency ω_{pe} and electron Larmor frequency Ω_e.

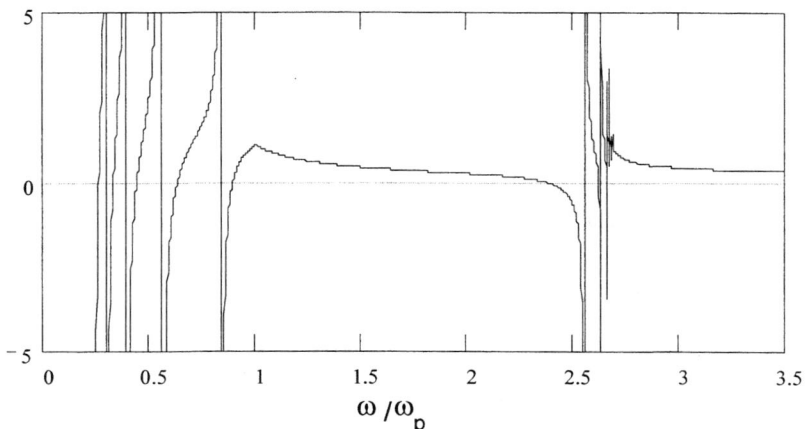

FIGURE 2. Function $f(k, \omega)$ of equation (4) as function ω for $ka = 4$, $b/a = 5$ and $\Omega_e/\omega_{pe} = 2.5$.

Equation (4) cannot be solved analytically and therefore its solutions were studied numerically. For every given k this equation has an infinite number of roots corresponding to a different number of potential variations as a function of radius. As will be seen below we are interested in the case of a sufficiently strong magnetic field and therefore, to simplify further analysis, we will consider below a plasma where $\_{e} \gtrsim \_{pe}$. In this case there are two well-separated groups of roots, as illustrated in figure 2 by a plot of function $f(k,\_)$. The first group is at low frequencies. Its roots belong to the solutions with primarily longitudinal motion of the electrons. They are grouped near the frequency where $\_{\parallel}(\_)$ approaches infinity, $\_=0$. The second group is at high frequencies. These roots belong to the solutions with primarily transverse motion of the electrons. The roots are grouped around the frequency where $\_{\perp}(\_)$ approaches zero, $\omega = \sqrt{\omega_{pe}^{2} + \Omega_{e}^{2}}$. To produce a clear picture only the first few roots from both groups are shown in figure 2. We denote roots using two numbers, like $\_{0,2}$. The first number, equal to 0 or 1, denotes the group number, and the second number denotes the number of potential variations – the number of zero crossings by potential dependence on radius. Note that when the sign of function $\_{\parallel}(\_) / \_{\perp}(\_)$ becomes positive the argument in Bessel functions becomes imaginary and they need to be replaced by the modified Bessel functions. This transformation was used for plotting the curve in figure 2.

The solution with zero number of variations in the first group has highest frequency among all other roots in the group. Its asymptotic for the case of long waves, $ka \ll 1$, is

$$\omega_{0,0}(k) = \omega_{pe} ka \sqrt{\frac{\ln(b/a)}{2}} \quad , \tag{7}$$

showing that this mode has linear dispersion and consequently constant phase velocity

$$v_{\phi 0} = \omega_{pe} a \sqrt{\frac{\ln(b/a)}{2}} \quad . \tag{8}$$

In the case of short waves, $ka \gg 1$, the asymptotic can be easily obtained for the cases of small and high magnetic field,

$$\omega_{0,0}(k) = \begin{array}{ll} \omega_{pe}/\sqrt{2} & , \quad \Omega_{e} \ll \omega_{pe} \quad , \\ \omega_{pe} & , \quad \Omega_{e} \gg \omega_{pe} \quad . \end{array} \tag{9}$$

Asymptotics for intermediate values of the magnetic field lie between these two values.

In the second group the solution with zero number of variations has the lowest frequency among all other roots in the group. Its asymptotics for the cases of long and short waves are

$$\omega_{1,0} = \begin{cases} \sqrt{\omega_{pe}^{\,2} + \Omega_e^{\,2}} & , \quad ka \ll 1 \ , \\[2mm] \sqrt{\dfrac{\omega_{pe}^{\,2} + \Omega_e^{\,2}}{2}} & , \quad kb \gg 1 \ . \end{cases} \tag{10}$$

Solutions with higher number of variations have the same asymptotic at small wavelength, while for short wavelength asymptotics, they are in the range of $\Omega_e, \sqrt{\Omega_e^{\,2} + \omega_{pe}^{\,2}}$.

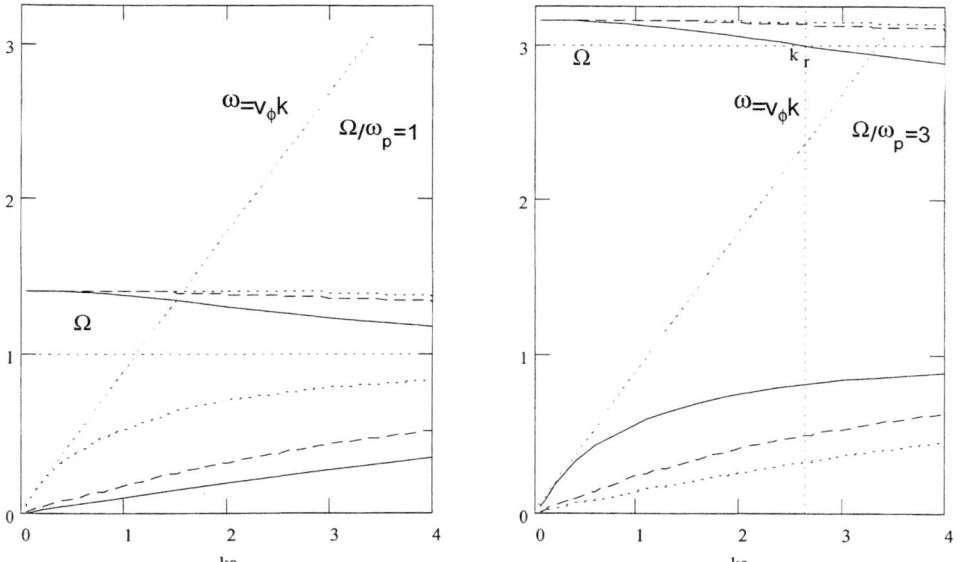

FIGURE 3. Dispersion curves for the first three modes of both low (bottom three curves in both pictures) and high (top three curves in both pictures) frequency groups; left picture - _ $e/_{pe}$=1.0, right picture - _ $e/_{pe}$=3.0.

Figure 3 illustrates behavior of the dispersion curves for the cases of small and high magnetic fields. Dispersion curves for solutions with zero, one, and two variations for both high and low frequency groups are shown. One can see that the magnetic field does not significantly affect curves belonging to the low frequency group, while it significantly changes the curves of the high frequency group.

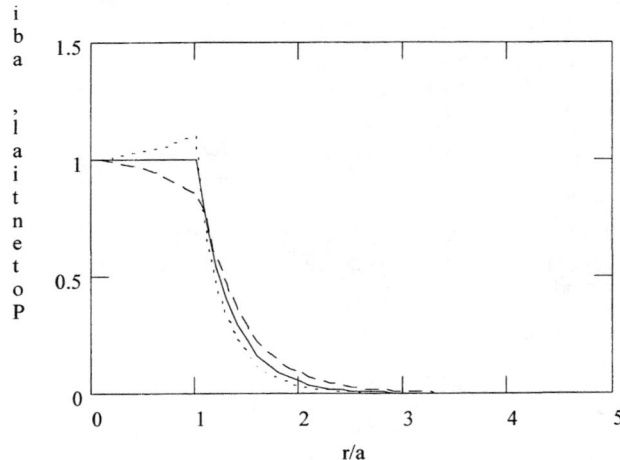

FIGURE 4. Dependence of potential on radius in vicinity and at resonance; $b/a = 5$ and $\_/\_p=3$; solid line - $\_/\_p = 3$, $ka = 2.64$, dotted line - $\_/\_p = 3.048$, $ka = 2.12$, dashed line - $\_/\_p = 2.953$, $ka = 2.12$

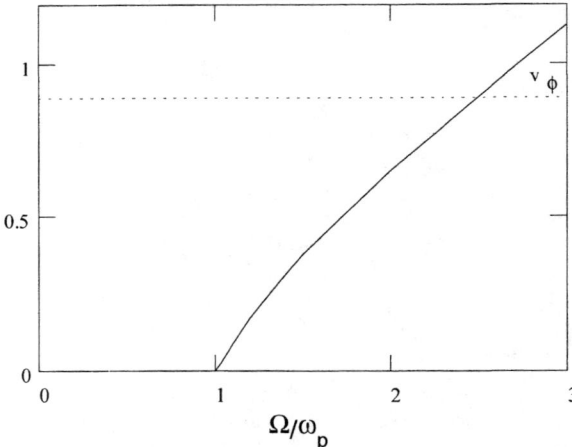

FIGURE 5. Dependence of the wave phase velocity (in units of $\_/(\_p ka)$) at resonance ($\_{1,0} = \_\_\_$ as a function of ratio $\_/\_p$; $b/a = 5$.

Dependence of the potential on the radius is determined by the value of $\varepsilon_\|(\omega)/\varepsilon_\perp(\omega)$ at corresponding eigen-frequency, $\_{m,n}(k)$. If $\varepsilon_\|(\omega)/\varepsilon_\perp(\omega)$ is negative the potential is

230

$$\varphi(r,z,t) = \varphi_0 \, J_0 \left(\overline{ka} \sqrt{-\frac{\varepsilon_\parallel(\omega_{m,n}(k))}{\varepsilon_\perp(\omega_{m,n}(k))}} \right) e^{i(\omega t - kz)} . \tag{11}$$

For the positive value the Bessel function in equation (11) has to be replaced by the modified Bessel function $I_0(x)$. As can be seen from figure 3, the high frequency zero variation root $\omega_{1,0}$ crosses frequency line $\omega = \omega_e$ where the transverse dielectric permittivity approaches infinity. At this point, $\omega_{1,0}(k_r) = \omega_e$, and the dependence of the potential on the radius inside the plasma vanishes, creating an ideal longitudinal accelerating wave, which does not have a transverse electric field and for which acceleration does not depend on radius. The dependencies of potential on radius for the resonance wave vector, k_r, and for 20% longer and shorter wavelengths are shown in figure 4. Figure 5 depicts the dependence of the phase velocity at resonance as a function of the ratio ω / ω_p. For frequency $\omega / \omega_p < 1$ the resonance condition cannot be fulfilled. Note that the phase velocity at resonance is higher than the phase velocity of the low frequency plasma wave (see equation 8) for $\omega / \omega_p > 2.51$.

2. TWO-BEAM INSTABILITY

The probe beam excites many small amplitude modes in the plasma. To create a good-quality accelerating wave from this seed excitation only one mode with correct structure and phase velocity has to be amplified. In our study we will analyze the two-beam instability as a candidate to amplify the beam acceleration in the plasma.

To simplify formulas we will consider a simple model where a non-relativistic electron beam propagates along the plasma column. The beam has the same radius as plasma and its velocity is v_0. The beam density, n_b, is uniform and is much smaller than the electron density in the plasma, n_e. The low density of the beam allows one to consider this system as a neutral plasma, and therefore the dispersion properties of such a system are described by the same equation (4) as properties of the plasma column. But the beam permittivities of equation (5) have to be corrected to take into account the electron beam motion,

$$\varepsilon_\parallel(\omega) = 1 - \frac{\omega_{pe}^2}{\omega^2} - \frac{\omega_{pb}^2}{(\omega - kv)^2} ,$$

$$\varepsilon_\perp(\omega) = 1 - \frac{\omega_{pe}^2}{\omega^2 - \Omega_e^2} - \frac{\omega_{pb}^2}{(\omega - kv)^2 - \Omega_e^2} , \tag{12}$$

where ω_{pe} and ω_{pb} are the plasma the frequencies related, correspondingly, to electrons of the plasma and the beam, and v is the electron beam velocity.

In the first approximation one can consider the plasma and the beam independent so that waves propagating in the plasma and in the beam do not interact with each other. The general instability criterion states that the instability can develop if the phase velocities of waves related to the plasma and to the beam are equal. Because of low electron density in the beam, the phase velocity of the plasma wave related to the beam is much smaller than the velocity of the beam and we can simplify the criterion comparing the wave phase velocity and the electron beam velocity. First, we want to avoid an instability at low frequencies. This implies that the electron beam velocity has to be higher than the phase velocities of the low frequency plasma waves, $v > v_0$. Second, we would like to excite a resonant wave considered above where the transverse dielectric permittivity approaches infinity and the transverse electric field vanishes. As can be seen from figure 5, this requires a strong magnetic field so that $\Omega/\omega_p > 2.51$.

The rest of this section will be devoted to study of properties of the resonant wave. To find the instability increment we will rewrite equation (4) in the following form

$$\frac{J_1\left(ka\sqrt{-\dfrac{\varepsilon_\|(\omega)}{\varepsilon_\perp(\omega)}}\right)}{J_0\left(ka\sqrt{-\dfrac{\varepsilon_\|(\omega)}{\varepsilon_\perp(\omega)}}\right)} = \frac{1}{\sqrt{-\varepsilon_\perp(\omega)\varepsilon_\|(\omega)}}\frac{I_1(ka)K_0(kb)+I_0(kb)K_1(ka)}{I_0(kb)K_0(ka)-I_0(ka)K_0(kb)} \quad . \tag{13}$$

One can see that if $\varepsilon_\perp(\omega)$ approaches infinity the expression in the right-hand side approaches zero and consequently Bessel functions at the left-hand side can be expended in Taylor series, $J_1(z)/J_0(z) = z/2 + \ldots$, yielding the following equation:

$$\varepsilon_\|(\omega) + \frac{1}{ka}\frac{I_1(ka)K_0(kb)+I_0(kb)K_1(ka)}{I_0(kb)K_0(ka)-I_0(ka)K_0(kb)} = 0 \quad . \tag{14}$$

Substituting $\varepsilon_\|(\omega)$ from equation (12) and expending the obtained equation in Taylor series near the resonance we obtain

$$\frac{\delta}{(x-y-u)^2} - \frac{2\omega_p}{\Omega^3}x - \frac{2\omega_p}{\Omega k_r a}\overline{f(k_r)} - \left.\frac{df}{dk}\right|_{k=k_r}k_r v = 0 \quad . \tag{15}$$

Here we took into account that at the resonance, $\omega = \Omega$,

$$\frac{\omega_p^2}{\Omega^2} - 1 = \frac{2}{k_r a}f(k_r) \quad , \tag{16}$$

and we denoted

232

$$f(k) = \frac{I_1(ka)K_0(kb) + I_0(kb)K_1(ka)}{I_0(kb)K_0(ka) - I_0(ka)K_0(kb)},$$

$$\omega = \Omega + \omega_p x \quad , \qquad k = k_r + \frac{\omega_p}{v_0} y \quad , \qquad v_0 = \frac{\Omega}{k_r} \tag{17}$$

$$\delta = \frac{\omega_b^{\,2}}{\omega_p^{\,2}} \cdot \frac{n_b}{n_e} \quad , \qquad v = v_0 + \frac{\omega_p}{k_r} u \quad .$$

Equation (15) is the cubic equation relative to the variable x (dimensionless frequency deviation from resonance value of _) as a function of y (dimensionless wave vector deviation from resonance value of k_r) and u (dimensionless deviation of the beam velocity from the resonance velocity v_0). At resonance, $y = u = 0$, and equation (15) yields the increment equal to

$$\lambda_{max} = \mathrm{Im}(\omega) = \omega_p\,\mathrm{Im}(x) = \frac{\sqrt{3}}{2\,?2^{1/3}}\Omega\delta^{1/3} \quad . \tag{18}$$

Numerical analysis of the roots of equation (15) exhibited that the maximum value of the instability increment is achieved at resonance. Dependence of the real and imaginary parts of equation 15 roots for _ = 0.001 is shown in figure 6. The instability increment is proportional to the imaginary part of a root, _ = _$_p$Im(x), while frequency is determined by its real part, _ = ___ _$_p$Re(x). The width of the amplification band is characterized by

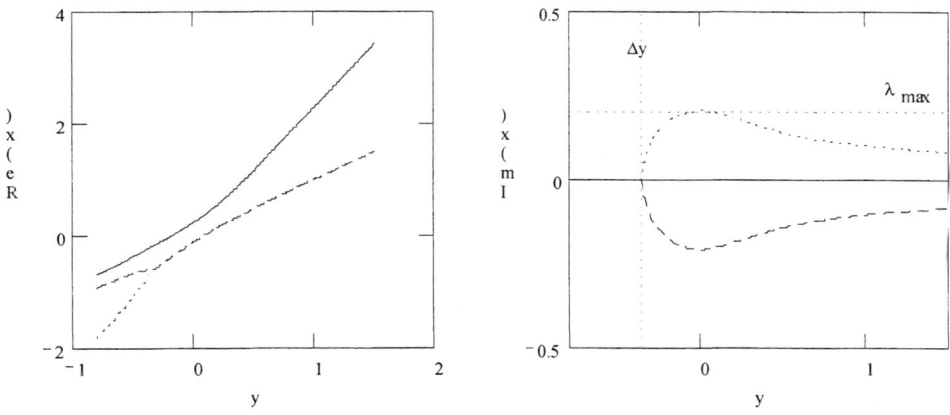

FIGURE 6. Dependence of the roots of equation (15) on the dimensionless wave vector y; _/ _$_p$ = 3, _ = 0.001, b/a = 5, u = 0.

233

$$\Delta k = \frac{\omega_p}{v_0} \Delta y \ldots \frac{3\Omega \delta^{1/3}}{2v_0 \overline{1 + \frac{\Omega^2}{\omega_p^2 k_r a}} \overline{f(k_r)} - \frac{df}{dk}\bigg|_{k=k_r} k_r} .$$

(19)

Numerical analysis also proved that the maximum increment and the bandwidth do not depend significantly on a small change of the electron beam velocity

It is important to note that presence of the electron beam changes also the group velocity of the plasma wave in the vicinity of resonance. Without the electron beam, as can be seen from figure 3, the group velocity, $d\_\_/dk \approx -0.1 \cdot v_0$, is negative, while the group velocity of the same mode is positive $d\_\_/dk \approx 0.5 \cdot v$ (see figure 6) in the presence of the beam. Positive group velocity in the whole amplification band implies that this instability is the convective instability, i.e. an initial perturbation grows with time but simultaneously moves downstream together with the electron beam, and therefore the amplitude of the wave does not grow to infinity at any given longitudinal coordinate.

3. ACCELERATOR SECTION PARAMETERS

Utilization of the resonant condition in plasma acceleration addresses three important issues. First, it removes the transverse electric field and thus prevents beam focusing by the accelerating field. Second, it suppresses the dependence of the accelerating field on the transverse coordinates. Third, it reduces the dependence of accelerating frequency on plasma density because the frequency is mainly determined by the magnetic field

$$\omega \cup \Omega \left(1 + \omega_{pe}^2 / \left(2\Omega^2\right)\right) .$$

(20)

But realization of this resonant condition requires high Larmor frequency, $\_/\_p >$ 2.51, and, consequently, high magnetic field. For acceleration gradients of about 1 GV/m the field value is beyond or on the boundary of today's state-of-art achievements. For an estimate of the accelerating section parameters (table 1) the magnetic field of 12 T is used.

The choice of instability increment is determined by the accelerating frequency and its achievable accuracy. Picking up an accuracy of plasma density of about 1%, and $\_/\_p \approx 3$, one obtains from equation (20) the relative accuracy of the accelerating frequency of about $5 \cdot 10^{-4}$, which implies that the plasma wave keeps correct phase for about 1000 oscillations. For an accelerating frequency of 340 GHz and amplification of 10 times, one obtains the instability growth time of about 1 ns. This time is significantly less than that required for an electron to pass the accelerating section.

To create such a beam-plasma system the following mechanism is suggested. An electron beam with pulse length of about 50 ns and rise and fall time less than about 10 ns is directed to a vacuum chamber with gas density equal to the required plasma

234

density. We will call this beam the excitation beam. At this time there is no plasma and the system is stable. To create the plasma, immediately after the beam current reaches its flat top a short laser pulse is aimed along the beam. A laser pulse with comparatively modest energy can ionize all gas on its way. Thus the pulse creates a plasma with the required density and makes the system unstable. After a small delay, which is mainly determined by the synchronization accuracy, the probe bunch follows. This bunch makes the seed excitation, which then is amplified by the instability. After the plasma wave amplitude reaches the required value (about 3 ns) the test bunch is accelerated by the plasma wave. This procedure does not require very good synchronization for the excitation and laser pulses, but requires sub-picosecond synchronization for the probe and test bunches, which can be comparatively easily achieved by accelerator means.

Main parameters of the accelerating section are shown in table 1. The parameters were estimated on the basis of the non-relativistic theory considered above with simple corrections taking into account an increase of the longitudinal mass for relativistic electrons. Because of the low intensity of the probe bunch and, consequently, the small energy loss on plasma excitation, the probe bunch can excite a plasma column of rather large length. The length of the acceleration section was chosen to be 10 m, which is mainly determined by engineering matters.

TABLE 1. Parameters of the accelerator section

Accelerating gradient	0.2 GeV/m
Wave length of the RF	0.89 mm
Section length	10 m
Plasma (electron beam) diameter	0.75 mm
Magnetic field	12 T
Plasma density	$1.6 \cdot 10^{14}$
Energy in plasma per unit length at max. field	74 mJ/m
Plasma frequency, $_p/2_$	112 GHz
Instability increment	$1 \cdot 10^9$ s
Energy of the excitation electron beam	2 MeV
Current of the excitation electron beam	200 A
Density of the excitation electron beam	$5.4 \cdot 10^{12}$
Energy in electron beam per unit length	1.4 J/m
Energy in ionization laser pulse	5 mJ
Ionization laser wave length	300 nm
Number of particles in excitation bunch	$1.1 \cdot 10^8$
Rms bunch length	50 m

CONCLUSION

The novel scheme of plasma acceleration considered above is based on the narrow band instability in a plasma. It allows one to choose and amplify only one of many modes excited by the probe beam in the plasma, and, consequently, it allows one to form a well-defined accelerating wave. The scheme has been illustrated by non-relativistic analysis of the two-beam instability. This analysis showed a possible way of carrying out such a scheme for non-ultra-relativistic particles. In particular it can be considered for acceleration of heavy ions in the energy range of 1–6 GeV/nucleon while more studies are required to make a realistic scheme. Further study is required for developing a similar scheme for acceleration of ultra-relativistic particles.

ACKNOWLEDGEMENTS

I would like to thank S. Corneliussen for his help in editing this article. This work was supported by the U.S. Department of Energy contract DE-AC05-84ER40150.

REFERENCES

1. Tajima, T. and Dawson, .J.M., *Phys. Rev. Lett,.* **43**, 267 (1979).
2. Kitogawa, Y. *et al.*, *Phys. Rev. Lett.*, **68**, 48 (1992).
3. *Advanced Accelerator Concepts*, edited by J. S. Wurtele, *AIP Conf. Proc,.* No. 279 (AIP, New York, 1993).
4. Chen, P., Dawson, J.M., Huff, R.W. and Katsouleas, T., *Phys. Rev. Lett.* **54**, 693 (1985).
5. Davidson, R.C., *Theory of Nonneutral Plasmas*, Addison-Wesley Publishing Company, Inc., 1988, ch. 2.7, pp. 45-49.

An X-ray Transition Radiation Beam Profile Detector for the LCLS[1]

Sergio Monteiro [a] Claudio Pellegrini [b]

(a) Department of Physics and Department of Electronics
Moorpark College
Moorpark, CA 93021
monteiro@sunny.moorpark.cc.ca.us

(b) Department of Physics
University of California, Los Angeles (UCLA)
405 Hilgard Ave.
Los Angeles, CA 90095-1547
pellegrini@physics.ucla.edu

[1] Project supported by DOE grant DE – FG03 – 98ER 45693

Abstract:

We discuss the characteristics of a transverse beam profile detector for a high energy charged particle accelerator, and propose a candidate detector based on transition radiation in the X-ray region of the spectrum. The detector is useful for low emittance, high energy beams, for example, the LCLS electron beam. We expect that it can resolve spatial details as small as a few microns. The advantages of this method over previous ones are that the measurements are linear with the beam density and the results are a point by point map of the beam density.

INTRODUCTION

Several methods to measure the transverse density distribution of a charged particle beam are based on electromagnetic radiation emitted by the beam under different conditions. One of the most popular methods is the wire scanner (1,2), which measures the Bremsstrahlung radiation produced by the interaction between the particle beam and a fine wire, and its associated optical form, the laserwire scanner (1), which measures the Compton scattering between the

CP468, *Nonlinear and Collective Phenomena in Beam Physics–1998 Workshop*,
edited by S. Chattopadhyay, M. Cornacchia, and C. Pellegrini
© 1999 The American Institute of Physics 1-56396-862-2/99/$15.00

charged particle beam and a finely focused laser beam. Another possibility is the transition radiation (3, 4, 5, 6, 7, 8, 9), which measures the radiation emitted by an electric charged particle that crosses the boundary between two media of different dielectric constants. Transition radiation was originally predicted in 1946 by V. L. Ginzburg and I. M. Frank (3). J. D. Jackson (4) besides an alternative approach to the subject and some useful observations, also presents some interesting order-of-magnitude calculations. Detailed analysis of transition radiation can be found in chapter 4 of M. L. Ter-Mikaelian (5), V. E. Pafomov (6), Ginzburg and Tsytovich (9), Boris Dolgoshein (8), and L. Wartski, S. Roland *et al.* (7). Transition radiation has been used in the visible as a method for measuring beam characteristics of low energy beams (~ 10's of MeV) (10), but its use in the X-ray part of the spectrum for high energy beams is still to be exploited.

Though wire scanners are currently more popular methods of transverse beam diagnostics, transition radiation based diagnostic devices are good alternatives for high energy beams, like the LCLS, and may offer advantages as the beam diameter decreases.

THEORETICAL CONSIDERATIONS

The beam characteristics of the LCLS in the undulator producing the X-rays are as follows (11):

energy	~ 15 GeV ($\gamma = 3 \times 10^4$)
bunch charge:	~ 1 nC (6×10^9 electrons)
emittance (rms)	~ 3×10^{-11} m-rad
bunch radius, 1σ	~ 25 μm
angular spread	~ 1.2 μrd
pulse duration, (rms)	~ 100 fs
pulse repetition rate	~ 120 Hz

The most important parameters for the use of the transition radiation detector we describe here are the beam energy and bunch radius; the beam energy controls the intrinsic divergence of the transition radiation, the bunch radius determines the spot size of each measurement, because the beam area must be resolved in no less than 50 points for acceptable resolution.

In first approximation, the transition radiation beam divergence (θ) is given by

$$\theta \approx \frac{1}{\gamma} \;, \qquad \text{for} \qquad \gamma \gg 1. \qquad (1)$$

For the LCLS case, at 15 GeV electron beam energy ($\gamma = 30{,}000$) the angle is

$$\theta \approx \gamma^{-1} \approx 3 \cdot 10^{-5} \approx 30 \mu rd \;. \qquad (2)$$

This intrinsic divergence, which is controlled by the relativistic deformation of the field, determines the maximum resolution that can be obtained with any wavelength. In the visible part of the transition radiation spectrum, the emitted light corresponds, if the object have been illuminated by an external source, only to the zeroth order diffraction. Yet, in order to create a proper image, light must be collected from as high a diffraction order as required to gather a substantial fraction of the radiation. As it was explained by Abbe (12, 13), the absence of the higher order diffraction orders precludes the formation of an image. In reality the image created by just the zeroth order diffraction is an even illuminated field; there is no image at all.

To create an image one needs to use such a wavelength that at least the first diffraction order is contained within the 30 µrd which contains all the emitted transition radiation light. This determines the wavelength that must be used. From standard diffraction theory,

$$d \sin \theta = m\lambda \qquad (3)$$

In Equation (3) the angle θ is fixed by the relativistic deformation of the field to ~ 30 µrd, d is the size of the object (beam diameter = 2 x R_{RMS} = 50 µm rms), and m must be at least 1 (first order of diffraction). This determines the minimum wavelength λ to be

$$\lambda = \frac{d \sin \theta}{m} \cong \frac{d\theta}{m} \qquad (4)$$

Which evaluates to 1.5 nm for $m = 1$, corresponding to a soft X-ray photon, energy $E \sim 0.7$ keV. A very poor image could be formed using this wavelength. Kurt Michel (12) shows, at page 254, a progression of better images created with the inclusion of two, three and five diffraction orders. These pictures, as well as the associated commentaries, well elucidate the point. A reasonable image may be created with the inclusion of the first five orders of diffraction. Using m = 5 in Equation (4), λ evaluates to 0.3 nm, corresponding to a 3 keV photon. Therefore 3 keV is the minimum photon energy that can be used to obtain an image of acceptable quality of a 50 µm point using electromagnetic energy restricted to a 30 µrd cone.

Having established that the constraints on the radiation divergence limit us to use photons with energy larger than 3 keV, one needs to determine the radiation intensity, which, together with considerations of the relative difficulty of detection, helps to determine the region of the spectrum to be used.

The intensity of the transition radiation is variable across the broadband spectrum emitted, and is given by Jackson and Ter-Mikaelian (4,5)

$$\frac{dI}{dw} = \frac{e^2}{\pi c}\left[\left(1 + 2\frac{w^2}{w_{cr}^2}\right)\ln\left(1 + \frac{w_{cr}^2}{w^2}\right) - 2\right].$$ (5)

After some modifications, equation (5) can be written also as

$$dI = \alpha\frac{1}{\pi}\left[\left(1 + 2\frac{w^2}{w_{cr}^2}\right)\ln\left(1 + \frac{w_{cr}^2}{w^2}\right) - 2\right](dE),$$ (6)

where α is the fine structure constant. Table 1 shows the number of photons produced from a 15 GeV electron beam, with 6×10^9 electrons (1 nC) in the bunch. Not all photons produced are captured by the detector system, mostly due to the lens transmissivity. Depending on the details of the final system, lens transmissivity is from 0.4 to 0.8 (14).

TABLE 1. Distribution of transition radiation photons created by a 15 GeV electron beam, 1 nC per bunch, in the X-ray spectral region.

Transition radiation photon energy (keV)	energy range (bin size) (keV)	energy emitted within range (keV)	number of transition radiation photons per electron	total number of transition radiation photons emitted from a 1 nC bunch
1	0.1	0.0025	0.0025	15×10^6
5	0.5	0.0088	0.0018	10×10^6
10	1	0.014	0.0014	9×10^6
50	5	0.035	0.00071	4×10^6
100	10	0.042	0.00042	2×10^6
600	60	0.011	0.000018	0.11×10^6

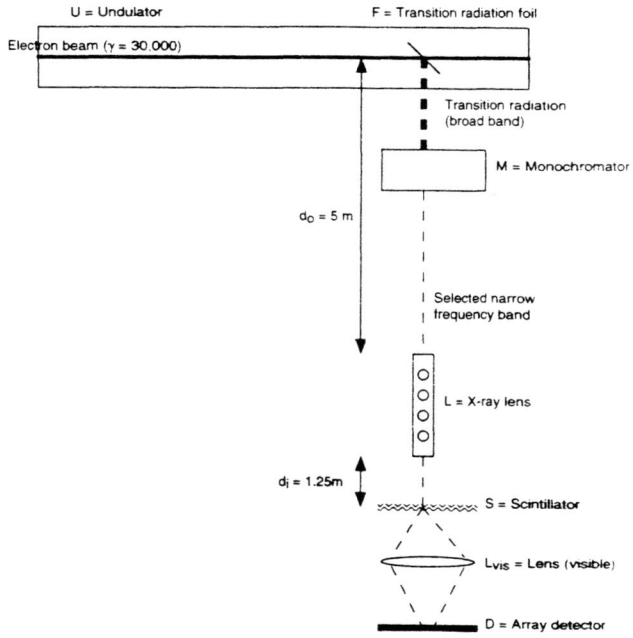

FIGURE 1. Schematic view of the detector system

DESCRIPTION OF THE DETECTOR

Figure 1 is a schematic diagram of the detector we propose. A is the undulator part of a high γ charged particle accelerator, for instance, the LCLS. F is a thin foil to generate transition radiation. We are analyzing different foil materials (mylar, aluminum, etc.) and thickness in view of the beam heating load on the foil, its possible deterioration with time and lifetime. F is at an acute angle, say $45°$, with the particle beam direction in order that the transition radiation is emitted out of the beam direction. M is a narrow band X-ray filter (monochromator) to narrow the bandwidth from the white spectrum emitted by the transition radiation foil. This step is necessary because the X-ray lens we intend to use suffers from chromatic aberration. L is a X-ray lens. Given the options available at the moment we believe that the best X-ray lens for this particular application is a diffractive lens (14). S is a scintillator, for example $CdWO_4$ as in (14). L_{vis} is a lens for the visible light emitted by the scintillator S. Finally D is a CCD or other type of surface array detector.

241

Based on the numbers from Table 1 and the minimum needed photon wavelength $\lambda = 0.3$ nm, as discussed before, we propose to use 20keV photons ($\lambda =\sim 0.05$ nm). A lens of the type described by Elleaume (14) with a focal length f = 1m, at a distance d_0 = 5m from the transition radiation foil would produce an inverted image at a distance d_i = 1.25 m beyond the lens, with a magnification M = -6.25. With such a system the LCLS electron beam (1 σ RMS diameter 50 μm) would produce an image of 300 μm diameter at the scintillator, with a further possible magnification by the visible lens L_{vis} that forms an image on the pixel detector, large enough for a good quality image with standard pixel detectors (resolution ~ 15-25 μm).

After narrowing the energy line width to 0.2 keV and accounting for the monochromator absorption we estimate some 10^5 photons reaching the X-ray lens with energy in the range (19.8 keV – 20.2 keV). The X-ray lens transmissivity is from 40 % to 80 %, leaving an estimated 50,000 X-ray photons reaching the scintillator. The projected image on the scintillator is approximately 300 μm in diameter (-6 x magnification), so its area is of the order of 70,000 μm². This means that about one X-ray photon hits every micrometer square on the scintillator – per bunch. If we assume 100 visible photons per each X-ray photon incident on the scintillator, that these visible photons are emitted isotropically, and that the visible optics collects 40% of these, with subsequent negligible absorption by the glass, there are approximately 40 visible photons per square micrometer of the beam image on the scintillator. With a visible optics system with a magnification equal to 2, the final beam image on the CCD detector has an area of the order or 280,000 μm². For a typical pixel detector of dimensions of the order of 20 μm there are 700 pixels to form the beam image, each of which collects 10 photons on the average. Naturally the central pixels collect more than this, while the peripheral collect a smaller number.

CONCLUSIONS

We have discussed the use of the X-ray part of the transition radiation spectrum for a transverse beam profile detector. In particular, the necessary X-ray optics is now available (14). Such a transition radiation detector has the advantages of being linear with the particle beam density, and the detector output is a direct map of the particle beam density. This detector can be built with available technology.

ACKNOWLEDGEMENTS

We acknowledge fruitful discussions with Nicholas Sereno (ANL), with Sandro Ruggiero (BNL), with Dinh Nguyen (LANL), and with the group 2 of the Nonlinear and Collective Phenomena in Beam Physics workshop held at Arcidosso, Italy, September 1-5, 1998.

BIBLIOGRAPHY

1. Marc Ross "Measurement Techniques for Short Bunches" Advanced Accelerator Concepts, Fontana, WI (1994), in Paul Schoessow (ed.) *AIP Conference Proceedings* **335** pg. 101-111

2. K. D. Jacobs *et al.* "Accelerator Beam Profile Measurements at the Bates LINAC" *1988 Linear Accelerator Conference*, CEBAF October 3-7, 1988, CEBAF Report 89-001 June 1987

3. V. L. Ginzburg and I. M. Frank "Radiation from a Uniformly Moving Electron Passing from One Medium to Another" *JETP* **16**, 15 (1946)

4. J. D. Jackson *Classical Electrodynamics* (2nd Edition) John Wiley (1975) pg 685 ff.

5. M. L. Ter-Mikaelian *High-Energy Electromagnetic Processes in Condensed Media* Wiley (1972)

6. V. E. Pafomov "Radiation of a Charged Particle in the Presence of a Separating Boundary" in D. V. Skobel'tsyn (Ed.) *Proceedings of the Lebedev Physics Institute* **44**, 25 - 157 (1969) English translation (1971)

7. L. Wartski, S. Roland *et al.* "Interference Phenomenon in Optical Transition Radiation and its Application to Particle Beam Diagnostics and Multiple-scattering measurements" *J. Appl Phys.* **46** (8), 3644-3653 (1975)

8. Boris Dolgoshein "Transition Radiation Detectors" *Nucl. Instr. Meth. In Phys. Res.* **A326**, 434-469 (1993) Comprehensive article

9. V. L. Ginzburg and V. N. Tsytovich *Transition Radiation and Transition Scatering*, Adam Hilger (1990).

10. Ralph Fiorito and Doanld Rule "Optical Transition Radiation Beam Emittance Diagnostics" in Robert Shafer (ed.) *AIP Conference Proceedings* **319,** 21-37 (1994)

11. LCLS Design Study Report SLAC-R-521, (1998)

12. Kurt Kichel *Die Grundzuege der Theorie des Mikroskops in Elementarer Darstellung*, Wissenschaftliche Verlagsgesellschaft m.b.H., Stuttgart (1964)

13. K. Kranjc "Simple Demonstration Experiments in the Abbe Theory of Image Formtion" *Am. J. Phys.* **30**, 342-47 (1962)

14. P. Elleaume "Optimization of Compound Refractive Lenses for X-rays" *Nucl. Instr. Meth. In Phys. Res.* **A412**, 483-506 (1998)

The Circular RFQ Storage Ring*

Alessandro G. Ruggiero

Brookhaven National Laboratory. AGS Department. Upton, NY 11973

Abstract. This paper presents a novel idea of storage ring for the accumulation of intense beams of light and heavy ions at low energy. The new concept is a natural development of the combined features used in a conventional storage ring and an ion trap, and is basically a linear RFQ bend on itself. In summary the advantages are: smaller beam dimensions, higher beam intensity, and a more compact storage device.

INTRODUCTION

There is need to develop compact storage rings for the accumulation of low-energy beams of ions for a variety of applications, namely: molecular and atomic physics, solid state, chemical-physics, astrophysics, and other more exotic applications like crystalline beams and ion fusion for energy production. Several experimental apparatuses can be conceived making use of these compact storage rings, for instance: colliding beams circulating in two intersecting storage rings, collision of a stored beam of ions with an internal target, and head-on collision with an electron beam or an X-ray beam from a synchrotron radiation source. Typical requirements are high beam intensity and density. Also, very small energy spreads are sought which can be achieved with cooling techniques like electron and laser cooling.

In this report we describe a novel idea of storage ring for the accumulation of intense and dense beams of light and heavy ions in a more compact structure. The concept takes advantage of established principles of operation of conventional low-energy storage rings, ion traps, and RFQ's. The proposed new storage ring is basically a circular RFQ bend on itself and closed mechanically. Instead of quadrupole magnets, focusing of the particles is provided by the rf field of the structure. Since electrically the structure is not closed on itself, it is expected that ion beams can be stored at intensities and densities higher than those achieved in conventional storage rings.

* Work performed under the auspices of the U.S. Department of Energy.

CP468, *Nonlinear and Collective Phenomena in Beam Physics–1998 Workshop*,
edited by S. Chattopadhyay, M. Cornacchia, and C. Pellegrini
© 1999 The American Institute of Physics 1-56396-862-2/99/$15.00

A CONVENTIONAL STORAGE RING

An example of a low-energy and small storage ring is ASTRID (1), used for the accumulation of light and heavy ions and for the demonstration of Laser Cooling. The ring magnetic rigidity is 1.87 T-m, from which it is then possible to estimate the maximum beam energy for different ion species. The circumference is about 40 m. As shown in Fig. 1, the ring is made of eight dipole magnets which bend the beam trajectory on a circular and closed orbit and sixteen quadrupole magnets for transverse focusing. The quadrupoles are arranged in eight doublets. There are four periods, each with a straight section of about 4-meter length, free of magnets. As an example, beam of ions $^7Li^+$ and $^{24}Mg^+$ have been stored at the energy of 100 keV for Laser Cooling experiments. Stored currents were in the 1-10 μA range.

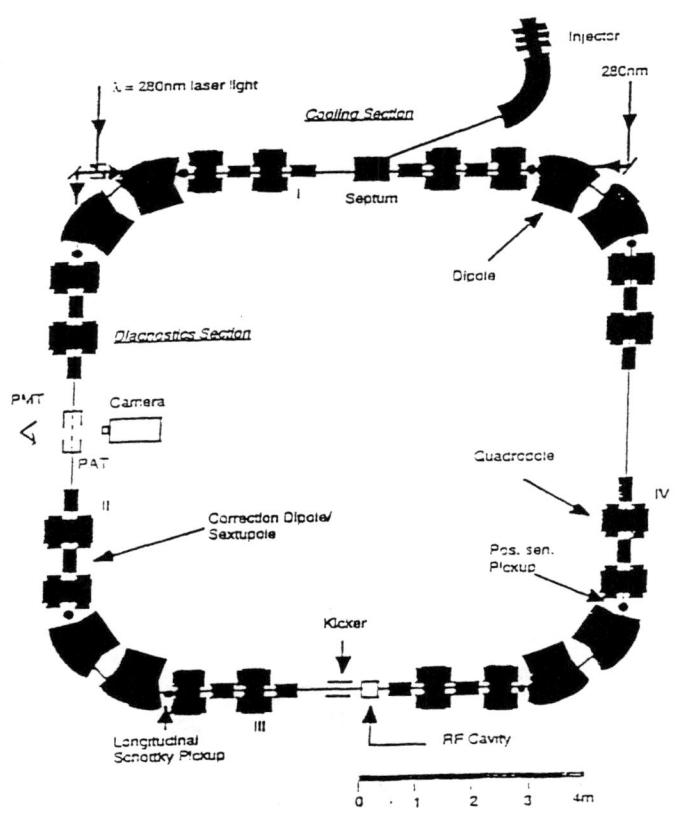

FIGURE 1. The ASTRID Storage Ring

There is a limit on the intensity that can be stored in a conventional storage ring like ASTRID. This is given by the space-charge tune-depression Δv, given by the formula

$$\Delta v \;=\; N Q^2 r_0 \, b \,/\, 2 A \, \beta^2 \gamma^3 \, \varepsilon, \tag{1}$$

where $r_0 = 1.535 \times 10^{-18}$ m is the classical proton radius, Q the ion charge state, A its mass number, β and γ respectively the velocity and energy relativistic factors, b a bunching factor (that is, the ratio of beam peak current to average current), N the number of ions stored, and, finally, ε the beam emittance, that is, the area occupied by the beam in the (x, x')-phase space (not forgetting to include the factor π). For practical purposes the tune-depression Δv cannot exceed a value at most as large as 0.5, but which more typically is taken at around 0.2. This limit is understood to be set by the presence of unavoidable random magnet imperfections which cause the lowering of the ring periodicity to essentially a unit, and the creation of stopbands around the half-integral tune values which cannot be crossed without substantial beam losses.

To simplify our concepts we shall consider below the case of beams of protons (Q = A = 1) completely debunched (b = 1) at the kinetic energy of 100 keV ($\beta = 0.0146$, $B\rho = 45.7$ kG-cm) and a tune-depression $\Delta v = 0.2$. From Equation (1) we derive the transverse phase-space density at the space-charge limit

$$D \;=\; N / \varepsilon \;<\; 2 \beta^2 \gamma^3 \, \Delta v \,/\, r_0. \tag{2}$$

For our case $D = 5.6 \times 10^{13}$ m^{-1}. It should be noticed that the phase-space density of Equation (2) does not depend on the dimension of the storage ring, but only on the beam energy.

A more relevant parameter to several experimental applications is the actual beam transverse density in the physical space $D_S = N / S$, where S is the beam cross-section area. For "circular" beams, $S \sim \pi a^2$, where a is the radius of the beam cross-section. There is a relation between the beam radius a and the emittance ε, so that

$$S \;=\; \beta_L \, \varepsilon \tag{3}$$

and

$$D_S \;=\; D \,/\, \beta_L. \tag{4}$$

The amplitude lattice function β_L for ASTRID is plotted in Fig. 2. The average value is about 5 meters, so that for our example the physical density that can be achieved at the space charge limit is $D_S = 1.1 \times 10^9$ cm^{-2}. In a conventional storage ring, the amplitude lattice function β_L is a measure of the strength of focusing. The stronger the focusing, the smaller the amplitude lattice function, and the smaller the beam cross-section. The focusing strength is increased by placing quadrupoles closer to each other. Actually

246

the average value of β_L is given by the length of the focusing period: the distance between doublets in ASTRID, or half of the cell length in a FODO structure. Unfortunately, in a conventional storage ring like ASTRID, the average value of β_L around the ring can hardly be less than a few meters; in fact, quadrupoles have a significant length, and space between quadrupoles is required to accommodate several functions, for instance, bending the beam trajectory.

In conclusion, the physical density that can be achieved in a conventional low-energy storage ring is limited, first, by the largest amount of space-charge tune-depression according to Equations (1 and 2) and, second, by the strength of focusing according to Equations (3 and 4).

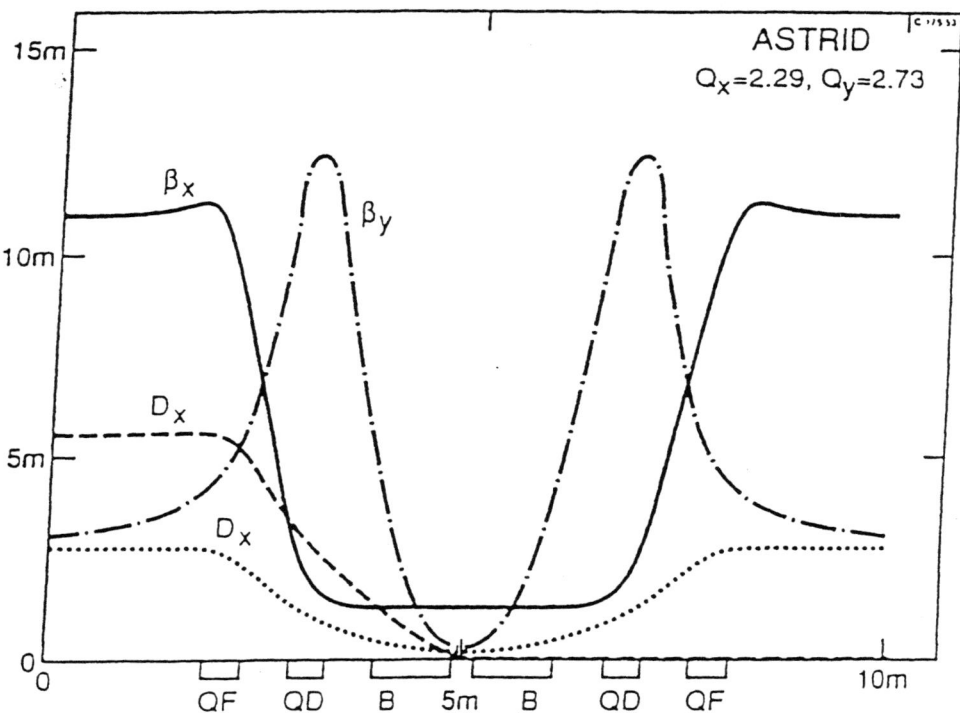

FIGURE 2. The Lattice Functions in ASTRID

THE LINEAR RADIO-FREQUENCY QUADRUPOLE

Another accelerator device is the Radio-Frequency Quadrupole (RFQ) which is used at the front end of a Linac. The device quickly accelerates the beam from an ion source to energies large enough when space-charge effects are considerably reduced, and provides simultaneous focusing of the transverse motion of the particles. The RFQ is not a magnetic device, but employs an alternating rf electric field for both the acceleration and the focusing of the particles. It is a straight waveguide, as shown in Fig. 3, with four internal metallic rods. An rf field at the frequency $f = \omega/2\pi$ is applied between diagonally opposite electrodes, as shown in Fig. 4, to generate in the opening a quadrupolar oscillating field of the same frequency. The motion of the particle will apparently modulate the actual field sequentially in time, creating the equivalent of an alternating focusing transverse field with periodicity $L = \beta\lambda$, where β is the particle velocity, and $\lambda = 2\pi c/\omega$ is the rf wavelength. Acceleration is provided by an axial oscillating electric field which is introduced by shaping the four electrodes with corrugations of the same length L, as shown in Fig. 3. As a consequence of acceleration, the beam, that is supposedly entering the RFQ with no time structure, will leave the RFQ at the other end bunched. If acceleration is not required, but only transverse focusing, which is the case of interest in our considerations here, the corrugation of the electrodes is not required, since the axial field is also not required, and the electrodes will appear just straight, as shown in Fig. 5. In this case the excitation is not an rf wave travelling down the waveguide structure, but just a stationary oscillating rf field between the pairs of electrodes.

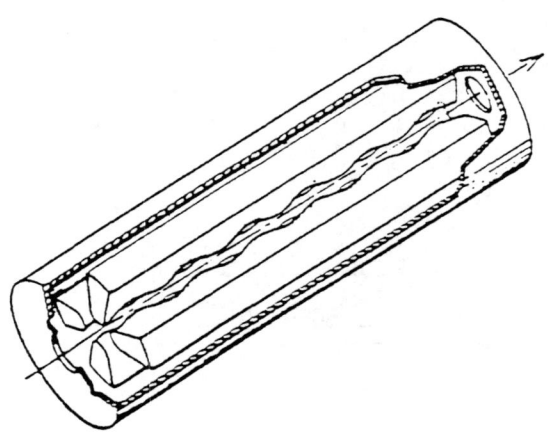

FIGURE 3.   Four-Vane RFQ (Linear) Corrugated Structure

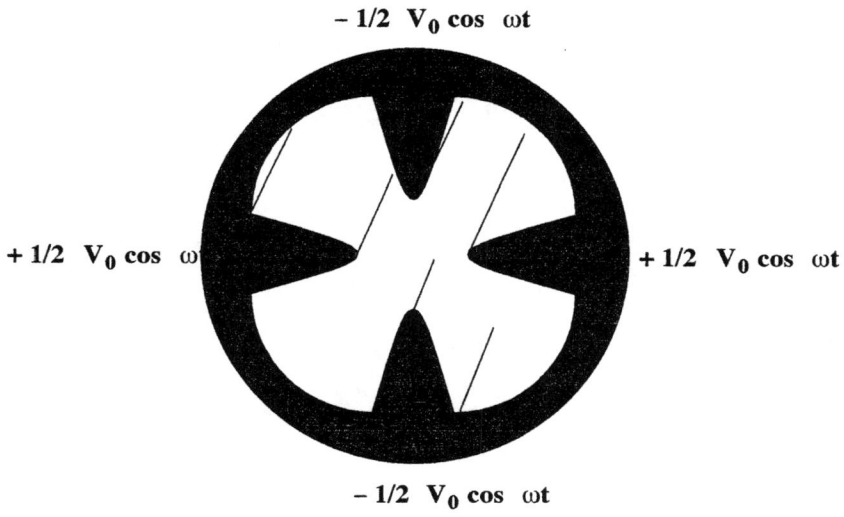

$- 1/2\ V_0 \cos\ \omega t$

$+ 1/2\ V_0 \cos\ \omega t$

$+ 1/2\ V_0 \cos\ \omega t$

$- 1/2\ V_0 \cos\ \omega t$

FIGURE 4.   Cross-section of a RFQ

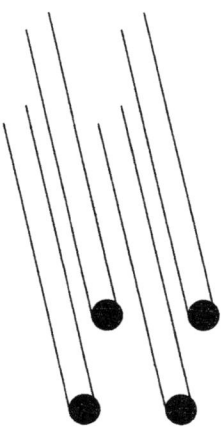

FIGURE 5.   Four-Vane RFQ without corrugation for Transverse Focusing only

In the case without acceleration, the equation of motion (2) of a particle of mass at rest m and electric charge e is

$$m \, d^2 x / dt^2 = (eV_0 / b^2) \, x \cos(\omega t), \tag{5}$$

where x is the lateral displacement (horizontal or vertical), 2b the internal diameter, and V_0 the peak rf voltage. Introducing the dimensionless RFQ parameter

$$B_0 = (eV_0 / mc^2)(\lambda / b)^2, \tag{6}$$

with the substitutions $B_0 = 2\pi^2 q$, $z = \beta ct$, and $\theta = \pi z/L$, Equation (5) transforms to

$$d^2 x / d\theta^2 - 2q \, x \cos(2\theta) = 0, \tag{7}$$

which clearly shows the alternating behavior of the focusing of the device with periodicity $L = \beta\lambda$. The phase advance per period depends only on the RFQ parameter B_0. For example, a phase advance of $90°$ per period L is obtained with $B_0 = 6.812$ or $q = 0.345$.

In our example of a proton beam with the kinetic energy of 100 keV, adopting an rf field at 200 MHz, that is $\lambda = 150$ cm, the periodicty of the focusing is $L = 2.2$ cm which is considerably shorter than the one that can be obtained in a conventional storage ring. Correspondingly, the average value of the amplitude lattice function β_L is significantly reduced, since $\beta_L \sim L$. Thus with the same beam emittance and energy, the beam size is also considerably smaller when compared to the beam dimension in a conventional storage ring.

There is another interesting feature of the RFQ. It is an open structure, both mechanically and electrically, where the beam enters one end and leaves the other. There is no global periodicity involved, as in the magnetic storage ring. There are no intrinsic imperfection resonances to worry about. It is thus legitimate to expect that considerably higher beam intensities can be transported since the conventional storage ring space-charge limitation equation (1) does not apply to a RFQ.

Space charge detracts of course from the focusing strength an amount Δ proportional to the beam density (2). The equation of motion modifies as follows:

$$d^2 x / d\theta^2 - 2q \, x \cos(2\theta) - \Delta x = 0, \tag{8}$$

where, for a uniform transverse and longitudinal distribution,

$$\Delta = 2 N r_0 \lambda^2 / (\pi^2 a^2 C), \tag{9}$$

with a the beam radius. There are N particles traversing at one time the length C of the RFQ. The maximum intensity that can be transported corresponds to a depression of the phase advance per period to about 45°. This yields Δ ~ 0.044. For our example, taking the RFQ length to be C = 40 m (the circumference of ASTRID), we have the physical beam density limit $D_S = N / \pi a^2 = 2.5 \times 10^{13}$ cm^{-2}, that is four orders of magnitude larger than what can be obtained in a conventional storage ring like ASTRID.

AN ION TRAP

This is shown in Fig. 6. It is also referred to as a quadrupole storage ring (3). The diameter of the device is only 12 cm. It is made of four annular electrodes with the cross-section as shown in the same figure. The internal diameter is 5 mm. An electrostatic voltage is applied between each pair of diagonally opposing electrodes. This generates a constant radial electric field which vanishes at the center and increases approximately linearly with the distance from the main axis. An atomic gas of the desired ion species is diffused in the region between the electrodes. An electron gun ionizes the atoms, and the resulting ions are trapped transversely in the small storage ring, oscillating around the circular main axis under the effect of the restoring forces of the electrostatic quadrupole. There is no beam in this configuration as in the previous more conventional magnetic storage ring, or in the linear Radio-Frequency Quadrupole, since the ions do not drift azimuthally along the main axis. The particles adjust their mutual longitudinal distance by Coulomb interaction, whereas the transverse interaction is compensated by the external quadrupolar forces. In a similar trap, ions of ^{24}Mg$^+$ were cooled transversely with a laser beam. A crystalline formation then observed were the ions arranged rigidly with respect to each other at a distance of about 20 μm (see Fig. 7).

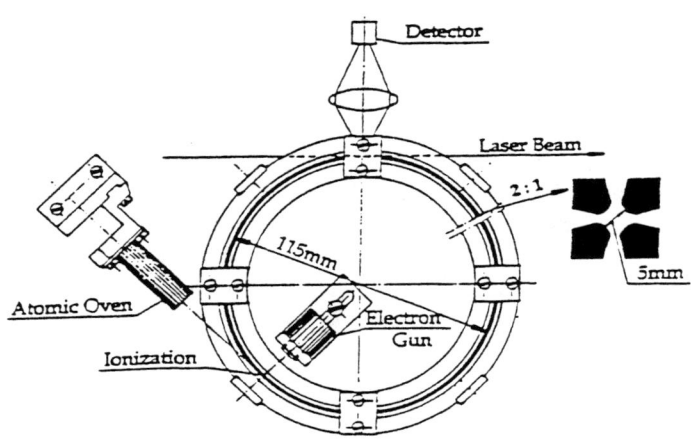

FIGURE 6.   The Ion Trap. (The Quadrupole Storage Ring)

251

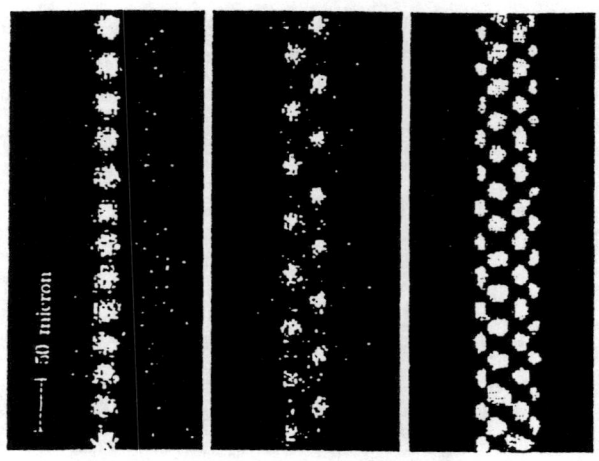

FIGURE 7. Observed Crystalline Structures in an Ion Trap

THE CIRCULAR RADIO-FREQUENCY QUADRUPOLE
STORAGE RING (CRFQ)

The Circular Radio-Frequency Quadrupole (CRFQ) is a compact storage device which includes features of all the devices described above. It is a storage ring where beams of light and heavy ions (including protons and negative ions) can circulate at constant speed corresponding to energies comparable to those used for ASTRID. As in the Ion Trap, focusing is provided by four annular electrodes. As in the linear RFQ, rf oscillating voltages are applied between the electrodes to provide transverse alternating focusing of motion over a short period of about few centimeters. From this point of view, the CRFQ resembles the linear RFQ without acceleration, and without, therefore, corrugated vanes. The CRFQ is actually a linear RFQ curved and closed mechanically on itself as shown in Figs. 8 and 9. The CRFQ can be used for a variety of applications where very intense and dense beams of ions are required. Some of these applications are in common with both the ASTRID storage ring and the Ion Trap. They range from molecular, atomic and ion physics, to Laser Cooling with the formation of Crystalline Beams, and to the demonstration of Intersecting and Colliding Beams for the study of Nuclear Fusion of light ions.

Below we shall describe typical parameters and performance of a CRFQ where a single beam of protons is circulating at the kinetic energy of 100 keV. In particular we shall explain how it is possible to store in this device ion beams with considerably higher intensity and density, well beyond the capability of conventional storage rings. We shall also describe how ions can be kept circulating in the device.

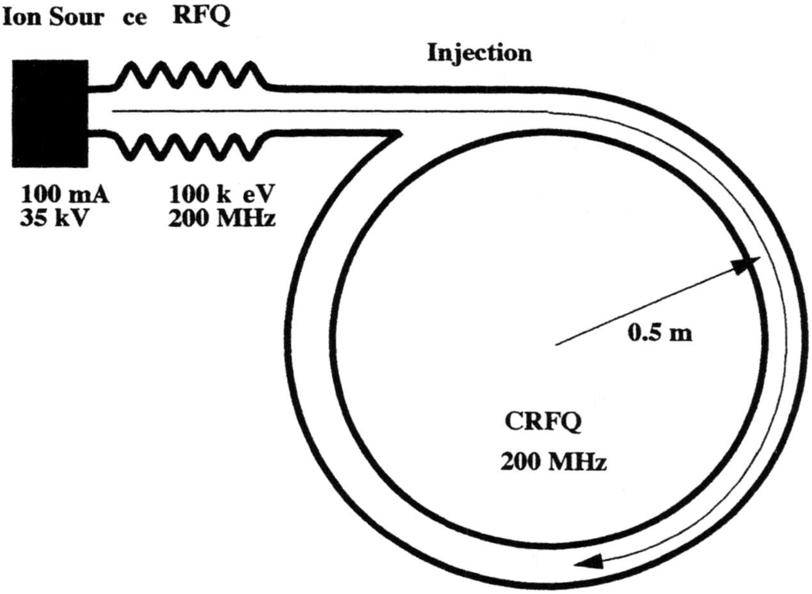

Ion Sour ce RFQ

Injection

100 mA **100 k eV**
35 kV **200 MHz**

0.5 m

CRFQ
200 MHz

FIGURE 8. Schematic layout of a Circular Radio-Frequency Storage Ring (CRFQ)

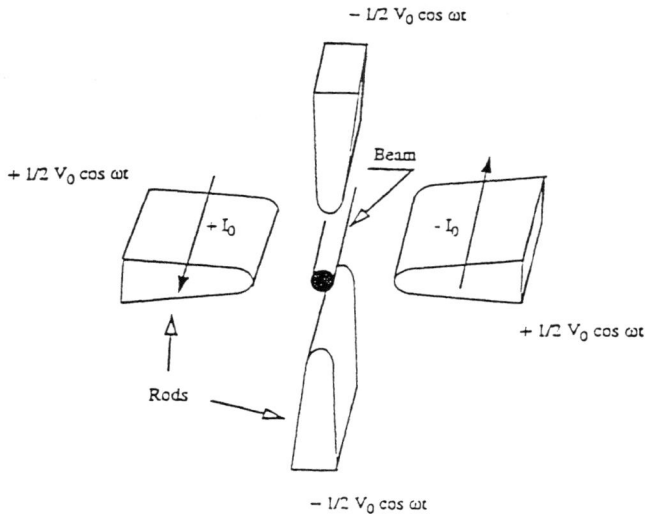

$- 1/2\, V_0 \cos \omega t$

$+ 1/2\, V_0 \cos \omega t$

Beam

$+ I_0$

$- I_0$

$+ 1/2\, V_0 \cos \omega t$

Rods

$- 1/2\, V_0 \cos \omega t$

FIGURE 9. Cross-section view of the CRFQ. A transverse rf field is excited between the electrodes. The rf excitation is as shown. DC current I_0 flows along the outer electrodes generating a magnetic field to bend the beam circulating in the region inside

253

EQUATIONS OF MOTION AND CONFINEMENT IN THE CRFQ

It is required to keep the ions on a circular orbit, centered with the azimuthal axis of the CRFQ. This can be accomplished in two ways. The CRFQ can be located inside and on the midplane of a pure $360°$ circular dipole magnet. Or, electric currents can be made to flow along and in opposite directions of the two horizontal electrodes as shown in Fig. 9. The electric currents then generate a bending field in the internal region of the CRFQ. In the following we shall adopt the latter solution. Obviously the bending field is to be adjusted to match a reference ion momentum p_0.

We shall assume in this section that the major radius of the CRFQ, that is the radius of curvature R, is much larger than the focusing period $L = \beta\lambda$, and that this is in turn larger than the internal diameter 2b. We can then disregard the contribution of the curvature to the rf focusing field. The equation of motion in vertical direction y is then still given by Equations (5 to 7) as in the linear RFQ,

$$d^2 y / d\theta^2 - 2q\, y \cos(2\theta) = 0. \tag{10}$$

There is in addition, in the radial plane, an extra focusing term due to the curvature of the trajectory, namely

$$d^2 x / d\theta^2 + 2q\, x \cos(2\theta) + x\, \beta^2\lambda^2 / \pi R^2 = \delta\, \beta^2\lambda^2 / \pi R, \tag{11}$$

where δ is the particle relative momentum deviation from the reference value p_0. The last term is also the source of dispersion. Both the contribution to the focusing of the curvature as well the dispersion are very small when compared to the focusing from the main q-factor, and, therefore, will be ignored in the following.

In case the beam is completely debunched and no longitudinal forces are applied, the azimuthal motion of the ions is an angular precession movement at the angular frequency

$$\omega_0 = 2\pi f_0 = \beta c / R. \tag{12}$$

All the considerations made for the linear RFQ apply also to the CRFQ. For instance, the alternating focusing period is $L = \beta\lambda$, and the q parameter, related to the RFQ parameter B_0 of Equation (6) by the relation $B_0 = 2\pi^2 q$, determines the phase advance per period and the amplitude lattice function $\beta_L \sim L$. As an example, the phase advance of $90°$ requires $q = 0.345$.

Thus, the Circular Radio-Frequency Storage Ring has the advantage of providing a very short focusing alternating period of only a few centimeters, considerably smaller than can be obtained in a conventional storage ring like ASTRID.

SPACE-CHARGE LIMITATIONS IN THE CRFQ

The other major advantage for the use of the CRFQ storage ring is the larger beam intensity that can be stored when compared to a conventional storage ring of the same energy. In a conventional storage ring, like ASTRID, we have seen that there is a limit caused by the Space-Charge Tune-Depression Δv that for practical purposes cannot exceed a value at most as large as 0.5. This limit is understood to be set by the presence of random magnet imperfections which cause the lowering of the ring periodicity to one unit, and the creation of stopbands around half-integral tune values which cannot be crossed without substantial beam losses.

In the CRFQ the situation is very different and we believe identical to that encountered in the linear RFQ. First of all, the focusing periodicity is very high and cannot be broken down by magnetic imperfections that do not exist. Possible errors on the focusing can of course be introduced because of the limited accuracy of the placement of the electrodes, but these errors are of a different nature and similar to those investigated in a linear RFQ. Secondly, the structure, though mechanically closed on itself, is electrically open. Turn after turn, the CRFQ is just like a long transport (unless the total betatron tune per revolution is exactly an integer, a situation that is to be avoided, as resonances are in this case created). One can then make the analogy with the linear RFQ where, as we have seen, the Space-Charge limit is caused by lowering the phase advance per period from 90° down to 45° (or less), below which the particle motion may become unstable.

Disregarding again the curvature of the ring, to take into account the finite beam intensity, the equations of motion are to be modified according to Equation (8) with the space-charge parameter Δ given again by Equation (9), where $C = 2\pi R$ is now the circumference of the ring, and N is the total number of particles stored. The same limit for space charge $\Delta \sim 0.044$ is assumed. If the CRFQ storage ring has the same circumference as ASTRID (40 m) the physical beam density that can be accepted is again

$$D_S = N / \pi a^2 = 2.5 \times 10^{13} \text{ cm}^{-2}.$$

AN EXAMPLE OF CRFQ

The main purpose of the following example of CRFQ is the construction of a prototype device to demonstrate the two basic principles: (i) that it is possible to achieve very short alternating focusing structure with a period of few centimeters, and (ii) that it is possible to store ion beams at intensities higher than those that can be achieved in conventional storage rings. We shall take again a proton beam at 100 keV and an rf of 200 MHz. The ring radius is $R = 50$ cm. The major parameters are listed in Table 1. The transverse electric field is around 21 MV/m, which at the chosen frequency is just below one kilpatrik unit, which is below the expected surface limit.

To bend the beam trajectory, one needs a bending magnetic field of 0.92 kG, which can be obtained by letting 1.7 kA current flow on the electrodes on the midplane of the CRFQ.

The physical density which corresponds to the space-charge limit (RFQ) is $N_S = 2 \times 10^{12}$ cm^{-2}. With a beam normalized emittance of 1 π cm mrad, the average beam radius at the space-charge limit is a = 6 mm, so that about 2×10^{12} protons can be stored in the ring. This is about 20 times larger than the conventional limit given by Equation (1) with Δv =0.2. Thus an experiment can easily be done to verify that indeed it is possible to store considerably more current in the CRFQ storage ring than in a conventional storage ring.

Table 1 shows that the circulating current at the RFQ space-charge limit is 460 mA, and only 20 mA if the conventional limit of Equation (1) should apply. An ion source, operating at a very low duty cycle, is certainly capable of producing a beam pulse of about 0.5 μs duration in excess of 100 mA.

The pulse duration is shorter than the revolution period so that only one turn needs to be injected in the CRFQ. The actual implementation of the single turn injection in the CRFQ device is a topic that still needs to be studied.

The ion source can be placed on a platform at 35 kVolt. The energy difference to 100 keV can be obtained by accelerating the beam in a short linear RFQ operating also at 200 MHz. At the end of the acceleration, the vanes of the linear RFQ are no longer corrugated, and only focusing is then provided. This linear section would then merge with and match to the entrance of the CRFQ.

The experiment can be performed at a very low duty cycle, for example, with a beam pulse injected and stored in the CRFQ every few minutes. There is thus not much beam power involved and the beam itself can directly be disposed of by turning off the bending field. It would also be very useful to learn how to extract the stored beam in a single turn and in a desired direction.

In conclusion, the experimental setup includes an Ion Source, a linear RFQ, a matching section between the RFQ and the CRFQ, an injection system to be developed, a 200-MHz rf source, a 2-kA dc-current source to bend the trajectory, and, of course, the CRFQ itself. The device is to be complemented with a vacuum, a beam diagnostic system, and control. One needs to study and to determine the construction tolerances of the four annular electrodes, the engineering of the grounded enclosure, and the rf and dc powering of the electrodes.

CONCLUSION

We have described a new concept of Storage Ring for low-energy ion beams. The principle of operation of the new device is similar to that of an ordinary RFQ, except that it is mechanically bent on itself. It is then possible to achieve very short alternating focusing periods and also to store considerable higher beam intensities well beyond the ordinary space-charge limit of conventional storage rings.

We have also outlined the design of a prototype that can be used for the demonstration of the new concept.

TABLE 1. Parameters of the proposed CRFQ

Kinetic Energy	100		keV
β	0.0146		
Magnetic Rigidity	45.7		kG - cm
Major Radius, R	50		cm
Minor Radius, b	7.5		mm
rf frequency, f	200		MHz
rf wavelength, λ	150		cm
RFQ parameter, B_0	6.812		
q	0.345		
peak rf Voltage, V_0	160		kVolt
Electric Field	21.33		MeV/m
Periodicity, $L = \beta\lambda$	2.2		cm
Space-Charge Δ	0.0	0.044	
Phase Advance / period	90°	45°	
No. of periods per turn	143.562		
Tune / revolution	35.9076	17.9806	
β - min	1.98	4.02	cm
β - max	2.85	5.69	cm
Bending Field	914		Gauss
d.c. current	1714		Amperes
Revolution Frequency, f_0	1.393		MHz
Revolution Period, T_0	0.718		μs
Density Limit, D_S	2.0×10^{12}		cm^{-2}
Normal.Emittance	1.0		π cm mrad
a-min	3.68	5.24	mm
a-max	4.42	6.24	mm
max No. of Protons, N		2.07×10^{12}	
max. Beam Current		460	mAmp

REFERENCES

1. Møller S. P., *Proc. of the 1991 IEEE Particle Accelerator Conference*, *IEEE 91CH3038-7*, page 2811.

2. Staples J. W., "RFQ's - An Introduction," *The Physics of Particle Accelerators*, *AIP Conference Proceedings*, 249 (1992), **Vol. II**, page 1483.

3. Walther H., "Spectroscopy of Trapped Ions," Crystalline Beams and Related Issues, page 149, *The Science and Culture Series - Physics 11*, (1996), Editors: D. M. Maletic and A.G. Ruggiero, World Scientific.

Radiative Interaction of Electrons in a Short Electron Bunch Moving in an Undulator

E.L. Saldin*, E.A. Schneidmiller* and M.V. Yurkov[†]

*Automatic Systems Corporation, 443050 Samara, Russia
[†]Joint Institute for Nuclear Research, Dubna, 141980 Moscow Region, Russia

Abstract. This paper presents investigations of the longitudinal radiative force in an electron bunch. The model of the electron bunch assumes line density distribution. General formulae are presented for the calculation of the radiative force in the bunch moving along an arbitrary small-angle trajectory. The case of a motion in an undulator (wiggler) has been studied in detail. Analytical solutions are obtained for a rectangular and for a Gaussian bunch shape. It is shown that the rate of the bunch energy loss due to the radiative interaction is equal to the power of the coherent radiation in the far zone. Numerical estimations presented in the paper show that the effects of induced energy spread due to the radiative interaction can be important for free electron lasers operating in the infrared wavelength range.

I   INTRODUCTION

The theory of the radiative interaction of electrons in an intensive microbunch traversing a curved trajectory is intensively developed nowadays. This is explained by the practical importance of the radiative effects for beam dynamics in linear colliders [1,2] and short-wavelength free electron lasers (FELs) [1,3,4]. When an intensive electron bunch passes bending magnets, bunch compressors, wigglers, etc, radiative interaction induces the energy spread in the electron beam and can lead to the transverse emittance dilution in dispersive regions [5,6]. Most of the previous theoretical studies of the radiative interaction (see, e.g., refs. [7,5,8]) describe periodical circular motion of a bunch. Recently, the problem connected with transient effects in a bending magnet of a finite length has been investigated [9,10]. From a practical point of view it is important to calculate radiative effects in more complicated magnetic systems such as a sequence of bending magnets, an undulator (wiggler), etc.

In this paper we present a universal algorithm for the calculation of the radiative interaction of the particles in a line-charge microbunch moving on an arbitrary curved trajectory. Furthermore, we use this general algorithm for the analysis of

CP468, *Nonlinear and Collective Phenomena in Beam Physics–1998 Workshop*,
edited by S. Chattopadhyay, M. Cornacchia, and C. Pellegrini

the radiative forces in a bunch moving in the undulator. The analytical solution is obtained using approximations of neglecting transient effects (when the bunch enters and leaves an undulator) and neglecting shielding effects (influence of a vacuum chamber on the radiative process). The models of a rectangular and Gaussian bunch profiles are studied in detail. The applicability region of obtained results is discussed and numerical estimations for several experiments are presented. The results obtained in this paper can be used for quick estimations of the considered effect and for testing numerical simulation codes.

II GENERAL ALGORITHM

In this section we present general expressions for calculation of the longitudinal (along a particle's velocity) radiative force in the ultrarelativistic electron beam. We consider the model of the electron microbunch with a line density distribution (no transverse dimensions) moving along a plane trajectory. All the particles in the bunch have the same module of velocity $v/c = \beta \simeq 1 - 1/(2\gamma^2)$. It is assumed that vectors of the velocities are always within a small cone on a part of the trajectory between a position of the bunch head and a retarded position of the tail. The motion of the particles is supposed to be given and we solve the electrodynamic problem only.

We consider here the curvature-dependent part of the bunch self-interaction which disappears when the curvature radius tends to infinity. The neglected part is the trivial longitudinal Coulomb self-interaction of the same bunch moving on a straight line (which can be calculated separately). The excluded part is nondissipative, in other words the total energy loss of the whole bunch is always zero. So, the dissipative self-interaction is connected with the curvature-dependent part, i.e., the calculated energy loss by the bunch must be equal to the energy of coherent radiation in the far zone. In ref. [11] the algorithm for the calculation of curvature-dependent interaction has been developed. The following general formula for calculation of the energy change rate of a particle in a rectangular bunch with linear density of particles $\lambda = $ const has been obtained (see Fig. 1):

$$\frac{d\mathcal{E}}{cdt} = e^2 \lambda \Phi(s, S) = e^2 \lambda \left(\frac{1}{\gamma^2 s} - \frac{2}{L} \frac{1 + \gamma^2 \phi \theta}{1 + \gamma^2 \theta^2} \right) . \tag{1}$$

The retardation condition is written as

$$s = S - S_{\text{tr}} - \beta L . \tag{2}$$

Formula (1) allows one to calculate the rate of the energy change as a function of the particle position along a rectangular bunch and of its position along a trajectory. For correct use of formulae (1) and (2) one should take into account the following. Let the test particle be placed at some distance s from the tail of the bunch. At a given moment of time it passes the point with coordinate S along the trajectory

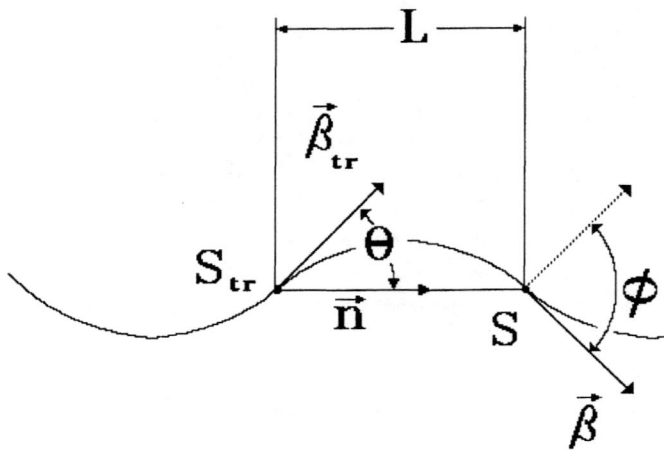

FIGURE 1. The scheme for calculation of the radiative force acting on a particle in a rectangular bunch.

with some definite direction of the velocity $\vec{\beta}$ (see Fig. 1). The retarded position of the bunch tail S_{tr} along the trajectory is defined by condition (2), where L is the retarded distance. Trigonometric functions which will appear in eq. (2) should be expanded in a small-angle approximation, keeping the main term and the next nonvanishing high-order term in angle. At the point S_{tr} of the trajectory the tail particle has the velocity $\vec{\beta}_{\mathrm{tr}}$ at the retarded moment of time, and the direction to the current position of the test particle is given by vector $\vec{n}$. The rate of the energy change of the test particle due to the curvature effects is given by formula (1), where $\theta = \arccos(\vec{n} \cdot \vec{\beta}_{\mathrm{tr}}/\beta)$ and $\phi = \arccos(\vec{\beta} \cdot \vec{\beta}_{\mathrm{tr}}/\beta^2)$. One should use the following rule for choosing the correct signs of these angles: if vectors $\vec{\beta}_{\mathrm{tr}}$, $\vec{n}$, and $\vec{\beta}$ have angles ψ_1, ψ_2 and ψ_3 in some polar coordinate system, then $\theta = \psi_1 - \psi_2$ and $\phi = \psi_1 - \psi_3$.

The solution for the rectangular bunch (1) with $\lambda(s) = \mathrm{const}$ can be generalized for the case of an arbitrary linear density $\lambda(s)$ by presenting the bunch as a sequence of rectangular bunches with a length of $(s-s')$ and a linear density of $ds' [d\lambda(s')/ds']$:

$$\frac{d\mathcal{E}}{c\,dt} = e^2 \int\limits_{-\infty}^{s} ds'\, \Phi(s - s', S)\frac{d\lambda(s')}{ds'} \ , \tag{3}$$

where function Φ is defined by eq. (1).

Now let us illustrate application of the obtained general formulae for the case of the periodical circular motion (see Fig. 2). In this case $\theta = \phi/2 \ll 1$, and the retardation condition (2) reduces to

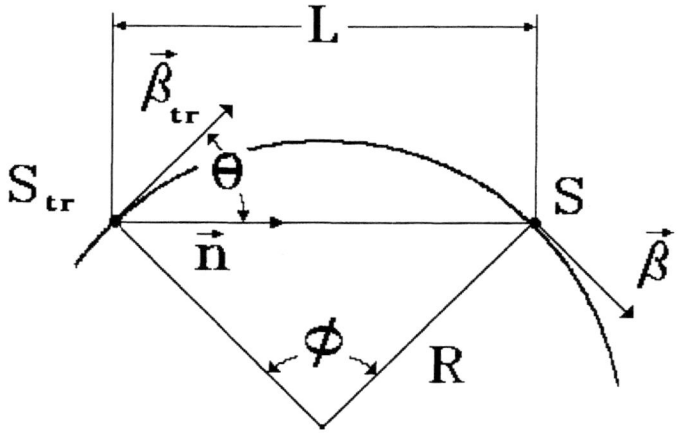

FIGURE 2. The scheme for the calculation of the radiative interaction in a rectangular bunch moving on a circle.

$$s = \frac{R\phi}{2\gamma^2} + \frac{R\phi^3}{24} \ , \tag{4}$$

and expression (1) takes the following form [9]:

$$\frac{d\mathcal{E}}{cdt} = e^2\lambda \left(\frac{1}{\gamma^2 s} - \frac{2}{R\phi} \frac{1 + \frac{\gamma^2\phi^2}{2}}{1 + \frac{\gamma^2\phi^2}{4}} \right) = -\frac{4e^2\lambda\gamma}{R} \frac{(\gamma\phi)(8 + \gamma^2\phi^2)}{(4 + \gamma^2\phi^2)(12 + \gamma^2\phi^2)} \ . \tag{5}$$

At large values of $\gamma\phi$ eq. (4) takes the simplified form $s \simeq R\phi^3/24$, and we obtain the well-known result [7,5]:

$$\frac{d\mathcal{E}}{cdt} = e^2\lambda\Phi(s) = -\frac{4e^2\lambda}{R\phi} = -\frac{2e^2\lambda}{3^{1/3}R^{2/3}s^{1/3}} \ , \tag{6}$$

This result describes the steady-state regime when the radiative interaction does not depend on the position of the bunch along the trajectory. Using eqs. (3) and (6), we can generalize it for the case of an arbitrary bunch profile [7,5,8]:

$$\frac{d\mathcal{E}}{cdt} = -\frac{2e^2}{3^{1/3}R^{2/3}} \int\limits_{-\infty}^{s} \frac{ds'}{(s - s')^{1/3}} \frac{d\lambda(s')}{ds'} \ . \tag{7}$$

261

III GENERAL SOLUTION FOR A BUNCH MOVING IN AN UNDULATOR

In this section we use the general algorithm presented in the previous section for the calculation of the radiative interaction of the electrons in a bunch moving in an undulator. The transient effects (when a bunch enters and leaves the undulator) are withdrawn from consideration.

Let us consider a rectangular electron bunch of a length l_b with a linear density $\lambda = N/l_b$ moving along the z direction in the undulator with magnetic field

$$H_x = H_w \cos(k_w z) .$$

The transverse and the longitudinal velocities of an electron can be approximated by

$$\beta_y = \frac{K}{\gamma} \sin(k_w z), \qquad \beta_z = \beta - \frac{K^2}{2\gamma^2} \sin^2(k_w z) ,$$

where $K = eH_w/k_w mc^2$ is the undulator parameter, $\gamma = \mathcal{E}/mc^2$ is the relativistic factor, $(1 + K^2/2)/\gamma^2 \ll 1$ and $\beta \simeq 1 - 1/2\gamma^2$. The transverse coordinate of the electron oscillates in accordance with

$$y = -\frac{K}{\gamma k_w} \cos(k_w z) .$$

We calculate the longitudinal radiative force assuming the motion of the particles to be given. The calculations are based on the general approach described in the previous section. Leaving the details of the calculations, we present the final result for the rate of the energy change as a function of the positions of the electron in the bunch and in the undulator, s and z, respectively:

$$\frac{d\mathcal{E}}{cdt} = e^2 k_w \lambda \, D(\hat{s}, K, \hat{z}) , \tag{8}$$

where

$$D(\hat{s}, K, \hat{z}) = \frac{1}{\hat{s}} - 2 \, \frac{\Delta - K^2 B(\Delta, \hat{z}) \left[\sin \Delta \cos \hat{z} + (1 - \cos \Delta) \sin \hat{z}\right]}{\Delta^2 + K^2 B^2(\Delta, \hat{z})} , \tag{9}$$

$$B(\Delta, \hat{z}) = (1 - \cos \Delta - \Delta \sin \Delta) \cos \hat{z} + (\Delta \cos \Delta - \sin \Delta) \sin \hat{z} , \tag{10}$$

and Δ is the solution of the transcendental equation:

$$\hat{s} = \frac{\Delta}{2} \left(1 + \frac{K^2}{2}\right) + \frac{K^2}{4\Delta} \{[2(1 - \cos \Delta) - \Delta \sin \Delta]$$
$$\times (\cos \Delta \cos 2\hat{z} + \sin \Delta \sin 2\hat{z}) - 2(1 - \cos \Delta)\} . \tag{11}$$

262

Here the following reduced variables are introduced: $\hat{s} = \gamma^2 k_w s$ and $\hat{z} = k_w z$. The physical sense of variable Δ becomes transparent when it is rewritten as $\Delta = k_w(z - z_{tr})$. Here z and z_{tr} are projections on the undulator axis of the current position of the reference particle and of the retarded position of the bunch tail, respectively. It follows from the geometry of the problem and from eqs. (9), (10) and (11) that function D is periodical with respect to the position of the reference particle z along the undulator. The period of function D is equal to half of the undulator period π/k_w (or, to π when using the normalized position $\hat{z}$).

Let us perform analysis of expression (8). First of all, it is seen that in the limit of $\hat{s} \to \infty$ function D tends to zero, i.e., at any point of infinitely long bunch the radiative interaction force is equal to zero as it should be. Now let the bunch be shorter than the characteristic wavelength of the spectrum of incoherent radiation (which is of the order of $(k_w\gamma^2)^{-1}$ for small values of K and of the order of $(Kk_w\gamma^2)^{-1}$ for large values of K). When

$$\hat{s} \ll 1, \qquad \hat{s}^2 K^2 \ll 1 , \tag{12}$$

function D is reduced to the following simple form:

$$D = -\frac{4}{3}K^2 \hat{s} \cos^2 \hat{z} . \tag{13}$$

The rate of the energy loss by the whole bunch is calculated as follows:

$$\frac{d\mathcal{E}_b}{cdt} = \int_0^{l_b} ds \, \lambda \frac{d\mathcal{E}}{cdt} . \tag{14}$$

Substituting expressions (8) and (13) into eq. (14) we obtain:

$$\frac{d\mathcal{E}_b}{cdt} = -\frac{2}{3} [NeKk_w\gamma \cos(k_w z)]^2 .$$

Using the definition of the parameter K we can rewrite this expression in the following form:

$$\frac{d\mathcal{E}_b}{cdt} = -\frac{2}{3} N^2 r_e^2 \gamma^2 H_w^2 \cos^2(k_w z) , \tag{15}$$

where $r_e = e^2/mc^2$ is the classical radius of the electron. One can see that in the limit of a "short" bunch the energy loss is identical to that of a single particle with the charge Ne. So, this limit formula (8) provides a correct and physically transparent result.

Now let us consider the limit of $\Delta \ll 1$. In this case the result should depend on the value of the local magnetic field only (or, on local curvature), and the solution (8) must be reduced to that corresponding to a circular motion. The approximate form of eq. (11) can be written as follows:

$$\hat{s} \simeq \frac{\Delta}{2} + \frac{K^2}{24}\Delta^3 \cos^2 \hat{z} \ . \tag{16}$$

The first term in the right hand side of this equation takes origin from the difference between the electron's velocity and the velocity of light, and the second one appears due to a local curvature of the trajectory. As a result, function D takes the form:

$$D = -\frac{4\Delta K^2 \cos^2 \hat{z}}{(4 + \Delta^2 K^2 \cos^2 \hat{z})} \frac{(8 + \Delta^2 K^2 \cos^2 \hat{z})}{(12 + \Delta^2 K^2 \cos^2 \hat{z})} \ . \tag{17}$$

Taking into account that

$$\frac{k_w^2 K^2 \cos^2 \hat{z}}{\gamma^2} = \frac{1}{R^2} \ ,$$

where R is the local radius of curvature, we obtain that solution (8) with the retarded condition given by (16) and function D given by (17) is identical to (5) where the retarded condition is given by (4). When conditions (12) are satisfied, expression (17) takes the simple form of eq. (13). Solution (17) is reduced to (6) at $(K \mid \cos \hat{z} \mid)^{-1} \ll \Delta \ll 1$. The latter asymptote means that the bunch is much shorter than the wavelength of the first harmonic of the undulator radiation, but is much longer than the characteristic wavelength R/γ^3 of the synchrotron radiation spectrum (radiated by a single electron at large values of the parameter K).

So, we have shown that solution (8) for a rectangular bunch has correct asymptotical behavior. In conclusion to this section we generalize this solution for the case of a bunch with an arbitrary linear density $\lambda(s)$. This general solution is given by the convolution of the solution for a rectangular bunch with the derivative of a linear density function:

$$\frac{d\mathcal{E}}{cdt} = e^2 k_w \int_{-\infty}^{s} ds' D(\hat{s} - \hat{s}', K, \hat{z}) \frac{d\lambda(s')}{ds'} \ . \tag{18}$$

In the subsequent sections we will study in detail the rate of the energy change averaged over the z coordinate:

$$\frac{d\bar{\mathcal{E}}}{cdt} = e^2 k_w \int_{-\infty}^{s} ds' \bar{D}(\hat{s} - \hat{s}', K) \frac{d\lambda(s')}{ds'} \ , \tag{19}$$

where

$$\bar{D}(\hat{s}, K) = \frac{1}{\pi} \int_{0}^{\pi} d\hat{z} D(\hat{s}, K, \hat{z}) \ . \tag{20}$$

IV   AVERAGED SOLUTION FOR A RECTANGULAR BUNCH

In this section we study the averaged solution (19) and (20) for the case of a rectangular bunch of a length l_b with a linear density $\lambda = N/l_b$. In this case eq. (20) can be written in the following form:

$$\frac{d\bar{\mathcal{E}}}{cdt} = \frac{e^2 k_w N}{l_b} \, \bar{D}(\hat{s}, K) \,, \tag{21}$$

where averaged function $\bar{D}(\hat{s}, K)$ should be calculated using the eqs. (20), (9), (10) and (11).

We start the investigation with small values of the undulator parameter $K \ll 1$. In this case function $\bar{D}(\hat{s}, K)$ takes the form:

$$\bar{D}(\hat{s}, K) = -K^2 \left(\frac{\sin^2 \hat{s}}{\hat{s}} + \frac{\sin 2\hat{s}}{2\hat{s}^2} - \frac{\sin^2 \hat{s}}{\hat{s}^3} \right) \,. \tag{22}$$

One can obtain that at small values of $\hat{s}$ this expression is reduced to that given by eq. (13) averaged over the z coordinate. The energy loss by the whole bunch is given by:

$$\frac{d\bar{\mathcal{E}}_b}{cdt} = \frac{e^2 N^2}{\gamma^2 l_b^2} \int_0^{\hat{l}_b} \bar{D}(\hat{s}, K) d\hat{s} \,, \tag{23}$$

where

$$\hat{l}_b = \gamma^2 k_w l_b \,.$$

Performing integration, we obtain:

$$\frac{d\bar{\mathcal{E}}_b}{dt} = -\frac{ce^2 N^2 K^2}{2\gamma^2 l_b^2} \left[\ln(2\hat{l}_b) - \mathrm{Ci}(2\hat{l}_b) + C + \left(\frac{1 - \cos(2\hat{l}_b)}{2\hat{l}_b^2} - 1 \right) \right] \,, \tag{24}$$

where $C = 0.577...$ is the Euler's constant and $\mathrm{Ci}(...)$ is the integral cosine [12].

Let us show that the rate of the energy loss by the electron bunch is equal to the power of coherent radiation in the far zone. The power of coherent radiation in the far zone is calculated as the integral of the power spectral density

$$\frac{dP_{coh}}{d\omega} = N^2 \eta(\omega) \frac{dP}{d\omega} \,, \tag{25}$$

where $\eta(\omega)$ is the form factor of the bunch equal to the squared module of the Fourier transform of a bunch shape. The form factor of the rectangular bunch of the length l_b is:

$$\eta(\omega) = \left(\sin\frac{\omega l_b}{2c}\right)^2 \left(\frac{\omega l_b}{2c}\right)^{-2} . \tag{26}$$

Function $dP/d\omega$ entering eq. (25) is the spectral density of the radiation power emitted by a single electron. For the first time it has been calculated in ref. [13]. In the case under consideration (small value of the undulator parameter K), this function is reduced to the following simple form:

$$\frac{dP}{d\omega} = \frac{e^2 K^2 \omega}{4c\gamma^2} \left(1 - \frac{\omega}{ck_w\gamma^2} + \frac{\omega^2}{2c^2 k_w^2 \gamma^4}\right) , \tag{27}$$

where ω extends from zero up to $2ck_w\gamma^2$. Substituting functions (26) and (27) into eq. (25) and integrating within these limits, one obtains that P_{coh} exactly coincides with $d\bar{\mathcal{E}}_b/dt$ taken with the opposite sign.

So, we have considered the case of small values of the undulator parameter K. In the case of an arbitrary value of K function $\bar{D}(\hat{s}, K)$ takes a much more complicated form than that given by eq. (22). Here we present asymptotical behaviour of this general averaged solution for a rectangular bunch at large distances from the tail ($\hat{s}/(1 + K^2/2) \gg 1$):

$$\bar{D}(\hat{s}, K) \simeq -\frac{(1 + K^2/2)}{\hat{s}} \left[2\left(1 - \frac{1}{\sqrt{1 + K^2}}\right)\sin^2\left(\frac{\hat{s}}{1 + K^2/2}\right)\right.$$
$$\left. +\frac{1}{\sqrt{1 + K^2}} - \frac{1}{1 + K^2/2}\right] . \tag{28}$$

At small values of K this solution is reduced to the first term in eq. (22).

V AVERAGED SOLUTION FOR A GAUSSIAN BUNCH

In this section we consider a bunch with a Gaussian distribution of linear density:

$$\lambda(s) = \frac{N}{\sqrt{2\pi}\sigma} \exp\left[-\frac{s^2}{2\sigma^2}\right] . \tag{29}$$

The averaged solution for the Gaussian bunch can be written in the form:

$$\frac{d\bar{\mathcal{E}}}{cdt} = \frac{e^2 N K^2}{\sqrt{2\pi}\sigma^2\gamma^2}\bar{G}(p, K, x) , \tag{30}$$

where $x = s/\sigma$ and p is the bunch length parameter:

$$p = \frac{\gamma^2 k_w \sigma}{1 + K^2/2} .$$

266

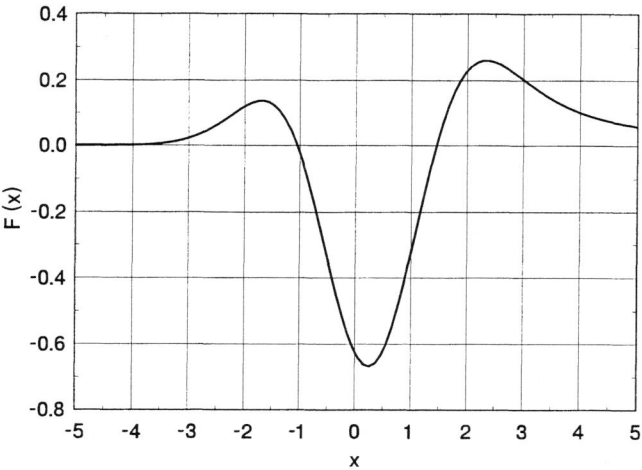

FIGURE 3. Function $F(x)$ given by eq. (32).

In the general case function $\bar{G}$ should be calculated by numerical integration of eq. (19). Nevertheless, in some region of parameters it can be expressed analytically. Let us study a practically important case of a long bunch, $p \gg 1$. As in the previous section, we start with the limit of small K. Under these conditions function $\bar{G}$ can be calculated analytically using eqs. (19) and (22):

$$\bar{G}(p,x) = \frac{x}{2}\exp\left(-\frac{x^2}{2}\right)\ln p + F(x) \ . \tag{31}$$

Here parameter p is reduced to $p \simeq \gamma^2 k_w \sigma$, and function $F(x)$ has the form:

$$F(x) = \frac{1}{4}(C + 3\ln 2 - 2)x\exp\left(-\frac{x^2}{2}\right) - \sqrt{\frac{\pi}{8}}\left[1 + \text{erf}\left(\frac{x}{\sqrt{2}}\right)\right.$$
$$\left. -x\exp\left(-\frac{x^2}{2}\right)\int_0^x dx'\exp\left(\frac{(x')^2}{2}\right)\left(1 + \text{erf}\left(\frac{x'}{\sqrt{2}}\right)\right)\right] , \tag{32}$$

where erf(...) is the error function [12]. The plot of function $F(x)$ is presented in Fig. 3. Figure 4 presents the plots of function $\bar{G}$ calculated at different values of parameter p.

In the case of an arbitrary value of the undulator parameter K, it is difficult to find an explicit analytical solution. Nevertheless, using the results of numerical integration of eq. (19), we can write function $\bar{G}$ in the following form ($p \gg 1$):

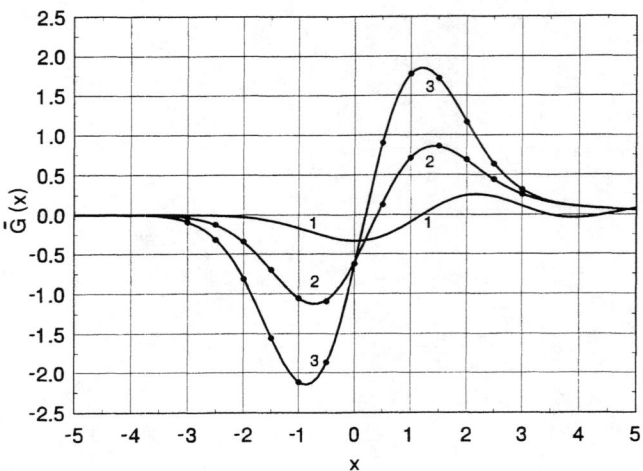

FIGURE 4. Function $\bar{G}$ for small value of the undulator parameter K and different values of the bunch length parameter p. Curve (1): $p = 1$, curve (2): $p = 30$, and curve (3): $p = 1000$. The curves are the results of numerical integration of eq. (19) and the circles are calculated with the help of analytical formula (31) for large values of parameter p.

$$\bar{G}(p, K, x) = \frac{x}{2} \exp\left(-\frac{x^2}{2}\right) [\ln p + g(K)] + F(x) , \tag{33}$$

where function $g(K)$ changes from 0 to 1 when K changes from small to large values. The plot of this function is presented in Fig. 5.

Let us show that the obtained solution is consistent with the conservation energy law, i.e., that the rate of the energy loss by the whole bunch is equal to the radiation power in the far zone. First, we calculate the rate of the energy loss by the whole bunch. Multiplying expression (30) with function $\bar{G}$ defined by eq. (33) by the function of linear density (29) and performing the integration, we get:

$$\frac{d\bar{\mathcal{E}}_{\mathrm{b}}}{dt} = -\frac{ce^2 N^2 K^2}{8\sigma^2 \gamma^2} . \tag{34}$$

Second, we use formula (25) to calculate the radiation power in the far zone. The form factor of the Gaussian bunch is given by

$$\eta(\omega) = \exp\left(-\frac{\sigma^2 \omega^2}{c^2}\right) . \tag{35}$$

It is seen that typical frequencies of coherent radiation are below c/σ. This means that in the case of a long bunch, $p \gg 1$, we can use asymptotical formula for the

268

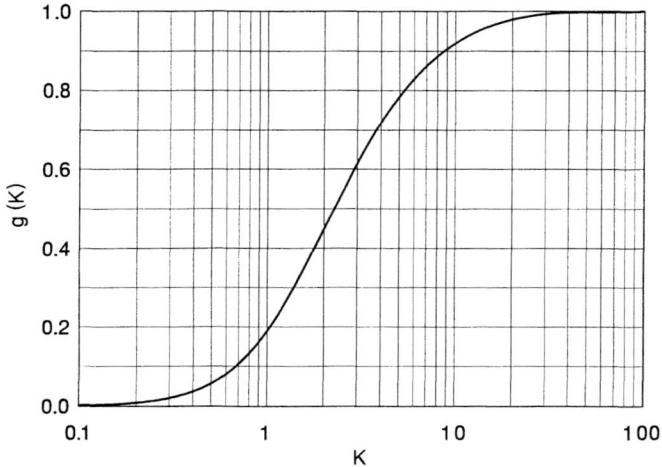

FIGURE 5. Function $g(K)$ entering eq. (33).

power spectral density of the radiation emitted by a single electron [13], assuming the frequency to be much less than the frequency of the first harmonic of the undulator radiation ($\omega \ll ck_w\gamma^2/(1 + K^2/2)$):

$$\frac{dP}{d\omega} \simeq \frac{e^2 K^2 \omega}{4c\gamma^2} \, . \tag{36}$$

Substituting eqs. (35) and (36) into eq. (25) and performing the integration, we obtain that the power of coherent radiation in the far zone exactly coincides with the bunch power loss (34) taken with the opposite sign.

In conclusion to this section we present the formula for the induced correlated energy spread in the Gaussian bunch due to the radiative interaction. Using expression (33) we write down this formula in the form convenient for practical calculations ($p \gg 1$):

$$\frac{d\sigma_\gamma}{cdt} = 0.219 \frac{IK^2}{I_A \sigma \gamma^2} \sqrt{[\ln p + g(K)]^2 + 0.933[\ln p + g(K)] - 0.786} \, , \tag{37}$$

where $I = Nec/\sqrt{2\pi}\sigma$ is the peak current, $I_A = 17$ kA is Alfven current, and

$$mc^2 \sigma_\gamma = \sqrt{\langle \bar{\mathcal{E}}^2 \rangle - \langle \bar{\mathcal{E}} \rangle^2} \, .$$

VI   DISCUSSION

Let us discuss the applicability region of the results obtained in the paper. The model approximations assume the line-charge bunch and disregarding the influence of the vacuum chamber on the process of radiative interaction. Besides, we have studied the steady-state regime, i.e., transient effects in an undulator of a finite length have been excluded out of consideration. Nevertheless, such a simple model provides correct physical description of many practical situations. Let us perform rough estimations of the applicability region of obtained results for the practically important case of a long bunch, $\sigma \gg (1 + K^2/2)/\gamma^2 k_w$. As a rule, the behaviour of the electron bunch after the undulator is not of interest for the FEL design and one should estimate the entrance transient effect only. The latter effect is not important when the undulator length, L_w, is sufficiently large,

$$L_w \gg \sigma\gamma_z^2 , \tag{38}$$

where $\sigma\gamma_z^2 = \sigma\gamma^2/(1 + K^2/2)$ is the typical formation length of the radiation.

Second, we estimate the region of parameters where we can disregard the influence of the bunch transverse size and of the vacuum pipe. It follows from simple geometrical consideration that a characteristic measure distinguishing these effects is the mean geometric value of the bunch length and the formation length of the radiation. Thus, we can roughly estimate the region where the considered effects can be disregarded:

$$\sigma_\perp \ll \sigma\gamma_z \ll b . \tag{39}$$

Here $\sigma_\perp$ and b are transverse dimensions of the bunch and of the vacuum chamber, respectively.

When the above limitations (38) and (39) are not satisfied, the transient and shielding effects begin to contribute to the process of radiative interaction. Namely, these effects suppress the process of the radiative interaction. So, we can conclude that the presented physical model of the radiative interaction gives the upper estimation of the effect. The obtained analytical results can be useful for quick estimation of the radiative interaction. Also, analytical results presented in the paper can serve as primary standards for testing numerical simulation codes.

In conclusion to this paper let us illustrate the practical application of the obtained results with two numerical examples. The first one is the 6 nm SASE FEL under construction at the TESLA Test Facility at DESY [4]. Parameters of the project are: the energy is 1 GeV, the rms bunch length 50 μm, the peak current 2.5 kA, the undulator period 2.73 cm, the undulator parameter K 1.27, and the undulator length 27 m. Substituting these values into formula (37), we find that the induced correlated energy spread, $\sigma_\gamma/\gamma = 4 \times 10^{-5}$, is much less than the FEL parameter $\rho \simeq 2 \times 10^{-3}$, and the radiative interaction does not influence the operation of the FEL amplifier. Besides, condition (38) and condition (39) for the shielding (the diameter of vacuum chamber is 1 cm) are not satisfied in the case under study. This will lead to the further significant reduction of the effect. It should be noticed that such a situation is typical for the projects of VUV and X-ray FELs.

The second example is the proposal by Duke University to construct a 1.4 μm SASE FEL using the PALADIN wiggler [14]. The energy is 200 MeV, the rms

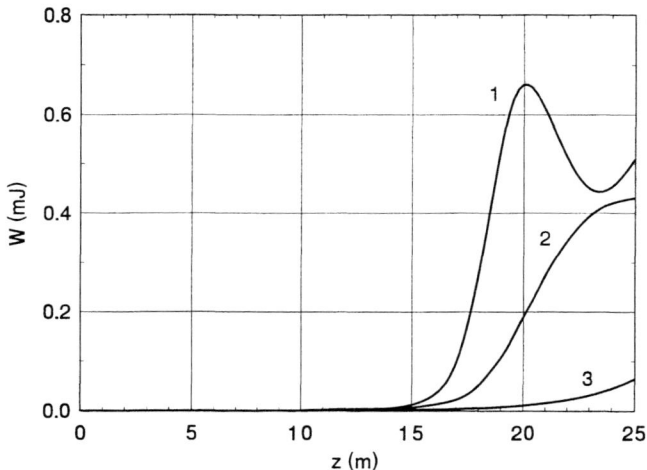

FIGURE 6. Energy in the radiation pulse versus the undulator length for a 1.4 μm SASE FEL project by Duke University. Curve 1 presents the results of simulations with steady-state code (the only diffraction effect is taken into account). Curve 2 presents the results of simulations taking into account diffraction and the slippage effect. Curve 3 is calculated taking into account all the effects (diffraction, slippage and the growth of the energy spread due to the radiative interaction).

bunch length 50 μm, the bunch radius 0.125 cm, the peak current 2.5 kA, the wiggler period 8 cm, the undulator parameter K equal to 3 and the wiggler length 15 – 25 m. Assuming the size of the vacuum chamber to be about 2 cm, we obtain that conditions (38) and (39) are fulfilled and our simple model provides correct estimation of the radiative interaction effect. Using formula (37), we obtain that the induced correlated energy spread will be $\sigma_\gamma/\gamma \simeq 8 \times 10^{-3}$ at the wiggler length of 15 m. Effective operation of the free electron laser requires $\sigma_\gamma/\gamma \ll \rho$, where ρ is the FEL parameter [15]. This condition is violated significantly in the case under study, since $\rho \simeq 5 \times 10^{-3}$. So, we can predict that the FEL process will be suppressed significantly by the radiative interaction effect. The actual power of the effect can be calculated only with an FEL simulation code taking into the growth of the energy spread due to the radiative interaction. We performed such simulations, gradually including important physical effects influencing the FEL amplifier operation (see Fig. 6). Curve 1 is calculated with three-dimensional, steady-state FEL simulation code FS2R [16], taking into account the only diffraction effects. In the case of the Duke SASE FEL, the bunch length is comparable with the co-operation length, so the slippage effect also leads to the degradation of the FEL process. The influence of the slippage effect is illustrated with curve 2 in Fig. 6.

271

Simulations have been performed with three-dimensional, time dependent simulation code FAST [17]. Finally, we include the growth of the energy spread in the electron beam due to the radiative interaction (see curve 3 in Fig. 6). It is seen that the latter effect is extremely strong and leads to significant degradation of the FEL performance.

ACKNOWLEDGMENTS

We wish to thank R. Brinkmann, C. Bohn, Ya. Derbenev, M. Dohlus, P. Emma, K. Flöttmann, D. Jaroszynski, J. Krzywinski, R. Li, T. Limberg, J. Rossbach and V. Shiltsev for useful discussions on the radiative interaction effects.

REFERENCES

1. *Conceptual Design of a 500 GeV e^+e^- Linear Collider with Integrated X-ray Laser Facility* (Editors Brinkmann R., Materlik G., Rossbach J., Wagner A.) **DESY 97-048, ECFA 1997-182**, Hamburg (1998).
2. *Zeroth-Order Design Report for the Next Linear Collider* **LBNL-PUB-5424, SLAC Report 474, UCRL-ID-124161** (1996).
3. Tatchyn R. et al., *Nucl. Instr. and Methods* **A375**, 274 (1996).
4. Rossbach J., *Nucl. Instr. and Methods* **A375**, 269 (1996).
5. Derbenev Ya.S., Rossbach J., Saldin E.L., and Shiltsev V.D., *DESY Print* TESLA-FEL 95-05, Hamburg (1995).
6. Dohlus M., and Limberg T., *Nucl. Instr. and Methods* **A393**, 494 (1997).
7. Iogansen L.V., and Rabinovich M.S., *Sov. Phys. JETP*, **37(10)**, 83 (1960).
8. Murphy J.B., Krinsky S., and Glukstern R.L., *Particle Accelerators* **57**, 9 (1997).
9. Saldin E.L., Schneidmiller E.A., and Yurkov M.V., *Nucl. Instr. and Methods* **A398**, 373 (1997).
10. Li R., Bohn C.L., and Bisognano J.J., *Shielded transient self-interaction of a bunch entering a circle from a straight path*, presented at SPIE's International Symposium on Optical Science, Engineering and Instrumentation, San-Diego, CA, July (1997).
11. Saldin E.L., Schneidmiller E.A., and Yurkov M.V., *Nucl. Instr. and Methods* **A417**, 158 (1998).
12. Abramowitz, and Stegun I.A., *Handbook of Mathematical Functions*, National Bureau of Standards (1964).
13. Alferov D.F., Bashmakov Yu.A., and Bessonov E.G., *Sov. Phys. Tech. Phys.* **18**, 1336 (1974).
14. O'Shea P.G., Neumann C.P., Madey J.M.J., and Freund H.P., *Nucl. Instr. and Methods* **A393**, 129 (1997).
15. Bonifacio R., Pellegrini C., and Narducci L., *Opt. Commun.* **50**, 373 (1984).
16. Saldin E.L., Schneidmiller E.A., and Yurkov M.V., *Phys. Rep.* **260**, 187 (1995).
17. Saldin E.L., Schneidmiller E.A., and Yurkov M.V., *FAST: Three dimensional, time-dependent FEL simulation code*, presented at the 20 th FEL Conference, Williamsburg (1998).

Possible Scheme of e-Beam Transverse Modulation and High Power CSR Production

A.A. Varfolomeev and T.V. Yarovoi

CRL, Russian Research Center "Kurchatov Institute"
123182 Moscow, Russia

Abstract. A scheme for e-beam transverse modulation is proposed and analyzed. It is shown that such a new type of modulation can be produced by laser fields in a special undulator. The transverse modulated electron beams can be used for coherent spontaneous radiation (CSR) production in another undulator at a frequency which is twice as high as that of the modulating laser field. It is shown by numerical calculations that CSR of some orders higher power than the primary laser beam power can be obtained.

INTRODUCTION

In our resent publication [1] we have shown that a new type of electron beam modulation (so called "transverse modulation," Fig. 1) is possible. This transverse electron velocity modulation can itself provide the coherence in the spontaneous radiation; i.e., the conventional e.b. spatial density modulation is not necessary for this. It was shown that such transverse modulated electron beams can be used for the CSR production in undulators. All characteristics of the radiation are similar to that of the coherent radiation produced by the conventional density modulated beams [2]. Requirements for e.b. quality are not more severe than those for the density modulated beams and, respectively, small transverse velocities amplitudes are required [1] what can make easier the problem of the adequate e.b. modulation. In this paper we describe one of the possible schemes for the transverse velocity modulation by an external laser field and present some calculation results of e-beam modulations.

TRANSVERSE MODULATION SCHEME

The principal scheme of the transverse bunching and CSR production device is shown in Fig. 2. It can be considered as a new version of the MOPA or muster oscillator – power amplifier scheme (see, for example, [3-5]). Transverse bunching is being produced in the modulator which is really an undulator U_1 inserted into a resonator of an external laser providing laser beam of frequency ω_0. Electrons are

CP468, *Nonlinear and Collective Phenomena in Beam Physics–1998 Workshop*,
edited by S. Chattopadhyay, M. Cornacchia, and C. Pellegrini
© 1999 The American Institute of Physics 1-56396-862-2/99/$15.00

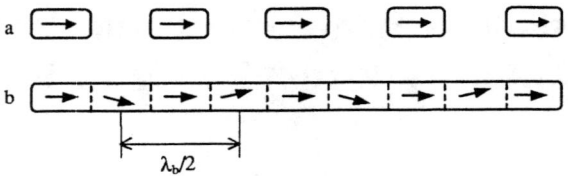

FIGURE 1. Schematic drawing of modulated beams. a)-Conventional density modulated beam. b)-Transverse velocity modulated beam.

passing through the modulator along its and the resonator axes. At the exit of the modulator the transversely modulated beam enters the generator undulator U_2 where it induces CSR.

The process of CSR production by the transversely modulated beam is analyzed in [1]. It was shown that the CSR of frequency ω_{s2} will be induced in the generator if the electron beam transverse velocities are modulated with some spatial period $\lambda_b = \dfrac{4\pi}{\omega_{s2}} \overline{v}_z$ where $\overline{v}_z$ is the mean electron velocity along the generator undulator axis. The required amplitude of this angular modulation is of the order $a_\perp \sim 1/\gamma\sqrt{N_{w2}}$, where γ is the electron energy in mc^2 units and N_{w2} is number of undulator periods in the undulator U_2. We are now going to show that the required modulation can principally be produced. Only one transverse modulation scheme will be considered, but some other schemes can also be suggested.

As was mentioned above, the modulator includes an undulator U_1 and an external laser source. The laser beam is passing through the undulator along its z-axis. The required undulator fields are stronger than those usually used in the conventional FEL schemes (with undulator field constant being 2-3 times higher). Another peculiarity is an unusual polarization of the laser beam in the undulator (Fig. 3). If the undulator provides electron oscillations in the xz plane, then the laser electric field E_s

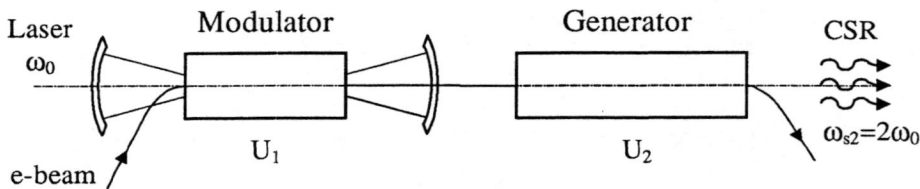

FIGURE 2. New MOPA scheme based on transverse velocity modulation. U_1 – undulator of the Modulator with normalized field strength K_1 and period λ_{w1}, U_2 – undulator of the Generator with normalized field strength K_2 and period λ_{w2}. Connection between parameters is given by $\lambda_{w1}(1+K_1^2/2)=4\cdot\lambda_{w2}(1+K_2^2/2)$, where ω_0 is frequency of the external laser field, $\omega_{s2}=2\omega_0$ is frequency of the CSR produced by the Generator.

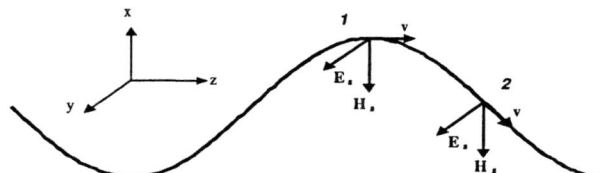

FIGURE 3. Laser magnetic field H_s deflects electrons in dependence of its z position (compare point 1 and point 2). Laser electric field E_s strength does not work on average because of slippage and does not change the electron energy.

should be directed along the y axis. In the first approximation the electric field remains orthogonal to the electron velocities at all time in the undulator. The magnetic field vector H_s, laying in the undulator oscillation plane, is directed at some angle with respect to the electron velocity. This angle, and therefore magnetic field strength acting on the electron, depend on the electron position z and relative electron phase in the laser wave field.

The equations of electron motion in the above undulator and the plane wave fields can be written in the form

$$\frac{dv_x}{dt}\gamma + v_x v_y \frac{eE_{sy}(z,t)}{mc^2} = \frac{e}{mc}v_z \cdot H_{w1y}(z), \tag{1}$$

$$\frac{dv_y}{dt}\gamma + v_y^2 \frac{eE_{sy}(z,t)}{mc^2} = \frac{eE_{sy}(z,t)}{m} - \frac{e}{mc}v_z \cdot H_{sx}(z,t), \tag{2}$$

$$\frac{dv_z}{dt}\gamma + v_z v_y \frac{eE_{sy}(z,t)}{mc^2} = \frac{e}{mc}v_y \cdot H_{sx}(z,t) + \frac{e}{mc}v_x \cdot H_{w1y}(z), \tag{3}$$

$$\frac{dv^2}{dt}\gamma^3 = v_y \frac{2eE_{sy}(z,t)}{m}. \tag{4}$$

Here $E_{sy}(z,t)=E_{s0}\cos(\omega_0\beta_z t - \omega_0 t + \varphi_0)$ is the e.m. wave electric field acting on the electron at the moment t, $\varphi_0 = \omega z_0/c$ the primary electron phase, $H_{sx}(z,t)=H_{s0}\cos(\omega_0\beta_z t - \omega_0 t + \varphi_0)$ the respective magnetic field, $v_x(z,t)$, $v_y(z,t)$, $v_z(z,t)$ the electron velocity components, γ - electron energy in mc^2 units. $H_{w1y}(z)=H_{w10}\cos(2\pi z/\lambda_{w1})$ is the undulator magnetic field. We will assume that the laser frequency ω_0 and the period of the undulator field λ_{w1} satisfy the equality

$$\omega_0 = (4\pi c/\lambda_{w1}) \cdot 2\gamma^2/(1+K_1^2/2), \tag{5}$$

where the normalized undulator field strength is given by

$$K_1 = \frac{eH_{w10}\lambda_{w1}}{2\pi mc^2}. \tag{6}$$

For usually obtained fields the following inequality is valid:

$$E_{s0}/\gamma^2 \ll H_{w10}. \tag{7}$$

It can be shown from the equation system (1-4) consideration that in the case (7) the undulator trajectory is defined mainly by the undulator fields. The condition (5) for this case means that the slippage of the electron of energy γ is equal to the wavelength λ_0 when the electron passes the distance $\lambda_{w1}/2$.

Let as consider eq. (4) for more detailed analysis. Because of slippage the function $E_{sy}(z,t)$ changes sign every time when the electron crosses the distance $\lambda_{w1}/2$. So the right side of eq. (4) alternates signs periodically. One can estimate from (2) that the transverse velocity provided by the laser field is relatively small ($\sim 1/\gamma$). As a result of these two circumstances the right side of (4) averaged over path length $\gg \lambda_{w1}$ is negligible $\langle v_y E_{sy}(z,t)\rangle = 0$ and we have

$$\frac{dv^2}{dt}\gamma^3 \cong 0. \tag{8}$$

Thus the absolute velocity value is not changing due to the electric field. All other fields are magnetic ones so they also change only the direction of electron velocity but not the energy. Therefore no transfer of energy between laser and electron beams takes place and absolute velocity of the electron remains constant.

With both (7) and (8) conditions being valid, the well known oscillating trajectory defined by the undulator field follows from equations (1) and (3), with disregarding the contribution due to the presence of the e.m. wave.

$$v_x = -\frac{K_1 c}{\gamma}\cos\left(\frac{2\pi}{\lambda_w}\overline{v}_z t\right) + const, \tag{9}$$

$$v_z = \overline{v}_z - \frac{K_1^2 c^2}{4\gamma^2 \overline{v}_z}\cos\left(\frac{4\pi}{\lambda_{w1}}\overline{v}_z t\right), \tag{10}$$

where $\overline{v}_z = v\left(1 - \frac{K_1^2 c^2}{4v^2\gamma^2}\right)$ is the averaged electron velocity in the undulator.

Now we can analyze the eq. (2) using the result (10):

$$\frac{dv_y}{dt}\gamma + v_y^2\frac{eE_{s0}}{mc^2}\cdot\cos\left[\omega_0\left(1-\frac{\overline{v}_z}{c}\right) + \frac{\omega_0}{c}z_0\right] = \frac{eE_{s0}}{m}\left(1-\frac{\overline{v}_z}{c}\right)\cdot\cos\left[\omega_0\left(1-\frac{\overline{v}_z}{c}\right) + \frac{\omega_0}{c}z_0\right] +$$

$$+\frac{eH_{s0}}{m}\cdot\cos\left[\omega_0\left(1-\frac{\overline{v}_z}{c}\right) + \frac{\omega_0}{c}z_0\right]\cdot\frac{K_1^2 c}{4\gamma^2 v}\cdot\cos\frac{2\pi}{\lambda_{w1}}\overline{v}_z t. \tag{11}$$

Supposing that velocity v_y is slowly varying at small distances comparable with λ_{w1}, we can again use the averaging approach by integrating both sides of (11) within distance Δz providing the slippage λ_0. It can be shown that this procedure using (5) transforms eq. (11) into the equation describing the evolution of the averaged velocity v_y as a function of the averaged strength:

$$\frac{dv_y}{dt}\gamma = \frac{eH_{s0}}{m} \cdot \frac{K_1^2 c}{8v\gamma^2} \cdot \cos\left(\frac{\omega_0}{c}z_0\right). \tag{12}$$

It is seen that the sign of the right side depends on the primary electron position z_0 (primary phase) and alternates periodically in space with the period $\lambda_0 = 2\pi c/\omega_0$. The averaged deflecting strength acting on an electron with given z_0 is nearly constant along the whole electron path length in the undulator. So the total vertical velocity induced by the modulator is given by

$$\frac{v_y}{c} = a_\perp \cdot \cos\left(\frac{2\pi}{\lambda_0}z_0\right); \quad a_\perp = \pi\frac{K_0 K_1^2 N_{w1}}{4\gamma^3} \cdot \frac{c^2}{vv_z}. \tag{13}$$

Here $K_0 = \frac{eH_{s0}\lambda_{w1}}{2\pi mc^2}$ is the normalized magnetic field strength of the e.m. wave. The result (13), giving the modulation amplitude $a_\perp$, describes the transverse bunching obtained. We see that electrons have been deflected at angles (13) but their absolute velocities (energies) are not changed. The spatial period of the transverse velocity modulation is equal to the wavelength λ_0 of the primary laser field used for the bunching.

RADIATION OF TRANSVERSELY MODULATED ELECTRON BEAM

Now we will describe briefly the process in the generator and present some results which will be used in the simulations. The undulator U_2 is supposed to be a routine planar one, providing e.b. oscillations in the xz plane with period

$$\lambda_{w2} = \lambda_{w1}(1 + K_1^2/2)/4(1 + K_2^2/2), \tag{14}$$

where $K_2 = \frac{eH_{w20}\lambda_{w2}}{2\pi mc^2}$ is the normalized undulator strength of the U_2 undulator. These oscillations induce spontaneous radiation with the electric field directed along the x-axis. The maximum of CSE intensity in the forward direction takes place at the radiation frequency $\omega_{s2} = 2\omega_0$. Frequency spread of CSE at the fixed observation angle is given by

277

$$\Delta\omega_{s2}/\omega_{s2}=1/(N_{w2}+n_b-1),\qquad(15)$$

where $n_b=2l_b/\lambda_0$ is the number of transverse microbunches in the electron bunch of length l_0. The total radiation intensity is proportional to $b^2(\omega_0)N_e^2$, where N_e is the total number of electrons in the bunch length l_b and $b(\omega_0)$ is the bunching Fourier component [1]. The angular spread of the electron beam should be limited for providing intensive CSE

$$\alpha_e<(1+K_{w2}^2/2)/(N_{w2}+n_b-1)^{1/2}\gamma.\qquad(16)$$

Peak intensity of radiation at zero observation angle per unit frequency range and unit solid angle is given by

$$\frac{dI}{d\omega_{s2}\,d\Omega}=\frac{e^2}{16c}\frac{\omega_{s2}^2K_2^2}{\omega_{w2}^2\gamma^2}\left[J_0(\xi_1)-J_1(\xi_1)\right]N_{w2}^2N_e\left[1+\frac{\pi^2N_{w2}^2N_e\gamma^4a_\perp^4}{16\cdot(1+K_{w2}^2/2+\gamma^2v_y^2/c^2)^2}\right],\qquad(17)$$

where J_0, J_1 – Bessel functions; $\xi_1=\dfrac{K_2^2}{8\gamma^2}\cdot\dfrac{\omega_{s2}}{\omega_{w2}}$; $\omega_{w2}=\dfrac{2\pi\bar{v}_z}{\lambda_{w2}}$.

The term proportional to $N_e^2a_\perp^4$ describes the CSR contribution. All the above results on CSR were obtained [1] for quasi monochromatic and well collimated beams with small angular divergence $\Delta\gamma/\gamma\ll1/N_{w2}$; $d_e\ll(\lambda_sN_{w2}\lambda_{w2})^{1/2}$; $v_{\perp0}/c\ll1/\gamma N_{w2}^{1/2}$. For this case rather small transverse modulation amplitudes $a_\perp$ are required, $a_\perp\leq1/\gamma N_{w2}^{1/2}$. The β-oscillation wavelength should be long enough not to decrease the angular modulation $\Lambda_\beta>4N_{w2}\lambda_{w2}$.

NUMERICAL SIMULATION RESULTS

Using both of the above results for the transverse bunching and radiation of the transversely bunched beam, the entire MOPA system can be considered. For an analysis of possible effects we have made simulations using the scheme parameters given in Table 1. The obtained results for induced radiation are given in Table 2. It is seen that rather high radiation power can be obtained if the quality of electron beams is high enough. More than one order higher power than the primary laser power is produced at twice shorter the wavelength. These results show that the described transverse MOPA scheme can be used for harmonic generation with high efficiency.

278

TABLE 1. Parameters used for numerical calculations.

Undulator U₁		Primary Laser	
Period	λ_{w1}=5.0 cm	Wave length	λ_0=10μm
Number of periods	N_{w1}=20	Intensity	I=5·10⁶ W/cm²
Field strength	$K_1 = 2\sqrt{2}$	Peak power	P_0=500 W
Undulator U₂		Electron Beam	
Period	λ_{w2}=1.923 cm	Number of electrons	N_e=10¹⁰ /bunch
Number of periods	N_{w2}=100	Energy of electrons	E_e=40 MeV
Field strength	$K_2 = 1.5\sqrt{2}$	Bunch length	l_b=0.15 cm
		Repetition rate	r.p.=1 bunch/ns
		Average current	I_{av}=0.6 A

TABLE 2. Radiation induced in the undulator U₂ by transversely bunched beam within observation solid angle $1/\gamma^2 N_{w2}$.

Resonance wavelength	λ_{s2}=5 μm
Coherence factor	F_{coh}=3.6·10⁸
Relative frequency range	$\Delta\omega_{s2}/\omega_{s2}$=2.5·10⁻³
Observation solid angle	$\Delta\Omega$=1.6·10⁻⁴
Radiation energy, radiated by one bunch within $\Delta\omega_{s2}$ and $\Delta\Omega$	ΔI_{s2}=1.2·10⁻⁵ J/bunch
Peak radiation power	P_{s2}=8 MW

LONGITUDINAL DENSITY MODULATION INDUCED BY THE TRANSVERSE VELOCITY MODULATION

Along with modulation of the transverse velocities, the longitudinal velocities are also modulating. The velocity v_z as a function the primary electron position z_0 can be found from equations (1-4). From the definition of the velocity components follows $v_z^2 = v^2 - v_x^2 - v_y^2$. In the approach we are using, the v_x component is defined by the undulator field. So one can find by averaging with respect to z over undulator period λ_{w2} that $\langle v^2 \rangle - \langle v_x^2 \rangle = \langle v_{z0}^2 \rangle$ where $\sqrt{\langle v_{z0}^2 \rangle} \equiv \overline{v}_{z0}$ is the r.m.s. value of the longitudinal velocity of the undulating electron given by (10). The velocity $\langle v_y^2 \rangle^{1/2} = v_y$ is given by (13). Finally we get for v_z, defined as the step averaged $\langle v_z^2 \rangle^{1/2}$,

$$v_z = \overline{v}_z - v \frac{a_\perp^2}{2} \cos^2 \left(\frac{2\pi}{\lambda_0} z_0 \right), \qquad (18)$$

where $\overline{v}_z = v \left(1 - \frac{K_2^2 c^2}{4\gamma^2 v^2} \right)$ and $a_\perp$ is the angular bunching amplitude (13).

279

An electron with the primary position z_0 after time of flight t will be positioned at the point

$$z = z_0 + \overline{v}_z t - vt \cdot \frac{a_\perp^2}{2}\cos^2\left(\frac{2\pi}{\lambda_0}z_0\right).$$ (19)

Here $vt \cong \overline{v}_z t$ is a drift path length L_{dr} which can be limited by the β-oscillation wavelength.

It can be shown that the primary density distribution of electrons given by dN_e/dz_0 transforms into the final distribution given by

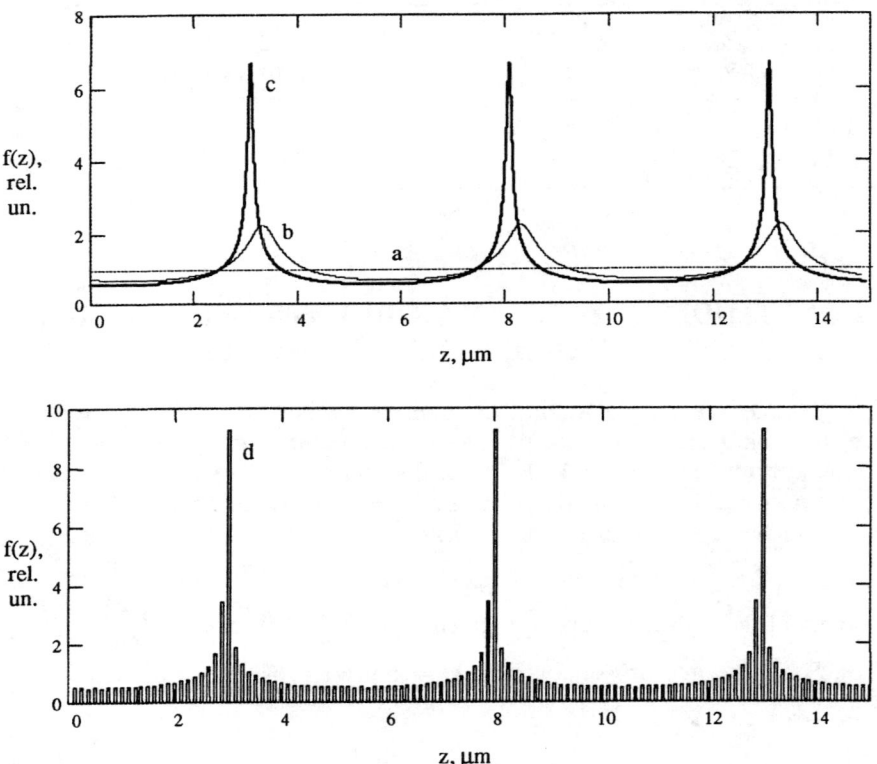

FIGURE 4. Density modulation induced by the transverse modulation. Simulation results obtained by using (19); (20) with the parameters $a_\perp=0.0014$, $\lambda_0=10\mu m$ and drift space distance L_{dr} giving the oscillating amplitudes $B_\perp=\pi L_{dr}a_\perp^2/\lambda_0$ respectively: (a) - $B_\perp=0$; (b) - $B_\perp=0.55$; (c) - $B_\perp=0.85$; (d) - $B_\perp=1.00$; The distribution (d) is obtained numerically by using (19).

$$f(z) = (dN_e / dz_0) \cdot \left[\frac{vt \cdot a_\perp^2 \pi}{\lambda_0} \cdot \sin\left(\frac{4\pi}{\lambda_0} z_0 \right) + 1 \right]^{-1}, \qquad (20)$$

where z_0 is defined by (19) for a given z. So (19) and (20) can be respected as parametric definition of the distribution function $f(z)$. It is seen from the structure of this function that even for the uniform primary distribution dN_e/dz_0, the final distribution is not uniform but periodical, with period $\lambda_0/2$. For illustration of this effect some simulation results are presented in Fig. 4. It is evident that the density modulation takes place and furthermore that this density distribution is not a single harmonic one. Rather not small higher harmonic contribution is evident. If the quality of electron beam is high enough to conserve this modulation, then this type of density modulation could be efficient for higher harmonic production.

CONCLUSION

Transverse velocity e.b. modulation can be produced by a strong undulator field and a laser beam with the magnetic field polarized along the undulator magnetic field. The transversely bunched beams induce both the routine spontaneous and the coherent spontaneous radiation and can be used for efficient production of the CSR in undulators or prebunched FEL devices. The respective requirements for the e.b. quality are not much more severe than that for the case of density modulated beams. Small transverse amplitudes are required ($<1/\gamma N_w^{1/2}$). The transverse velocity modulation transforms into density modulation. The transverse modulation has some advantages. It can be more easily produced, makes the space charge effect less destructive and gives opportunity for radiation frequency upconversion. So it can provide good opportunities for high intensity superradiant sources.

REFERENCES

1. Varfolomeev, A.A., and Yarovoi, T.V., "Transverse velocity modulated e-beam propagating through an undulator as a source of coherent spontaneous radiation," presented at the 20-th International FEL Conference, Williamsburg, VA, Aug. 16-21, 1998.
2. Alferov, D.F., Bashmakov, Yu.A., and Bessonov, E.G., *J. of Tech. Phys.* **48**, 1592; 1598, (1978), in Russian.
3. Coisson, R., and De Martini, F., in *Phys. of Quant. Electr.*, ed. by Jacobs, S.F. and Sargent, M., Addison–Wesley, Reading, Mass. 1982, vol. 9, ch. 42.
4. Baccaro, S., De Martini, F., and Ghigo, A., *Opt. Lett.* **7**, 174, (1982).
5. Varfolomeev, A.A., *Sov. Phys. JETP* **58** N1, 24, (1983).

PHYSICS OF, AND PHYSICS WITH,

HIGH ENERGY DENSITY BEAMS

Density Effects on Quantum Fluctuation of Radiation in Synchrotrons

S.V. Koutin and A.N. Lebedev

P.N. Lebedev Physical Institute of RAS. 117924, Leninsky prosp. 53, Moscow, Russia

Abstract. We discuss an influence of short-time electron correlation on quantum excitation of particle oscillations by synchrotron radiation. The effect should decrease the "natural" beam emittance and can be observable at reasonable intensities at 1-2 GeV energy.

INTRODUCTION

The excitation of synchrotron and betatron oscillations in the cyclic electron accelerators and storage rings due to the quantum nature of radiation significantly affects their characteristics and operating mode. Together with radiation cooling, which is a pure classical effect, this effect defines the equilibrium emittance of the beam. The emittance has a macroscopic value, although it is proportional to the Compton wavelength of the electron $\dot{E} = \hbar/mc$.

The effect of the quantum excitation of oscillations, similar to the diffusion increase of beam phase volume, has been calculated theoretically both in quantum and semi-classical theory and was confirmed in experiments. On the one hand, the semi-classical approach seems to be natural since the quantum numbers that correspond to macroscopic betatron and synchrotron oscillations are enormously high. On the other hand, it is based on a "feasible", semi-intuitive statistical point of view, which has to be justified by experiments and quantum theory inside the bounds of the last. (The existing quantum theory uses the single particle wave functions approach.) The semi-classical approach is based on single-particle theory as well, i.e. it ignores all correlation effects, connected to the large number of radiating particles. Since the decrease of quantum excitation is of large practical importance, we will try to study this approach and apply it to the case of a high-density beam, when the correlation effects cannot be neglected. The quantum solution of this problem seems to be quite complicated but even the semi-classical approach leads us to qualitatively new effects.

Main assumptions of single particle theory are:

• Quantum excitations are defined mostly by quanta with energy $\hbar\omega_0\gamma^3$, which is in the higher range of synchrotron radiation spectra. Here γ is the electron Lorentz factor

CP468, Nonlinear and Collective Phenomena in Beam Physics–1998 Workshop,
edited by S. Chattopadhyay, M. Cornacchia, and C. Pellegrini
© 1999 The American Institute of Physics 1-56396-862-2/99/$15.00

and $\omega_0 = c/R$ its angular frequency. The characteristic time of quantum radiation transition (radiation of one quantum) is $\tau_{rad} \cup 2/\omega_0 \gamma$. During this time the electron passes a distance approximately given by $R/\gamma >> \lambda$, where $\lambda \cup R/\gamma^3$ is the wavelength of the characteristic quanta. Time τ_{rad} is significantly less than any characteristic times of electron motion are. The recoil looks like a one-time kick, and sequence of such kicks is the white noise, that leads to diffusion of particles in the phase space.

• The average number of quanta radiated during the time τ_{rad} is of the order of the fine structure constant $\alpha \cup 1/137 << 1$. The minuteness of this number means that consecutive quanta are emitted statistically independently.

• The phases of the fields radiated by the electron in the given direction in two consequent turns are random. This means that the spectra in the short wave range in reality will be continuous, like the one of random process.

• The radiation of each quantum corresponds to loss of energy (and momentum) for one electron. For this to be perfectly true the distance between electrons should be significantly larger than the wavelength, i.e. the density should be rather low. Otherwise, the quantum is radiated by a system of few particles, which accepts the recoil momentum of the quantum.

The violation of the last assumption means the coherence of synchrotron radiation and lead to changes in its spectra and total intensity. However, this spectra is coherent only if the coherent position of two or more electrons is kept during several turns.

Let us consider a simple one-dimensional chain of N electrons, that are dispensed by the normal dispersion law on an average distance Δ/N from each other with uncertainty of position δ, completely random on consequent turns. In a far-field zone, the field from such a system is a pack of randomly distributed similar short pulses with a length of $\cup 1/\omega_0 \gamma^3$, which repeats (with other realization of distribution) after the time of $2\pi/\omega_0$. Simple calculation gives the following spectral intensity of synchrotron radiation of such a system:

$$\frac{W(\omega)}{W_0(\omega)} = N + [\frac{\sin^2(\Delta\omega/2c)}{\sin^2(\Delta\omega/2Nc)} - N]\exp(-\frac{\omega^2\delta^2}{2c^2}), \tag{1}$$

where $W_0(0)$ is the spectral radiation intensity of one electron (See Fig. 1). The first term corresponds to fully incoherent radiation, and the second describes coherent effects. The term in square bracket gives interference modulation of spectra. However, in the most significant high-frequency range of spectra, wavelengths are smaller than the position uncertainty δ and this modulation decreases exponentially; that means that radiation is completely incoherent. Note that the number of particles in the radiation zone of length λ at certain incidental moments can be larger than one.

The last can significantly change the recoil momentum of each particular electron during radiation of one quantum. Really, n particles, which find themselves simultaneously in near-field zone (radiation zone) and keep their relative position at least during the time τ_{rad}, behave like an organic whole, and share recoil momentum and energy loss.

286

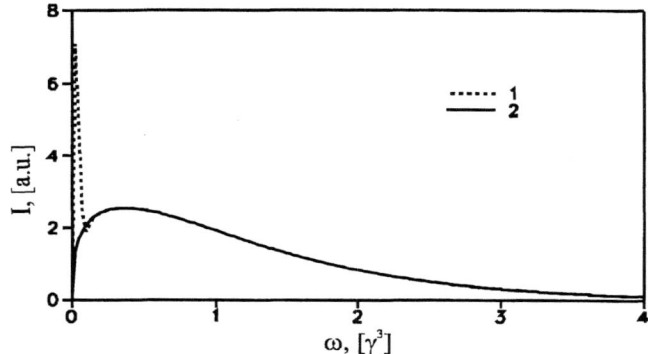

FIGURE 1. 1 - Spectrum intensity per one electron of an one-dimensional chain of electrons with fixed average separation and uncertainty of position; 2 - spectra of single electron.

The fact that the number of quanta is increased proportionally to n^2 does not play a significant role, since this happens, as shown above, due to the soft coherent part of spectra, when the number of high-energy incoherent quanta is proportional to n only. For assurance we consider n to be less than 137, so the probability of radiation of two coherent consequent high-energy quanta in the same direction is negligible.

EXCITATION OF OSCILLATIONS BY RADIATION FLUCTUATION

Let us, for example, consider linear synchrotron oscillation of value u — deviation of energy from its equilibrium value E_s [2].

During the radiation du/dt does not change and u jumps down on the value ε — quantum energy, that gives the change of amplitude:

$$\Delta A = -\frac{u}{A}\varepsilon + \frac{\varepsilon^2}{2A}(1 - \frac{u^2}{A^2}); \; (\Delta A)^2 = \frac{u^2}{A^2}\varepsilon^2 . \quad (2)$$

Let us average this expression with probability of radiation of quantum $\hbar\omega$ per unit time $P(u,\omega)$, which in the quasi-classical approximation is equal to:

$$P(u,\omega) = \frac{1}{\hbar\omega}[W_{coh}^n(\omega) + W_{incoh}^n(\omega)] . \quad (3)$$

The coherent part of the spectral intensity $W_{coh}^n(\omega)$ should not depend on single particle energy, and the incoherent part is equal to:

287

$$W_{incoh}^{n} = np_n W_0(u, \omega) = (nW_s(\omega) + n \, ?u \frac{fW_s(\omega)}{fE_s}) \, ?p_n, \qquad (4)$$

where n is the number of electrons participating in the act of emission, p_n — probability of such realization. By averaging we get:

$$V = < \Delta A \, ?P(u, \omega) >= -\frac{A}{2} \frac{fW_s}{fE_s} \frac{\varepsilon n}{\hbar \omega} + \frac{1}{4A} \frac{\varepsilon^2}{\hbar \omega}(W_{coh}(\omega) + nW_s(\omega)); \qquad (5)$$

$$D = \frac{1}{2} < (\Delta A)^2 \, ?P(u, \omega) >= \frac{1}{4} \frac{\varepsilon^2}{\hbar \omega}(W_{coh}(\omega) + nW_s(\omega)). \qquad (6)$$

For an estimation we can put $\varepsilon = \hbar \omega / n$ and neglect the terms with $W_{coh}(\omega)$ which are significant only at low frequencies, where ε^2 is very small. Then by averaging over possible realizations (number n) and spectra, for the diffusion coefficients we get:

$$< V >= -\frac{A}{2} \Gamma_s + \frac{< \varepsilon >}{4A} W_s; \quad < D >= \frac{< \varepsilon >}{4} W_s, \qquad (7)$$

where W_s is the total intensity of synchrotron radiation of a single particle, $< \varepsilon > \cup < \hbar \omega / n >$ is the average loss of energy per one particle, and $\Gamma_s = fW_s / fE_s$ is a well-known constant of radiation dumping of synchrotron oscillation [3].

Calculated values of diffusion coefficients give the evolution in time of the distribution function by amplitude F(A,t) defined by Fokker-Plank-like equation:

$$\frac{fF}{ft} + \frac{1}{A} \frac{f}{fA} A(< V > - < D > \frac{f}{fA})F = 0. \qquad (8)$$

The stabilized distribution is: $F_{st} = a \, ?\exp(-a^2 / 2)$; $a = \sqrt{2\Gamma_s / W_s \varepsilon}$ and gives the mean-square stabilized amplitude:

$$A_{st}^2 = W_s \varepsilon / \Gamma_s. \qquad (9)$$

CONGREGATE RADIATION ZONE

By congregate radiation zone, we will understand space volume, so that when several electrons appear in this zone for a short time τ_{rad} they behave as one congregate radiator that accepts the total recoil momentum. One can get an idea about size and configuration of this zone, considering near zone fields. Let us consider one electron, moving along the circumference; the observation point is on fixed angular distance μ. (For simplicity we assume the motion to be plane). From the general equation for retarded fields, we found force acting at point μ [1].

$$F_\tau(\mu) = \frac{e^2}{R^2} \frac{\sin\mu' - 2\beta^2\cos\mu'\sin\frac{\mu'}{2} + \gamma^2\beta^2(\sin\mu' - 2\beta^2\sin\frac{\mu'}{2})(\cos\mu' - 1)}{\gamma^2(2\sin\frac{\mu'}{2} - \beta\sin\mu')^3} \quad (10)$$

where

$$\mu' - \mu = 2\left|\sin\frac{\mu'}{2}\right|. \quad (11)$$

At small μ' $\mu \cup \mu' - \beta|\mu'| + \mu'^3/24$, and

$$\frac{R^2}{e^2}F_\tau = \frac{|\mu|}{\gamma^2\mu^3} + \begin{cases} -4\gamma^4/3 & \text{for} \quad \mu > 0 \\ 0 & \text{for} \quad \mu < 0 \end{cases}. \quad (12)$$

Anti-symmetrical and diverging at $\mu \bullet 0$ part of this force we identify as Coulomb interaction of two electrons. It is not relevant to considered effect, at least since it does not change the average momentum of interacting particles. We will come back to this question later.

The radiation part of the force $F_\tau(\mu)$ is strongly asymmetric and in a relativistic case it is practically equal to zero behind the radiating particle. At the point where the particle is situated ($\mu \bullet 0$) the self-interaction force is equal to $4\gamma^4 e^2/3R^2$, the well known value of the recoil momentum per unit time transferred by radiation. The radiation part of force $F_\tau(\mu)$ vs. angle μ is shown in Fig. 2. One can see that this force became practically equal to zero when μ is about $(1.5\text{-}2)\cdot\gamma^3$. Therefore, we might conclude that the electrons radiate independently, when distance between them is larger than $2\gamma^3$.

When the pair radiates collectively, back and front electrons share recoil momentum in some proportion. However, since to the moment of the next photon radiation ($\cup 137 R/c\gamma$) the given pair will decay with large probability, the "back" electron might become a "front" one in the other pair; i.e. in average the electron in pair radiation will lose half of the photon momentum. If there are n electrons simultaneously in the zone of congregate radiation, each of them will get one nth of photon momentum. The transverse dimension of the congregate radiation zone can be obtained from the similar but more complicated calculations. However, it is quite easy to see that when radiation concentrated in a small angle of approximately γ^{-1} one can consider the wave as a plane one. Hence the transverse dimensions are γ times lager than the longitudinal one and have an order of γ^2 in units of circumference radius. Thus, taking into consideration the qualitative character of all relations above we will consider the volume of the congregated radiation zone to be equal to $R^3\gamma^7$. It is interesting that the zone of

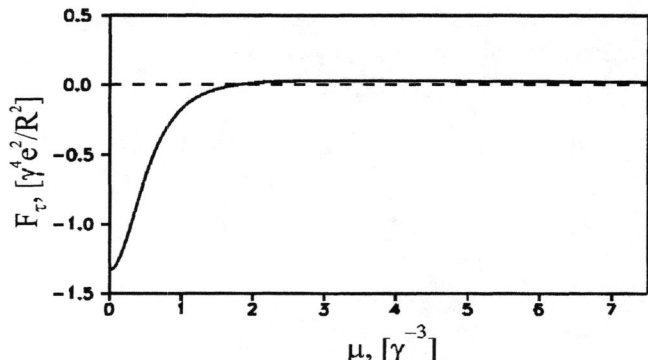

FIGURE 2. Tangential electromagnetic force acting from back electron to forward one without Coulomb part vs. angular distance between electrons.

congregate radiation defined this way does not depends on radiation frequency and is defined only by the upper bound of the radiation spectrum.

The average number of particles n, which are simultaneously in the near-field zone at the moment of quantum emission, depends on their distribution over the bunch. For simplicity, we can use the Poisson distribution with:

$$p_n = \exp(\bar{n})\bar{n}^n / n!; \ < n^{-1} > = (1 - \exp(\bar{n})) / \bar{n} . \tag{13}$$

Here $\bar{n} = \nu N$ (ν is the ratio of this zone volume to the volume of the whole bunch, N is number of particles in bunch). The announced effect consists of a decrease of value n^{-1} (averaged over spectra).

EQUILIBRIUM EMITTANCE

We will estimate the effect in a case when bunch dimensions are defined only by quantum fluctuation. Although the existing accelerators do not satisfy this condition, there are certain experiments on direct measurement of quantum limits. The equilibrium mean-square dimensions of the bunch can be presented as follows:

$$\begin{aligned} A_x^2 &= C_x \alpha^2 R\Lambda\gamma^2 / n \\ A_z^2 &= C_z R\Lambda / Q^2 n \\ A_\tau^2 &= C_\tau 137\alpha \ ?ctg\phi_s / q\gamma n \end{aligned} \qquad , \tag{14}$$

where Λ is the Compton wavelength, α the momentum compaction factor, q the harmonic number, Q the betatron tune and ϕ_s the equilibrium phase. Numeric coefficients

290

C are of the order of unity and are defined by structural functions and distribution of radiation damping decrements [3]. Since our calculations are qualitative only, one need not define such coefficients with more precision.

In accordance with the consideration above for an average number of electrons in the congregate radiation zone, we get:

$$n = 1 + \begin{array}{ll} R^3 N n^{3/2}/\tilde{a}^7 A_x A_z A_\hat{o} & \text{for} \quad A_z^2 > nR^2/\tilde{a}^4 \\ R^3 N n/\tilde{a}^5 A_x A_\hat{o} & \text{for} \quad A_z^2 < nR^2/\tilde{a}^4 \end{array} \quad . \tag{15}$$

The last condition appears because the vertical size of the bunch can be smaller than the congregate zone. For the other degrees of freedom this seems unrealistic (A_z has the lowest value due to the specific of vertical oscillations excitations by quanta coming away from the orbit plane). For simplicity we replace the above equation by a single one,

$$N^2 = \frac{A_x^2 A_\tau^2}{R^6}(A_z^2 + \frac{R^2 n}{\gamma^4})\frac{(n-1)^2}{n^3}\gamma^{14} \quad . \tag{16}$$

By substituting, we get a relation between the number of particles in the bunch and a value of n that characterizes compression of the bunch due to the density effect:

$$N^2 = N_0^2 \frac{(n-1)^2}{n^3}(n+n_0), \tag{17}$$

where

$$N_0^2 = 137 C_x C_\tau \frac{\Lambda \alpha^2 ctg\phi_s}{Rq}\gamma^{11};$$

$$n_0 = C_z \frac{\Lambda\gamma^4}{RQ^2}. \tag{18}$$

In absence of other perturbations, the density effect amplifies itself: the smaller the bunch size for a fixed number of particles is, the higher density and the lower quantum excitations are. Formally, it leads to radiation collapse at $N>N_0$ i.e. $n \blacklozenge \times$ (See Fig. 3), although at large n many of the assumptions made above are violated. Note the interesting hysteresis-like behavior of the model at $n_0>2$. It can provide additional possibilities for achieving higher bunch density by optimal choosing of function N(t) and γ(t). The threshold value of particle number N_0 for the energy of about 1GeV is practically obtainable and has an order of 10^{11}-10^{12}. Unfortunately, for the higher energy the number N_0 is unachievably high.

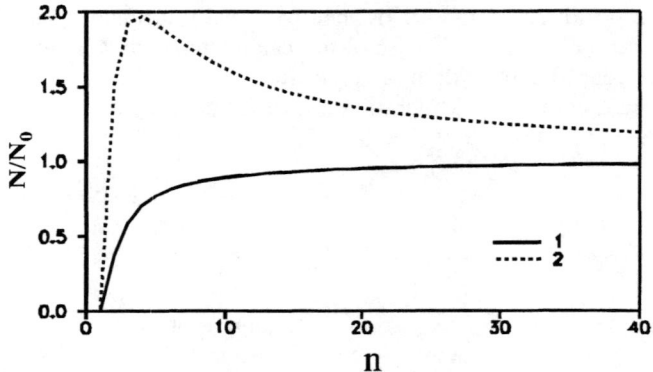

FIGURE 3. Normalized number of electrons in bunch vs. average number of electrons in congregate radiation zone. **1** — for $n_0=1$; **2** — for $n_0=10$

CONCLUSION

The principal possibility of decreasing a bunch size below limits, defined by quantum fluctuations, will open interesting perspectives for cooler rings, for producing super-short electron bunches and for synchrotron radiation sources (possibly coherent). Therefore, despite the qualitative character of our arguments, they give a basis for more detailed study of density effect: quantum theoretical and experimental. Even in semi-classical considerations, some moments need additional study. For example, the Coulomb field in the congregate zone is comparable or exceeds the radiation field that requires consideration of intra-beam scattering. (We are grateful to A.N. Skrinsky.) Concerning this question, we can note the somewhat different physical nature of these two effects, one of which is quantum one and another the purely classical, and depending on velocity distribution of particles. In addition, at $n \gg 1$ intra-beam scattering cannot be considered as a two-particle effect and needs a special consideration.

ACKNOWLEDGMENTS

The work has been supported by the Russian program "Physics of Microwaves."

REFERENCES

1. Landau, L.D., Lifshitz, E.M., *The Theory of Fields*. M, Science, 1988.
2. Sands M., *Phys. Rev.*, **97**, 470 (1955).
3. Kolomensky, A.A., Lebedev, A.N., *Theory of Cyclic Accelerators*. North-Holland, 1966.

Modeling of Saw Tooth Instability in Storage Rings

G. Dattoli*, L. Mezi*, M. Migliorati† and L. Palumbo†

*ENEA, Dipartimento Innovazione, Frascati, Rome, Italy
† Università di Roma LA SAPIENZA, Dip. di Energetica, and INFN - LNF

Abstract. Assuming the validity of the Boussard criterion for the determination of the microwave instability threshold, we derive two coupled non linear differential equations which describe the time evolution of the energy spread and of the instability growth rate under different conditions. The equations reproduce the characteristic features of the saw tooth instability and are in agreement with the results of a time domain simulation code.

The performance of the new generation of storage rings can be seriously limited by the microwave instability, an effect which causes an increase of energy spread and consequent bunch lengthening. Although it is commonly present in storage rings, it still remains not completely understood. One of its aspects is the so called saw tooth instability [1], which manifests itself as a periodic oscillations of relaxation type of the energy spread as shown in Fig. 1. To study the phenomenology of the saw tooth instability, we have developed a time domain simulation code which uses the single particle equations of motion turn by turn, and takes into account the self induced wake fields [2]. From the simulations with 3×10^5 macroparticles, we have observed that a pure inductive impedance produces a clear saw tooth behavior (Fig. 1), while a pure resistive impedance gives a very noise energy spread vs time, as shown in Fig. 2, without any apparent saw tooth instability.

The phase space distribution obtained with the simulation code in the saw tooth instability regime has been used to study the mechanisms responsible of the growth of the energy spread and of its damping. When the energy spread is close to its minimum (see Fig. 1), the phase space distribution shows a perturbation superimposed to the stationary distribution which starts to increase (Fig. 3). It produces a growth of the energy spread, the distribution becomes more chaotic (Fig. 4), and the local phase space density decreases. When it is sufficiently low, the Landau damping becomes able to counteract the instability, and the energy spread tends to become constant. Eventually the natural radiation damping effect reduces again the phase space distribution which becomes more uniform (Fig. 5) until coming back to the starting conditions.

CP468, *Nonlinear and Collective Phenomena in Beam Physics–1998 Workshop*,
edited by S. Chattopadhyay, M. Cornacchia, and C. Pellegrini
© 1999 The American Institute of Physics 1-56396-862-2/99/$15.00

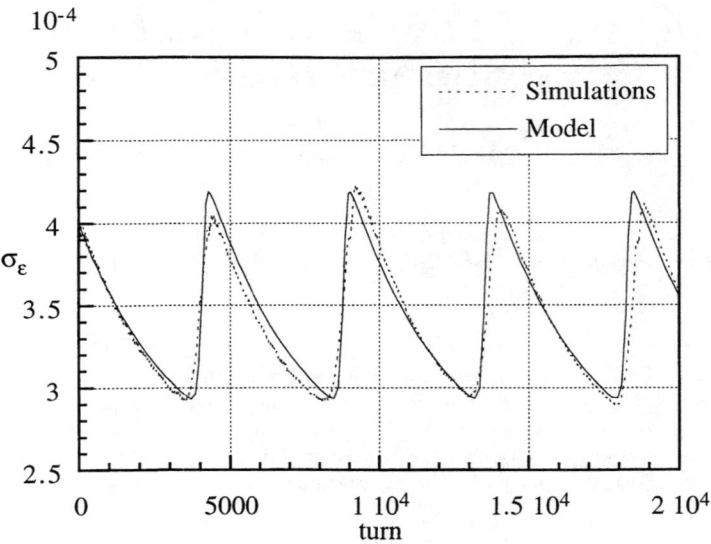

FIGURE 1. Energy spread vs number of turns. Comparison between simulations and coupled equations above microwave instability with pure inductive impedance.

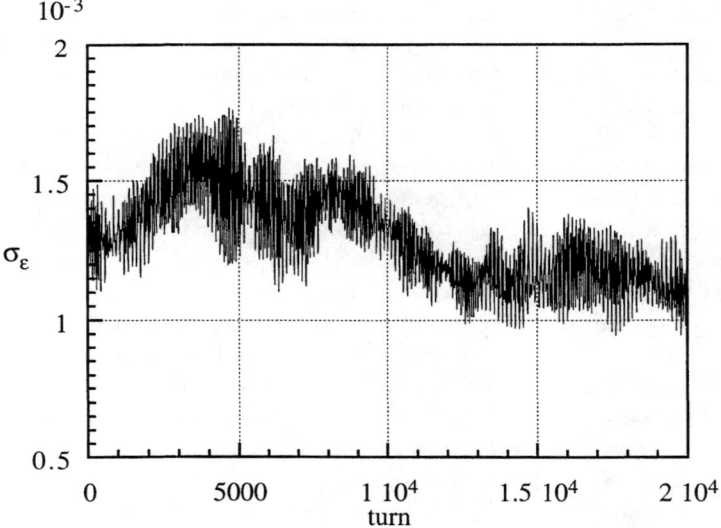

FIGURE 2. Energy spread vs number of turns for pure resistive impedance.

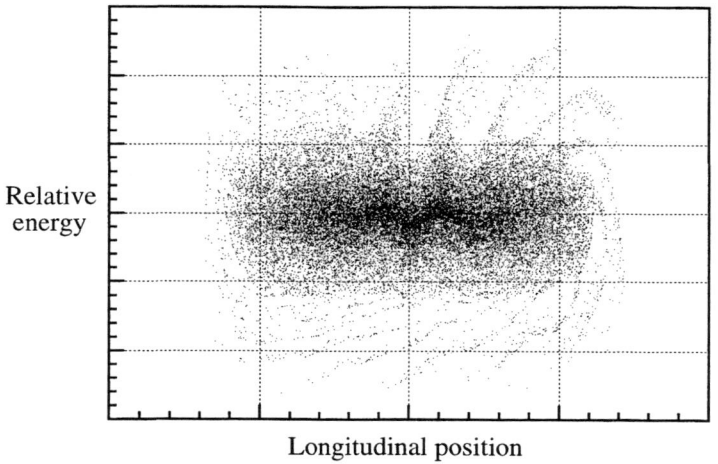

Relative
energy

Longitudinal position

FIGURE 3. Phase space distribution in the microwave instability regime.

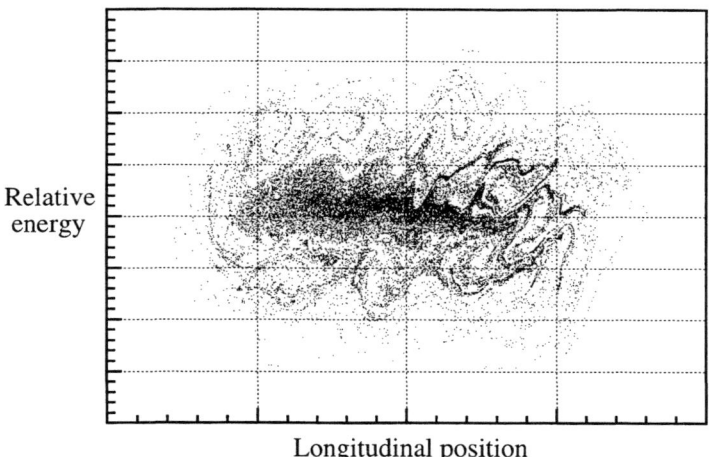

Relative
energy

Longitudinal position

FIGURE 4. Phase space distribution in the microwave instability regime.

295

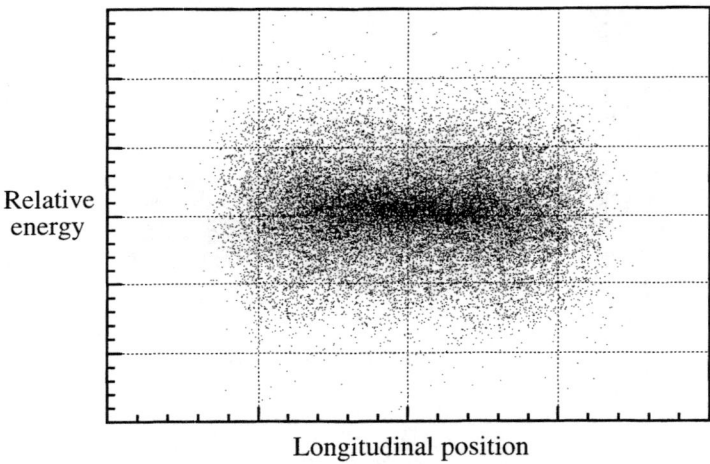

Relative energy

Longitudinal position

FIGURE 5. Phase space distribution in the microwave instability regime.

To model the saw tooth instability, we start by considering the coasting beam case, for which the theory predicts an instability caused by a competition between two effects: a growth of a perturbation produced by the wake fields, and a damping deriving from a spread in the oscillation frequencies (Landau damping). Under the simplifying assumption that the oscillation frequencies are distributed as a Lorentz spectrum [3] such a growth rate can be expressed as

$$\alpha_{CB} = \frac{n}{T_0} \sqrt{\frac{2\pi \alpha_c I_0 |Z/n|}{(E_0/e)}} - n\frac{2\pi}{T_0}\alpha_c\sigma_\varepsilon, \qquad (1)$$

where n is an harmonic of the revolution frequency, α_c the momentum compaction, T_0 the revolution period, I_0 the average beam current, Z/n the broad band impedance at the nth harmonic of the revolution frequency, E_0 the beam energy, e the electron charge, and σ_ε the energy spread.

An extension of Eq. (1) to the bunched beam case has been derived by Boussard [4], who investigated the threshold conditions for the microwave instability as an equilibrium between the two mentioned competing effects ($\alpha_{CB} = 0$). From Eq. (1), by substituting the average beam current with the bunch peak current, and considering a gaussian distribution, we obtain the important condition for the onset of the microwave instability

$$\frac{\sqrt{2\pi}I_0\nu_s |Z/n|}{(E_0/e)\,\alpha_c\sigma_\varepsilon} = 2\pi\alpha_c\sigma_\varepsilon^2, \qquad (2)$$

known as Boussard criterion. Here ν_s is the synchrotron tune.

We suppose now that the energy spread σ_ε is a quadratic combination of the natural $(\sigma_{\varepsilon n})$ and instability induced (σ_i) parts. We write, indeed

$$\sigma_\varepsilon = \sigma_{\varepsilon n}\left(1 + \sigma_r^2\right)^{\frac{1}{2}}, \qquad \sigma_r = \frac{\sigma_i}{\sigma_{\varepsilon n}}, \tag{3}$$

By introducing the relations (2) and (3) into Eq. (1), we can rewrite the instability growth rate as

$$\frac{n}{T_0}\sqrt{\frac{(2\pi)^{3/2} I_0 \nu_s |Z/n|}{(E_0/e)\,\sigma_{\varepsilon n}}}\,\frac{1}{(1 + \sigma_r^2)^{\frac{1}{4}}} = \frac{A}{(1 + \sigma_r^2)^{\frac{1}{4}}}, \tag{4}$$

and the Landau term as

$$\frac{n}{T_0} 2\pi\alpha_c \sigma_{\varepsilon n}\left(1 + \sigma_r^2\right)^{\frac{1}{2}} = B\left(1 + \sigma_r^2\right)^{\frac{1}{2}}. \tag{5}$$

The introduction of the above parameters A and B is particularly useful to state whether the instability may grow or not: the condition $A > B$ ensures that we are above threshold.

It is worth mentioning that the Boussard criterion and therefore the characteristic rates given by the relations (4) and (5) have been obtained with a linear theory which considers the growth rate of the instability independent on time. Under this simplifying hypothesis, the saw tooth behavior of Fig. 1 can not be explained.

If we denote by α the growth rate of the instability, the equation controlling the evolution of σ_r^2 can be written as

$$\frac{1}{\sigma_r^2}\frac{d\sigma_r^2}{dt} = \left(\alpha - \frac{2}{\tau_s}\right), \tag{6}$$

where τ_s is the longitudinal damping time.

Since the growth rate may have an exponential behavior, here we make the ansatz that α satisfies an equation of the type

$$\frac{1}{\alpha}\frac{d\alpha}{dt} = \frac{A}{(1 + \sigma_r^2)^{\frac{1}{4}}} - B\left(1 + \sigma_r^2\right)^{\frac{1}{2}}. \tag{7}$$

Eqs. (6) and (7) provide our coupled equations. They give stationary solutions with $\sigma_\varepsilon = \sigma_{\varepsilon n}$ in any situation for which $A < B$. The case $A = B$ is exactly the Boussard criterion which is therefore satisfied.

In Fig. 6, we compare the simulations and the results of Eqs. (6) and (7) under the condition of stability, that is for $A < B$. The coefficients A and B in this case differ from those obtained with Eqs. (4) and (5) by a factor 2.9 and .4 respectively.

The case of $A > B$ is shown in Fig. 1 where we report the results of Eqs. (6) and (7), represented by a solid line, compared with numerical simulations. Also in

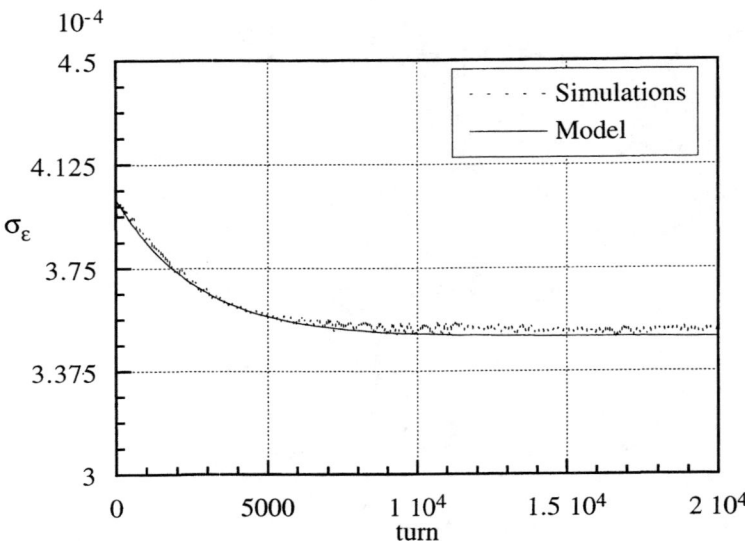

FIGURE 6. Energy spread vs number of turns. Comparison between simulations and coupled equations below microwave instability.

this case the coefficients A and B differ from those obtained with Eqs. (4) and (5) by the same factors as in Fig. 6.

It must be said that the coupled equations rely on simple assumptions which are not necessarily fulfilled in the beam dynamics. For example, in case of the A parameter, the actual bunch shape is distorted by the potential well and is not gaussian. Furthermore in the relation (2) we have used the average bunch current I_0 while a local density perturbation may be responsible of the microwave instability.

For what concern the B parameter, the major approximations are in the linear relationship assumed between the spectrum width producing Landau damping and the energy spread σ_ε, and the assumption of a Lorentz spectrum. Actually the oscillation frequency spectrum depends on the bunch distribution and on the non linearities of the wake fields. A last remark on both A and B is related to the choice of the harmonic number n. In the coasting beam theory it represents the unstable mode index, while in the extension to bunched beam case, the value

$$n = \frac{L_0}{2\pi\sigma_z} \tag{8}$$

is generally used [5]. But if we interpret n as the unstable mode index applied to the bunched beam case, and observe the phase space distribution of Fig. 3, it is easy to see that the choice of Eq. (8) could be quite arbitrary.

If we increase the damping time τ_s and eventually it tends to infinity, the saw

298

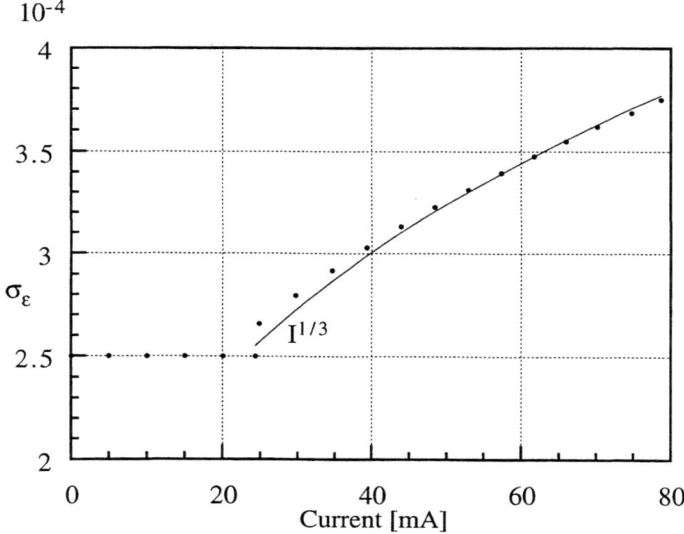

FIGURE 7. Energy spread vs current for $\tau_s \to \infty$.

tooth behavior disappears, thus giving a stationary energy spread which depends on the initial conditions and on the coefficients A and B, that is on the intensity of the instability and of Landau damping. Such a stationary energy spread has a dependence on the current of the kind $I_0^{1/3}$ as shown in Fig. 7.

In conclusion it is almost surprising that such a complex dynamics can be described under different conditions (stationary solution, Boussard criterion, saw tooth instability, $I_0^{1/3}$ behavior) by Eqs. (6) and (7) which contain only averages values.

REFERENCES

1. Heifets S., Proceedings of 14th ICFA Workshop, Frascati (1998). To be published.
2. Bane K., and Oide K., Contributed to 1993 Particle Accelerator Conference (PAC 93), Washington DC (1993) p. 3339.
3. Chao A., Physics of Collective Beam Instabilities in High Energy Accelerators, John Wiley & Sons (1993).
4. Boussard D., CERN LABII/RF/INT/75-2 (1975).
5. Hofmann A., and Maidment J. R., LEP note 168 (1979).

Particle-Beam Approach to Collective Instabilities — Application to Space-Charge Dominated Beams

K.Y. Ng

Fermi National Accelerator Laboratory,* P.O. Box 500, Batavia, IL 60510

S.Y. Lee

Physics Department, Indiana University, Bloomington, IN 47405

Abstract. Nonlinear dynamics deals with parametric resonances and diffusion. The phenomena are usually beam-intensity independent and rely on a particle Hamiltonian. Collective instabilities deal with beam coherent motion, where the Vlasov equation is frequently used in conjunction with a beam-intensity dependent Hamiltonian. We address the questions: Are the two descriptions the same? Are collective instabilities the results of encountering parametric resonances whose driving force is intensity dependent? We study here the example of a space-charge dominated beam governed by the Kapchinskij-Vladimirskij (K-V) envelope equation [1]. The stability and instability regions as functions of tune depression and envelope mismatch are compared in the two approaches. The study has been restricted to the simple example of a uniformly focusing channel.

I INTRODUCTION

Traditionally, the thresholds of collective instabilities are obtained by solving the Vlasov equation for collective motion. The modes of instabilities are described by the set of orthonormal eigenfunctions of the characteristic equation and the corresponding complex eigenvalues give the initial growth rates.

The beam dynamics of the Vlasov equation derives from a Hamiltonian that includes wakefields. The unperturbed beam distribution function is computed under the influence of the mean field. The perturbative distribution function is obtained by solving the Vlasov equation, which is often linearized so that the collective eigenmodes can be determined.

The nonlinear Hamiltonian can generally be decomposed into an unperturbed part, H_0, and a perturbation H_1. The unperturbed Hamiltonian is derived from the external forces such as the quadrupoles, rf focusing potential, the space-charge mean-field potential, and the potential-well distortion due to low-frequency components of the wakefields.

The perturbation H_1 can arise from nonlinear magnetic fields, high-frequency wakefields, etc. It may have a time independent component, for example, the part

---

*) Operated by the Universities Research Association, under contracts with the US Department of Energy.

CP468, *Nonlinear and Collective Phenomena in Beam Physics–1998 Workshop*,
edited by S. Chattopadhyay, M. Cornacchia, and C. Pellegrini

involving the nonlinear magnetic fields, that gives rise to the dynamical aperture limitation of the dynamical system. On the other hand, it may also have a time dependent component, which includes the effects of wakefields and produces coherent motion of beam particles. The harmonic content of the wakefields depends on the structure of accelerator components. If one of the resonant frequencies of the wakefields is equal to a fractional multiple of the unperturbed tune of H_0, including the mean-field potential, a resonance is encountered and coherent particle motion is introduced. This may result in a runaway situation such that collective instability is induced.

Experimental measurements indicate that a small time dependent perturbation can create resonance islands in the longitudinal or transverse phase space and profoundly change the bunch structure [2]. For example, a modulating transverse dipole field close to the synchrotron frequency can split up a well-behaved bunch into beamlets. Although these phenomena are driven by a beam-intensity independent source, they can also be driven by the space-charge force and/or the wakefields of the beam which are also intensity dependent. Once perturbed, the new bunch structure can further enhance the wakefields inducing even more perturbation to the circulating beam. Experimental observation of hysteresis in collective beam instabilities seems to indicate that resonance islands have been generated by the wakefields.

For example, the Keil-Schnell criterion [3] of longitudinal microwave instability can be derived from the concept of bunching buckets, or islands, created by the perturbing wakefields. Every particle of the beam will execute *synchrotron* motion inside these buckets leading to growth in the momentum spread of the beam. In fact, the collective growth rate is exactly the angular synchrotron frequency inside these buckets. If the momentum spread of the beam is much larger than the bucket height, only a small fraction of the particles in the beam will be affected and collective instabilities will not occur. This mechanism has been called Landau damping.

As a result, we believe that the collective instabilities of a beam can also be tackled from a particle-beam nonlinear dynamics approach, with collective instabilities occurring when the beam particles are either trapped in resonance islands or diffuse away from the beam core because of the existence of a sea of chaoticity. The advantage of the particle-beam nonlinear dynamics approach is its ability to understand the hysteresis effects and to calculate the beam distribution beyond the threshold condition. Such a procedure may be able to unify our understanding of collective instabilities and nonlinear beam dynamics.

In this work, the stability issues of a space-charge dominated beam in a uniformly focusing channel are considered as an example. In Sect. II, the Hamiltonian formulation of the envelope equation is reviewed. In Sect. III, we present a brief account of the collective-motion approach as studied by Gluckstern, Cheng, Kurennoy, and Ye [4], using the Vlasov equation. For the nonlinear dynamic approach, we start with the particle Hamiltonian in Sect. IV, where the particle tune is computed in the presence of beam-envelope oscillations. Section V is devoted to parametric resonances, where beam stability is investigated as a function of the space-charge perveance and beam envelope mismatch. In Sec. VI, beam particles having nonzero angular momentum are included. Finally, in Sec. VII, conclusions are given.

II ENVELOPE HAMILTONIAN

First, the envelope Hamiltonian is normalized to unit emittance and unit period. In terms of the normalized and dimensionless envelope radius R, together with its conjugate momentum P, the Hamiltonian for the beam envelope in a uniformly focusing channel can be written as [5,6]

$$H_e = \frac{1}{4\pi}P^2 + V(R) \ , \tag{2.1}$$

with the potential

$$V(R) = \frac{\mu^2}{4\pi}R^2 - \frac{\mu\kappa}{\pi}\ln\frac{R}{R_0} + \frac{1}{4\pi R^2} \ , \tag{2.2}$$

where $\mu/(2\pi)$ is the *unperturbed* particle tune and $\kappa = Nr_{\rm cl}/(\mu\beta^2\gamma^3)$ plays the role of the *normalized* space-charge perveance, N is the number of particles per unit length having the classical radius $r_{\rm cl}$, and β and γ are the relativistic factors of the beam. The *normalized* K-V equation then reads

$$\frac{d^2R}{d\theta^2} + \left(\frac{\mu}{2\pi}\right)^2 R = \frac{2\mu\kappa}{4\pi^2 R} + \frac{1}{4\pi^2 R^3} \ , \tag{2.3}$$

The radius R_0 of the matched beam envelope or core is determined by the lowest point of the potential; i.e., $V'(R_0) = 0$, or

$$\mu R_0^2 = \sqrt{\kappa^2 + 1} + \kappa = \frac{1}{\sqrt{\kappa^2 + 1} - \kappa} \ . \tag{2.4}$$

From the second derivative of the potential, the small amplitude tune for envelope oscillations is therefore

$$\nu_e = \frac{2\mu}{2\pi}\left[1 - \kappa\left(\sqrt{\kappa^2 + 1} - \kappa\right)\right]^{1/2} \longrightarrow \begin{cases} 2\dfrac{\mu}{2\pi} & \kappa \to 0 \\[2mm] \sqrt{2}\dfrac{\mu}{2\pi} & \kappa \to \infty \end{cases} \tag{2.5}$$

For a mismatched beam, R varies between $R_{\min}$ and $R_{\max}$. To derive the tune of the mismatched envelope, it is best to go to the action-angle variables. The envelope action can be computed from the envelope Hamiltonian via

$$J_e = \frac{1}{2\pi}\oint P dR \ . \tag{2.6}$$

302

The envelope tune is then

$$Q_e = \frac{dE_e}{dJ_e} = \nu_e + \alpha_e J_e + \cdots , \qquad (2.7)$$

where E_e is the Hamiltonian value of the beam envelope, and the detuning α_e, defined by

$$H_e = \nu_e J_e + \tfrac{1}{2}\alpha_e J_e^2 + \cdots , \qquad (2.8)$$

is computed to be

$$\alpha_e = \frac{3}{16\pi^3 R_0^4 \nu_e^2} \left(\mu\kappa + \frac{5}{R_0^2} \right) - \frac{5}{48\pi^5 R_0^6 \nu_e^4} \left(\mu\kappa + \frac{3}{R_0^2} \right)^2 + \cdots . \qquad (2.9)$$

To obtain the envelope tune for large mismatch, one must compute numerically the action integral to obtain

$$Q_e^{-1} = \frac{dJ_e}{dE_e} = \frac{1}{2\pi} \oint \frac{\partial P}{\partial E_e} dR , \qquad (2.10)$$

The envelope tune is plotted in Fig. 1 as a function of the maximum envelope radius $R_{\max}$, which for small mismatch is related to the envelope action J_e by

$$R = R_0 + \left(\frac{J_e}{\pi\nu_e} \right)^{1/2} \cos Q_e \theta . \qquad (2.11)$$

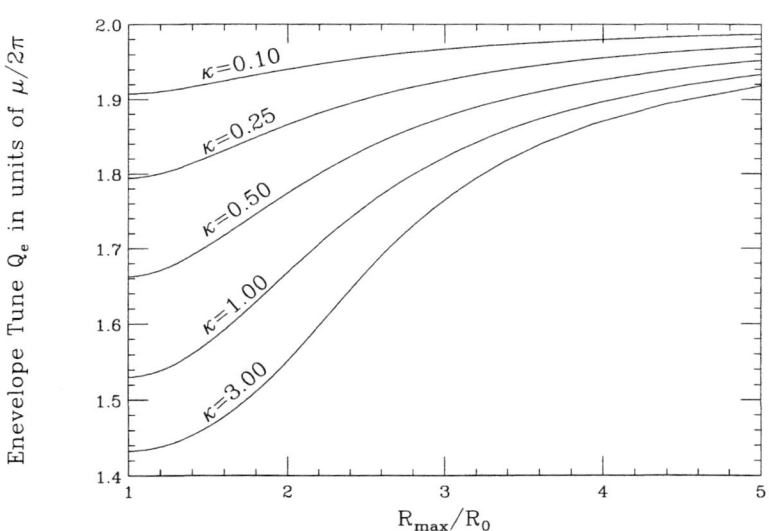

FIGURE 1. Envelope tune Q_e versus envelope mismatch $R_{\max}/R_0$ for various space-charge perveance κ. Notice that Q_e is represented by ν_e at $R_{\max}/R_0 = 1$ when the beam envelope is matched.

303

III COLLECTIVE MOTION APPROACH

Gluckstern, Cheng, Kurennoy, and Ye [4] have studied the collective beam stabilities of a space-charge dominated K-V beam in a uniformly focusing channel. The particle distribution f is separated into the unperturbed distribution f_0 and the perturbation f_1:

$$f(u,v,\dot{u},\dot{v};\theta) = f_0(u,v,\dot{u},\dot{v}) + f_1(u,v,\dot{u},\dot{v};\theta) , \qquad (3.1)$$

where u and v are the normalized transverse coordinates which are functions of the 'time' variable θ. Their derivatives are denoted by $\dot{u}$ and $\dot{v}$. The unperturbed distribution

$$f_0(u,v,\dot{u},\dot{v}) = \frac{I_0}{v_0\pi^2}\delta(u^2 + v^2 + \dot{u}^2 + \dot{v}^2 - 1) \qquad (3.2)$$

is the steady-state solution of the K-V equation (2.3) and is therefore time-independent. In the notation of Gluckstern, Cheng, Kurennoy, and Ye, I_0 is the average beam current and v_0 the longitudinal velocity of the beam particles. The perturbed distribution generates an electric potential G, which is given by the Poisson's equation

$$\nabla^2 G(u,v,\theta) = -\frac{1}{\epsilon_0}\int d\dot{u}\int d\dot{v} f_1(u,v,\dot{u},\dot{v};\theta) , \qquad (3.3)$$

so that the Hill's equations become

$$\ddot{u} + u = -\frac{e\beta}{m_0 v_0^2 \epsilon}\frac{\partial G}{\partial u} , \qquad \ddot{v} + v = -\frac{e\beta}{m_0 v_0^2 \epsilon}\frac{\partial G}{\partial v} , \qquad (3.4)$$

where ϵ stands for the transverse emittance of the beam and m_0 the rest mass of the beam particle.

For small perturbation, the perturbation distribution is proportional to the derivative of the unperturbed distribution. This enables us to write

$$f_1(u,v,\dot{u},\dot{v};\theta) = g(u,v,\dot{u},\dot{v};\theta)f_0'(u^2 + v^2 + \dot{u}^2 + \dot{v}^2) . \qquad (3.5)$$

Substituting into the linearized Vlasov equation, we obtain

$$\frac{\partial g}{\partial \theta} + \dot{u}\frac{\partial g}{\partial u} + \dot{v}\frac{\partial g}{\partial v} - u\frac{\partial g}{\partial \dot{u}} - v\frac{\partial g}{\partial \dot{v}} = \frac{2e\beta}{m_0 v_0^2 \epsilon}\left[\dot{u}\frac{\partial G}{\partial u} + \dot{v}\frac{\partial G}{\partial v}\right] . \qquad (3.6)$$

Noting that the potential G is a polynomial, Gluckstern, et. al. are able to solve for g and G consistently in terms of hypergeometric functions. Thus a series of orthonormal eigenmodes are obtained for the perturbed distribution with their corresponding eigenfrequencies. These modes are characterized by (j,m), where j is the radial eigennumber and m the azimuthal eigennumber.

For the azimuthally symmetric modes, $m=0$. The $(1,0)$ is the breathing mode of uniform density at a particular time. The $(2,0)$ mode oscillates with a radial

304

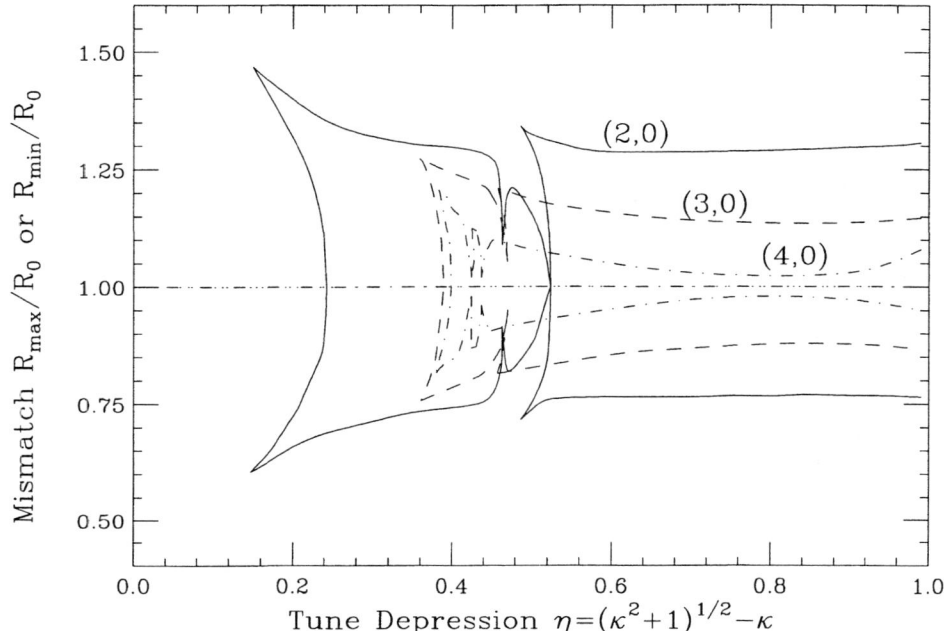

FIGURE 2. Beam stability plot versus particle tune depression η and beam envelope mismatch. The stability region for the (2,0) mode is enclosed by the solid curve, that for the (3,0) mode by the dashed curve, and that for the (4,0) mode by the dot-dashed curve. (Reproduced from Ref. 4).

node between $R = 0$ and $R = R_0$ so that the density becomes nonuniform. The higher modes are similar, with the $(j,0)$ having $j-1$ radial nodes. When the eigenfrequency of a mode is complex, the mode becomes unstable with a collective growth rate. Stability is studied in terms of tune depression $\eta = \sqrt{\kappa^2 + 1} - \kappa$ and the amount of envelope mismatch. The former is defined as the ratio of the particle tune with space charge to the the particle tune without space charge for a *matched* beam. Thus η ranges from 0 to 1; $\eta = 1$ implies zero space charge while $\eta = 0$ implies infinite space charge.

Gluckstern, *et. al.* showed that the (1,0) mode is stable for any amount of mismatch and tune depression. The (2,0) mode becomes unstable at zero mismatch when the tune depression $\eta < 1/\sqrt{17} = 0.2435$. It is also unstable when the mismatch is large. This is plotted in Fig. 2 with the stable region enclosed by the solid curve, a reproduction of Ref. 4. The stability regions of the (3,0) and (4,0) modes, enclosed by dashes and dot-dashes, respectively, are also shown. These latter two modes become unstable at zero mismatch when the tune depressions are less than 0.3859 and 0.3985, respectively. They found that the modes become more unstable as the number of radial nodes increases. Among all the azimuthals, they also noticed that the azimuthally symmetric modes $(m=0)$ are the most unstable.

IV NONLINEAR APPROACH

A Particle Hamiltonian

We want to investigate whether the instability regions in the plane of tune depression and mismatch can be explained by nonlinear parametric resonances. First let us study the transverse motion of a particle having zero angular momentum. The situation of finite momentum will be discussed later in Sec. VI. We choose y as the particle's transverse coordinate with canonical angular momentum p. Its motion is perturbed by an azimuthally symmetric oscillating beam core of radius R. The particle Hamiltonian is [6]

$$H_p = \frac{1}{4\pi}p^2 + \frac{\mu^2}{4\pi}y^2 - \frac{2\mu\kappa}{4\pi R^2}\,y^2\,\Theta(R - |y|) - \frac{2\mu\kappa}{4\pi}\left(1 + 2\ln\frac{|y|}{R}\right)\Theta(|y| - R) \;, \quad (4.1)$$

giving the equation of motion for y,

$$\frac{d^2y}{d\theta^2} + \left(\frac{\mu}{2\pi}\right)^2 y = \frac{\mu\kappa}{2\pi^2 R^2}y\,\Theta(R - |y|) + \frac{\mu\kappa}{2\pi^2|y|}\,\Theta(|y| - R) \;. \quad (4.2)$$

For a weakly mismatched beam, the envelope radius can be written as

$$R = R_0 + \Delta R \cos Q_e\theta \;. \quad (4.3)$$

The particle Hamiltonian can also be expanded in terms of the equilibrium envelope radius R_0, resulting

$$H_p = H_{p0} + \Delta H_p \;. \quad (4.4)$$

The unperturbed Hamiltonian is

$$H_{p0} = \frac{1}{4\pi}p^2 + \frac{\mu^2}{4\pi}y^2 - \frac{2\mu\kappa}{4\pi R_0^2}\,y^2\,\Theta(R_0 - |y|) - \frac{2\mu\kappa}{4\pi}\left(1 + 2\ln\frac{|y|}{R_0}\right)\Theta(|y| - R_0) \;,$$

$$(4.5)$$

and the perturbation

$$\Delta H_p \approx -\frac{\mu\kappa}{\pi R_0^2}\left[\frac{\Delta R}{R_0}(y^2 - R_0^2) + \frac{3\Delta R^2}{2R_0^2}\left(y^2 - \tfrac{1}{3}R_0^2\right) + \cdots\right]\Theta(R_0 - |y|) \;. \quad (4.6)$$

Note that many non-contributing terms, like the ones involving the δ-function and δ'-function, have been dropped. Additionally, envelope oscillations do not perturb particle motion outside envelope radius; thus the perturbing potential in Eq. (4.6) exists only inside the envelope.

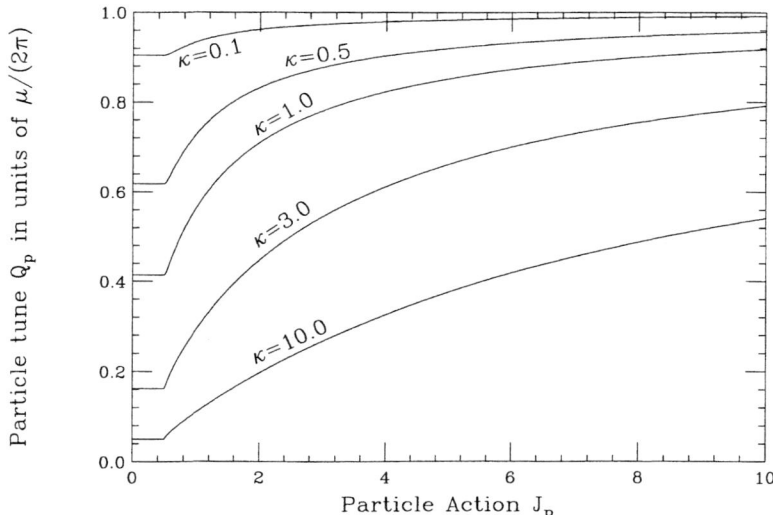

FIGURE 3. Particle tune Q_p as function of particle action J_p and space-charge perveance κ for a matched beam.

For a matched beam, $\Delta H_p = 0$. Inside the core of uniform distribution, the particle motion is linear and its tune can be readily obtained:

$$\nu_p = \frac{\mu}{2\pi}\left(1 - \frac{2\kappa}{\mu R_0^2}\right)^{1/2} = \frac{\mu}{2\pi}\left(\sqrt{\kappa^2 + 1} - \kappa\right) .$$

Thus, $\eta = \sqrt{\kappa^2 + 1} - \kappa$ is the tune depression.

When the particle spends time oscillating outside the beam envelope, its tune has to be computed numerically. First the particle action is defined as

$$J_p = \frac{1}{2\pi}\oint p\, dy . \tag{4.7}$$

The particle tune Q_p is then given by

$$Q_p^{-1} = \frac{dJ_p}{dE_p} = \frac{1}{4\pi}\oint \frac{\partial p}{\partial E_p}dy , \tag{4.8}$$

where E_p is the Hamiltonian value of the beam particle. The result is shown in Fig. 3 for various space-charge perveance κ. We see that when the particle motion is completely inside the beam envelope ($J_p < \frac{1}{2}$), the particle tune is a constant and is given by ν_p depending on κ only. As the particle spends more and more time outside the beam envelope, its tune increases because the space-charge force decreases as y^{-1} outside the envelope.

307

B Particle Tune Inside a Mismatched Beam

To simplify the algebra, it is advisable to scale away the unperturbed particle tune $\mu/(2\pi)$ through the transformation:

$$\mu R^2 \longrightarrow R^2 \ , \qquad \left(\frac{\mu}{2\pi}\right)\theta \longrightarrow \theta \ . \tag{4.9}$$

The envelope equation becomes

$$\frac{d^2 R}{d\theta^2} + R = \frac{2\kappa}{R} + \frac{1}{R^3} \tag{4.10}$$

The particle equation is, after the same transformation,

$$\frac{d^2 y}{d\theta^2} + y - \frac{2\kappa}{R^2} y\,\Theta(R - |y|) - \frac{2\kappa}{y}\,\Theta(|y| - R) = 0 \ , \tag{4.11}$$

where the replacement $\mu y^2 \to y^2$ has been made. For one envelope oscillation period, the envelope radius R is periodic and Eq. (4.11) *inside* the envelope core becomes a Hill's equation with effective 'field gradient' $K(\theta) = 1 - 2\kappa/R^2(\theta)$. The solution is then exactly the same as the Floquet transformation by choosing $y = aw(\theta)\cos[\psi(\theta) + \delta]$. It is easy to show that the differential equation for w is exactly the envelope equation of Eq. (4.10). Thus we can replace w by R and R^2 which is the effective betatron function. Since the particle makes Q_p/Q_e betatron oscillations during one envelope fluctuation period, where Q_p is the particle tune, we have

$$\frac{Q_p}{Q_e} = \frac{\Delta\psi}{2\pi} = \frac{1}{2\pi}\oint \frac{d\theta}{R^2(\theta)} \ . \tag{4.12}$$

In Floquet's notation, we define $\hat{y} = y/R$. Then Eq. (4.2) describing the motion of a particle modulated by a beam envelope becomes

$$\frac{d^2\hat{y}}{d\psi^2} + \hat{y} + 2\kappa R^2 \left[\frac{\hat{y}^2 - 1}{\hat{y}}\right]\Theta\left(|\hat{y}| - 1\right) \ . = 0 \tag{4.13}$$

Thus, all particles inside the beam envelope have a fixed tune depending on the amount of space charge and envelope mismatch. Particles spending part of the time outside the beam envelope will have larger tunes.

For a small mismatch core fluctuation, we can write $R = R_0(1 - M\cos Q_e\theta)$, where M can be interpreted as the mismatch parameter. The integral can be performed analytically to give

$$Q_p = \frac{\nu_p}{(1 - M^2)^{3/2}} \ , \tag{4.14}$$

where $\nu_p = R_0^{-2} = \sqrt{\kappa^2 + 1} - \kappa$, in the presence of the transformation (4.9), is the particle tune when the envelope is matched. The analytic formula of Eq. (4.14),

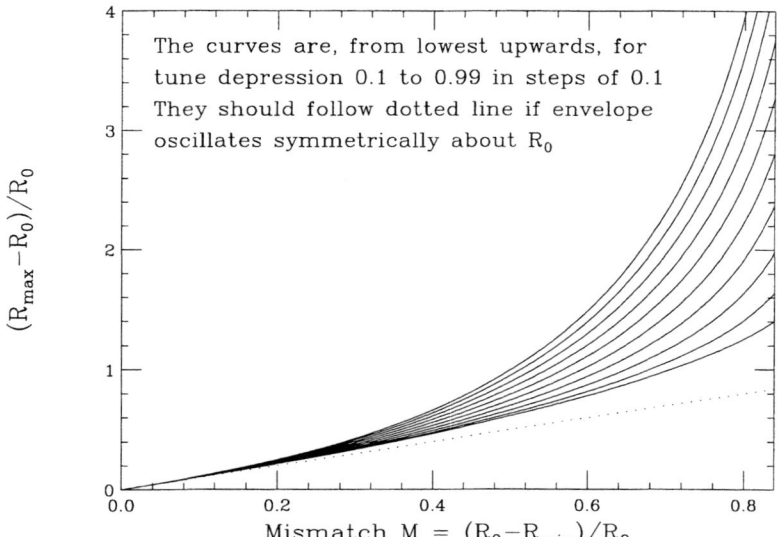

The curves are, from lowest upwards, for
tune depression 0.1 to 0.99 in steps of 0.1
They should follow dotted line if envelope
oscillates symmetrically about R_0

FIGURE 4. Plot of $(R_{\max} - R_0)/R_0$ versus $M = (R_0 - R_{\max})/R_0$ showing that the envelope oscillation is very asymmetric about the equilibrium radius R_0 when both the mismatch and tune depression are large.

however, is only valid when the mismatch parameter $M \lesssim 0.2$. The reason is that the envelope equation is nonlinear in the presence of space charge. In other words, while minimum envelope radius is given by $R_{\min} = (1 - M)R_0$, the maximum envelope radius is always $R_{\max} > (1 + M)R_0$. In fact, when $M \to 1$, $R_{\min} \to 0$, but $R_{\max} \to \infty$. This can be seen in Fig. 4 when $(R_{\max} - R_0)/R_0$ is plotted against $M = (R_0 - R_{\max})/R_0$. If the envelope oscillations were asymmetric about R_0, the plot should follow the $45°$ dotted line instead. We see that the deviation is large when the mismatch and tune depression are large. When the approximation $R = R_0(1 - M \cos Q_e \theta)$ breaks down, the particle tune can still be easily evaluated by performing the integral in Eq. (4.12) numerically. Figure 5 shows the deviation of the actual particle tune Q_p from its analytic formula of Eq. (4.14).

V   PARAMETRIC RESONANCES

Particle motion is modulated by the oscillating beam envelope. Therefore, to study the resonance effect, we need to include the perturbation part ΔH_p of the particle Hamiltonian. We expand it as a Fourier series in the angle variable ψ_p yielding, for example,

$$(y^2 - R_0^2)\,\Theta(R_0 - |y|) = \sum_{n=-\infty}^{\infty} G_n(J_y)e^{in\psi_y} \ . \tag{5.1}$$

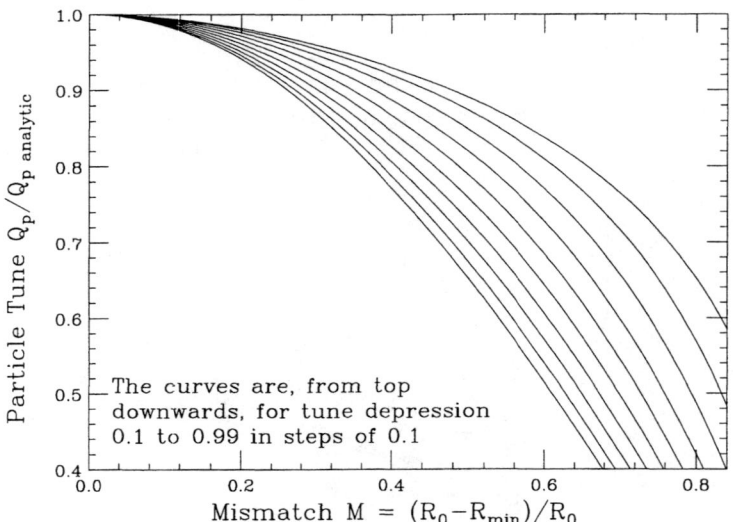

The curves are, from top
downwards, for tune depression
0.1 to 0.99 in steps of 0.1

FIGURE 5. Plot showing the deviation of the actual particle tune Q_p from the value of the analytic formula of Eq. (4.14) in the presence of mismatch.

Since ΔH_p is even in y, only even n harmonics survive. The particle Hamiltonian then becomes

$$H_p = H_{p0} + \frac{\mu\kappa}{2\pi R_0^2} \sum_{m=1}^{\infty} \sum_{\substack{n>0 \\ \text{even}}} (m+1)M^m|G_{n,m}| \\ \times \left[\cos(n\psi_y - mQ_e\theta + \gamma_n) + \cos(n\psi_y + mQ_e\theta + \gamma_n)\right] + \cdots \tag{5.2}$$

where γ_n are some phases and use has been made of $R_0 - R = MR_0\cos Q_e\theta$, the approximation for small mismatch.

Focusing on the $n{:}m$ resonance, we perform a canonical transformation to the resonance rotating frame to get

$$\langle H_p \rangle = E_p(I_y) - \frac{m}{n}Q_eI_y + h_{nm}(I_y)\cos n\phi_y , \tag{5.3}$$

with the effective κ-dependent resonance strength given by

$$h_{nm} = \frac{(m+1)M^m\mu\kappa}{2\pi R_0^2} |G_{nm}(I_y)| . \tag{5.4}$$

As usual, there are n stable and n unstable fixed points which can be found easily. Since ΔH_p is a polynomial up to y^2 only and $y \propto \sin\psi_y$, we have, inside the envelope,

$$G_{nm} = \frac{1}{4\pi Q_e}J_y\delta_{n2} , \tag{5.5}$$

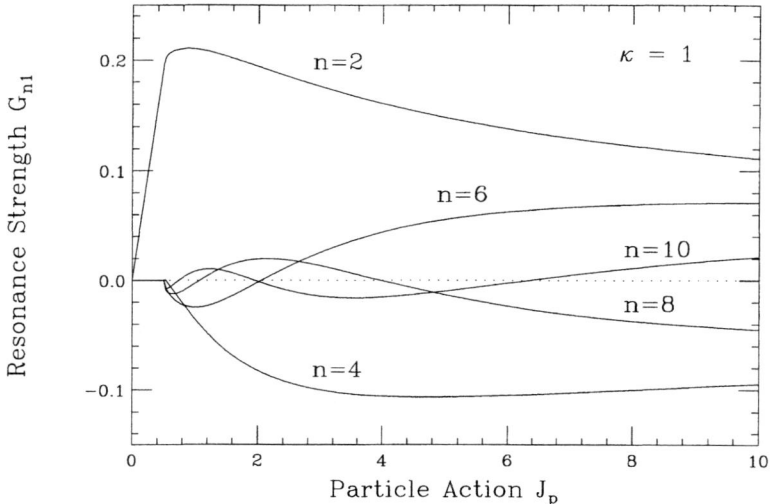

FIGURE 6. Plot of driving strengths of the first-order resonance, G_{n1}, versus the particle action J_p. Inside the envelope ($J_p < \frac{1}{2}$), only G_{21} is nonzero. Once outside the envelope, however, $|G_{n1}|$ for $n \geq 2$ increases rapidly from zero.

implying that only $2{:}m$ resonances are possible. Outside the envelope the resonance driving strength can also be computed. These are plotted in Fig. 6. We see that although the driving strengths G_{n1} for $n > 2$ vanish inside the envelope ($J_p < \frac{1}{2}$), they increase rapidly once outside. Including noises of all types, particles inside the K-V beam envelope can leak out. This situation is particularly true when the particle tune is equal to a fractional multiple of the envelope tune. A small perturbation may drive particles outside the beam envelope. Once outside, because of the nonvanishing driving strengths, these particles may be trapped into resonance islands or diffuse into resonances farther away. Once trapped or diffused, these particles cannot wander back into the envelope core to stabilize the core distribution. As more and more envelope particles leak out, this is viewed as an instability.

Our job is, therefore, to map out the location of parametric resonances in the plane of mismatch and tune depression. Because particles are affected only by resonances when they are just outside the envelope core, their tunes are essentially the tune inside the beam envelope. At zero mismatch, the threshold for the $n{:}m$ resonance can therefore be derived by equating ν_p/ν_e to m/n. Thus

$$\frac{\nu_p}{\nu_e} = \frac{\sqrt{\kappa^2 + 1} - \kappa}{2\left[1 - \kappa\left(\sqrt{\kappa^2 + 1} - \kappa\right)\right]^{1/2}} \leq \frac{m}{n}, \quad \text{or} \quad \kappa \geq \frac{\left(\frac{n}{m}\right)^2 - 4}{\sqrt{8\left[\left(\frac{n}{m}\right)^2 - 2\right]}}. \tag{5.6}$$

In particular, for the 6:1 resonance, $\kappa \geq 8/\sqrt{17} = 1.9403$, or the tune depression is

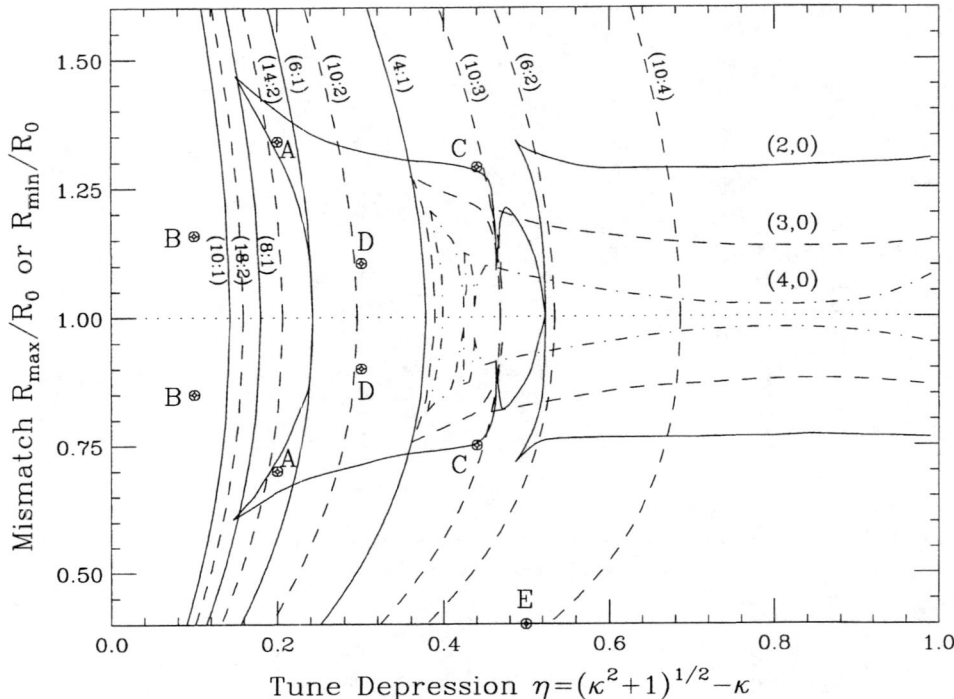

FIGURE 7. Plot of parametric resonance locations in the plane of tune-depression beam envelope mismatch. First-order resonances are shown as solid while second- and higher-order resonances as dashes. Overlaid on top are the instability boundaries of modes (2,0), (3,0), and (4,0) derived by Gluckstern, *et. al.*

$\eta \leq 1/\sqrt{17} = 0.2425$, which agrees with Gluckstern's instability threshold for the (2,0) excitation.

For a mismatched beam, the threshold for the $n{:}m$ resonance is obtained by equating Q_p/Q_e at that mismatch to m/n. These resonances are labeled in Fig. 7 in the plane of tune depression and mismatch. The locus of the 2:1 resonance is the vertical line $\eta = 1$. This is obvious, because at zero space charge the particle tune is exactly two times the envelope tune regardless of mismatch. Also, it is clear from Eq. (4.13) that there will not be any Mathieu instability or half-integer stopband. Thus it appears that the 2:1 resonance would not influence the stability of a space-charge dominated beam. This is, in fact, not true. The stable fixed points of the 2:1 resonance are usually far away from the beam envelope. Thus particles can diffuse towards the 2:1 resonance to form the halo of the beam. As more and more particles diffuse from the beam core into the 2:1 resonance, the beam becomes unstable.

Trackings have been performed for particles outside the envelope core using the fourth-order symplectic integrator. The Poincaré surface of section is shown in

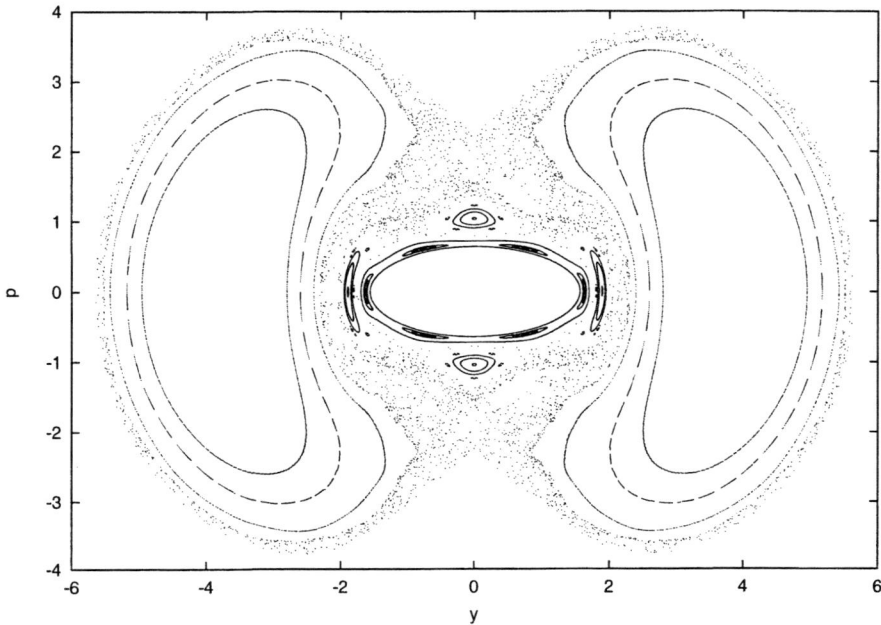

FIGURE 8. Poincaré surface of section in particle phase space (y, p) with $\eta = 0.20$ ($\kappa = 2.4$) and $M = 0.3$ corresponding to Points A in Fig. 7.

Fig. 8 for the situation $\eta = 0.20$ ($\kappa = 2.4$) and $M = 0.3$. This corresponds to Points A in Fig. 7. The innermost torus is the beam envelope. The sections are taken every envelope oscillation period when the envelope radius is at a minimum. For each envelope oscillation period, 500 to more than 1000 time steps have been used. We see that as soon as particles diffuse outside the beam envelope, they will encounter the 6:1 resonance, which is bounded by tori. This explains the front stability boundary of the (2,0) mode of Gluckstern, *et. al.* Since the 4:1 resonance is a strong one, its locus explains the front stability boundaries of Gluckstern's (3,0) and (4,0) modes also.

The Poincaré surface of section corresponding to Points B of Fig. 7 with $\eta = 0.10$ ($\kappa = 4.95$) $M = 0.15$ is shown in Fig. 9. This is a close-up view showing only the region near the beam envelope; the 2:1 resonance and its separatrices are not shown because they look similar to those depicted in Fig. 8. We see resonances like 14:2, 8:1, 16:2, 9:1, 10:1, etc, which are so closely spaced that they overlap to form a chaotic region. Particles that diffuse outward from the beam envelope will wander easily towards the 2:1 resonance along its separatrix. This region, where $\eta \lesssim 0.2$, is therefore very unstable.

Figure 10 shows the close-up Poincaré surface of section of Points C in Fig. 7 with $\eta = 0.44$ ($\kappa = 0.916$) and $M = 0.25$. Here the particles see many parametric resonances when they are outside the beam envelope. First the 10:3, followed by the 6:2, 8:3, 10:4, and then a chaotic layer going towards the 2:1 resonance. The

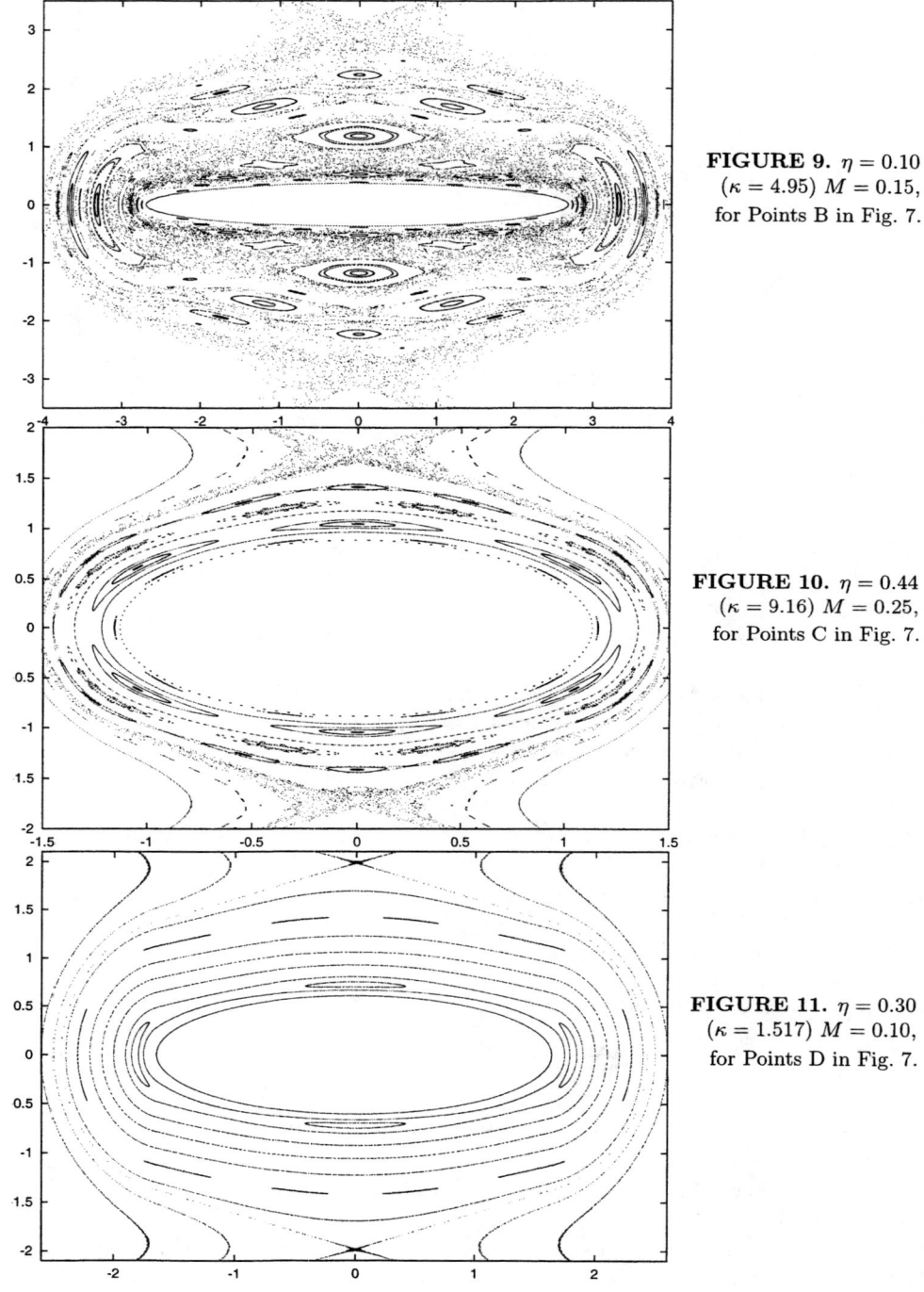

FIGURE 9. $\eta = 0.10$ ($\kappa = 4.95$) $M = 0.15$, for Points B in Fig. 7.

FIGURE 10. $\eta = 0.44$ ($\kappa = 9.16$) $M = 0.25$, for Points C in Fig. 7.

FIGURE 11. $\eta = 0.30$ ($\kappa = 1.517$) $M = 0.10$, for Points D in Fig. 7.

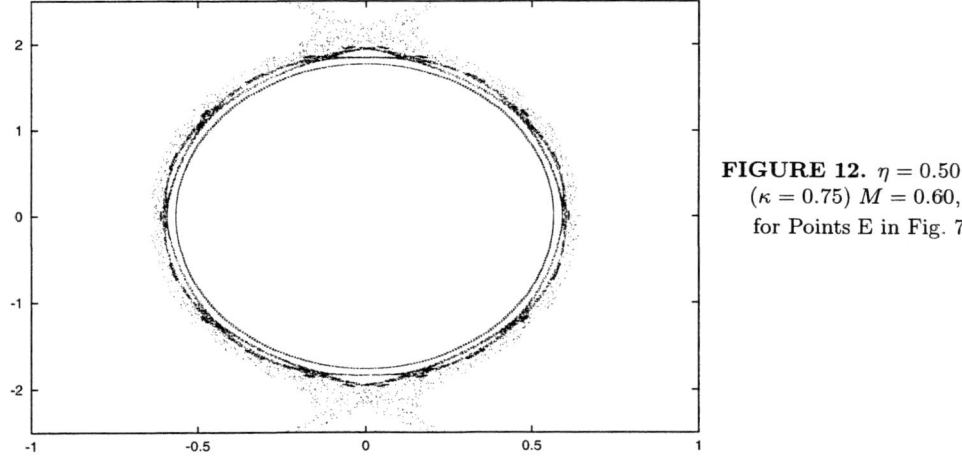

FIGURE 12. $\eta = 0.50$
$(\kappa = 0.75)$ $M = 0.60$,
for Points E in Fig. 7.

resonances are separated by good tori and the instability growth rate should be small. Thus, this is the region on the edge of instability.

On the other hand, the Poincaré surface of section in Fig. 11 corresponding to Points D of Fig. 7 with $\eta = 0.30$ ($\kappa = 1.517$) and $M = 0.10$ shows the 6:2 resonance well separated from the 10:4 resonance with a wide area of good tori. Also the width of the 10:4 resonance is extremely narrow and so that particles can hardly be trapped there. Unlike the situation in Figs. 9 and 10, there is no chaotic region at the unstable fixed points and inner separatrices of the 2:1 resonance, making diffusion towards this resonance impossible. This region will be relatively stable.

Next let us consider the region with very large beam envelope mismatch like Points E of Fig. 7 with $\eta = 0.50$ ($\kappa = 0.75$) and $M = 0.60$. (The other Point E is at $R_{\max}/R_0 = 2.067$ and is therefore not visible in Fig. 7). The close-up Poincaré surface of section in Fig. 12 shows the beam envelope radius at $y = 0.566$ when $p = 0$. We can see that the unstable fixed points and the inner separatrices of 2:1 resonance are very close by and are very chaotic. As soon as a particle diffuses out to $y = 0.62$, it reaches the chaotic sea and wanders towards the 2:1 resonance. Because the chaotic region is so close to the beam envelope, this region of large mismatch is also unstable. As a result, we can interpret Gluckstern's region of instability at large mismatch as chaotic region leading to the 2:1 resonance being too close to the beam envelope.

The deep fissures of the (2,0) mode near $\eta = 4.7$ and 5.3 are probably the result of encountering the 10:3 and 6:2 parametric resonances. The width of the fissures should be related to the width of the resonance islands.

We tried very hard to examine the region between the 4:1 and 10:3 resonances with a moderate amount of mismatch. We found this region very stable unless it is close to the 10:3 resonance. We could not reproduce the slits that appear in Gluckstern's (4,0) mode.

315

VI ANGULAR MOMENTUM

When angular momentum is included, it is preferable to use the circular coordinates (r, φ) as independent variables; their canonical momenta are, respectively, p_r and p_φ. The particle Hamiltonian becomes

$$
\begin{aligned}
H_p =& \frac{1}{4\pi} \left(p_r^2 + \frac{p_\varphi^2}{r^2} \right) + \frac{\mu^2}{4\pi} r^2 + \frac{\mu\kappa}{2\pi R^2} r^2 \Theta(R - r) \\
&+ \frac{\mu\kappa}{2\pi} \left[\frac{r^2}{R^2} - \left(1 + 2\ln\frac{r}{R} \right) \right] \Theta(r - R) \ .
\end{aligned}
\tag{6.1}
$$

Since H_p is φ-independent, the angular momentum p_φ is therefore a constant of motion. To understand the parametric resonances created by the modulation of the beam envelope, we need to derive the radial tune inside the beam envelope. We simplify the Hamiltonian by performing the transformation of Eq. (4.9) and let $\mu r^2 \to r^2$ and $p_r^2/\mu \to p_r^2$. We then have, inside the beam envelope,

$$
H_p = \frac{1}{2} \left(p_r^2 + \frac{p_\varphi^2}{r^2} \right) + \frac{r^2}{2} \left(1 - \frac{2\kappa}{R^2} \right) \ .
\tag{6.2}
$$

Let us look at the the time evolution of a K-V particle. For a given nonzero angular momentum, what is the initial radial position of the K-V particle when $p_r = 0$? Notice that when $p_r = 0$ and $p_\varphi = 0$, a K-V particle is characterized by $r = R$ or $E_p = \frac{1}{2}R^2 - \kappa$. Thus when $p_\varphi \neq 0$, a K-V particle should have a radial amplitude r or initial radial position (at $p_r = 0$) given by

$$
R^2 \left(1 - \frac{2\kappa}{R^2} \right) = \frac{p_\varphi^2}{r^2} + r^2 \left(1 - \frac{2\kappa}{R^2} \right) \ ,
\tag{6.3}
$$

or

$$
\left(\frac{r}{R} \right)^2 = \frac{1}{2} + \left[\frac{1}{4} - \frac{p_\varphi^2}{R^2(R^2 - 2\kappa)} \right]^{1/2} \ ,
\tag{6.4}
$$

where the positive sign has been chosen because $r \to R$ as $p_\varphi \to 0$. For a matched beam, substituting for R_0 from Eq. (2.4) leads to the restriction of $|p_\varphi| \leq \frac{1}{2}$. This agrees with the result of Riabko [6] that $2J_r + |p_\varphi| = \frac{1}{2}$ for a K-V particle inside a matched beam envelope, where J_r is the radial particle action obtained from the Hamiltonian (6.2). In fact, inside a matched beam envelope, the equation of radial motion for a particle with finite angular momentum is identical to the envelope equation (2.3) with the space-charge perveance κ set to zero. The tune of the radial oscillation can be shown to be exactly two times the particle tune with zero angular momentum.

For a mismatched beam, because of the additional p_φ^2/r^2 term, the particle equation is no longer a linear Hill's equation even when the particle is inside the beam envelope, and the radial tune is difficult to compute. For a K-V particle, Eq. (6.4)

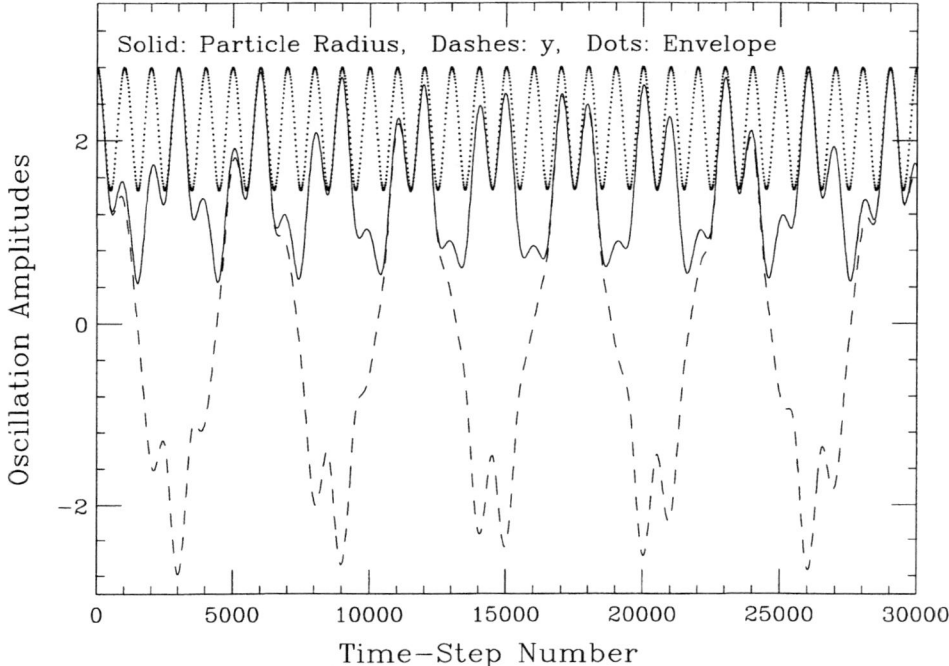

FIGURE 13. Plot showing the time evolution of the radial position r of a K-V particle with angular-momentum $p_\varphi = 0.3$ inside a beam envelope with mismatch $M = 0.30$ and space-charge perveance $\kappa = 2.059$ ($\eta = 0.23$). The time evolution y of a zero angular-momentum particle is also shown. Notice that the former oscillates two times as fast as the latter, thus having essentially twice the tune of the latter. The simulation has been performed at the 6:1 resonance for the nonzero angular-momentum particle.

shows that $|p_\varphi|$ can be larger than $\frac{1}{2}$ in the presence of mismatch. Simulations have been performed for the time evolution of the radial motion of such a particle and then compared with the time evolution of the transverse motion of a particle with zero momentum. One of the simulations is shown in Fig. 13. The particle is a K-V particle with angular momentum $p_\varphi = 0.3$ in a mismatched beam envelope with $M = 0.30$ having a tune depression of $\eta = 0.23$. We see that the shape of oscillations of r shown as solid is very similar to that of y with zero angular momentum shown as dashes. Since r does not go negative, its tune appears to be twice the tune of a zero angular-momentum particle. This plot was performed near a 6:1 resonance for a zero angular-momentum particle and it therefore translates into a 3:1 resonance for a nonzero angular-momentum particle.

It is interesting to point out that a particle inside the beam envelope remains inside the envelope if $|p_\varphi| \leq 1$. However, as $|p_\varphi|$ approaches 1, the humps that exhibit in the time evolution of the radial motion become more pronounced, and eventually coincide with oscillations of the envelope radius R. This is because the equation of motion of a $|p_\varphi| = 1$ particle is exactly the K-V equation for envelope oscillations. Even with $|p_\varphi| < 1$, the equation of motion can be scaled to the envelope equation. Thus $r(\theta) = \sqrt{|p_\varphi|}R(\theta)$ is a solution. Now the radial tune is equal to the envelope tune and is *not* two times the particle tune with zero angular momentum as in the matched envelope situation. These particles, however, are not K-V particles.

Fortunately, we are interested in particles that escape from the beam envelope and they must be originally K-V particles or close to K-V particles. For this reason, it is safe to conclude that the nonzero angular-momentum particles that we study here have radial tunes two times the tune of the zero angular-momentum particles, even in the presence of beam-envelope mismatch. For all the $n{:}m$ parametric resonances that we studied in the previous section, they translate into the $\frac{n}{2}{:}m$ resonances when the angular momentum is finite. As a result, the stability investigation in the previous section should hold even when particles with finite angular momentum are included.

VII   CONCLUSION

(a) Collective instabilities of a K-V beam in a uniformly focusing channel have been explained using the particle-beam nonlinear-dynamics approach. First, the loci of the first- and higher-order parametric resonances were mapped in the plane of tune depression and beam envelope mismatch. Different regions in this plane were discussed in terms of their potential for unstable motion. Because the K-V equation is far from realistic, and because of the existence of noises of all types in the accelerators (as well as in the numerical simulations), some particles will diffuse away from the K-V distribution. If the beam envelope ellipse is bounded by invariant tori, and if the diffusion rate is small, the beam will eventually reach equilibrium and we consider this situation stable.

When the particle tune is equal to some fractional multiples of the envelope tune, particle motion encounters parametric resonances in the presence of the space-charge force. As shown in Fig. 6, the resonance strengths for K-V particle are zero inside the beam envelope. However, if the diffusion process is included, K-V particles can diffuse towards the resonances outside the K-V envelope. These particle can move along the separatrices of the parametric resonances, and eventually escape from the K-V core.

As particles escape from the K-V core, density fluctuations will occur in the K-V core. Furthermore, this may enhance the envelope oscillations, driving more particles to the outside of the K-V core. Such a process can induce collective beam instabilities.

It may happen that the island chains outside the beam envelope are so close together that they overlap to form a chaotic sea. In the case where the last invariant torus breaks up, particles leaking out diffuse towards the 2:1 resonance, which is usually much farther away from the beam envelope. These particles form a beam halo. Since the rms radius of the beam core decreases, more particles will find themselves outside the envelope radius. As this process continues, the beam eventually becomes unstable.

Another severe instability occurs when the envelope mismatch is large even though the space-charge perveance is moderate. In this case, high-order parametric resonances exist near beam envelope. Since the slope of the tune versus the particle action is small, high-order resonances overlap causing more local chaos. Also the unstable fixed points of the 2:1 resonance are nearby. Particles can diffuse easily from the beam envelope towards the 2:1 resonance and, again become halo particles.

(b) We see that the region of stability is similar to but not exactly the same as the results of Gluckstern, *et. al.* There can be many reasons:

1. Only particles outside the envelope core encounter parametric resonances. So far we have been using particle tune Q_p *inside* the envelope core. The tune outside the core is larger. This may produce more curvature of the resonance loci in Fig. 7.

2. Gluckstern, *et. al.* introduced an envelope tune that depends on the space-charge perveance κ only, but is independent of the amount of envelope mismatch. We used an envelope tune that varies with the amount of mismatch.

3. To the lowest order, the Vlasov equation studied by Gluckstern, *et. al.* does involve the perturbation force induced by the perturbation distribution, f_1, or Poisson's equation of (3.3). In our nonlinear dynamics approach, the particle that escapes from the beam envelope core, always sees the Coulomb force of the *entire unperturbed beam core*, independent of any variation of the core distribution due to the leakage of particles. This is a shortcoming in our approach, which we need to improve in the future.

4. Although the results of Gluckstern, *et. al.* had been checked by simulations using a particle-in-cell code, we suspect the simulations do not have sufficient precision. For example, in one envelope period, the code [7] uses only 100 times steps and only first-order symplectic tracking. In our simulations, we used a fourth-order symplectic integrator developed by Forest and Berz [8], and found that at least 500 and sometimes more than 1000 time steps are required in many circumstances. Also, the projections of the perturbed distribution onto the various excitation modes exhibit large fluctuations. Therefore, the growth rates as evaluated might have not been accurate.

(c) It is possible that many collective instabilities can be explained by the particle-beam nonlinear dynamics approach. The wakefields of the beam interacting with the particle distribution produce parametric resonances and chaotic regions. Collective instabilities will be the result of particles trapped inside these resonance islands. The perturbed bunch structure further enhances the wakefields to induce these collective particle instabilities.

REFERENCES

1. Kapchinskij, I.M., and Vladimirskij, V.V., *Proceedings of the International Conference on High Energy Accelerators*, p. 2724 (CERN, Geneva, 1959).

2. Caussyn, D.D., *Phy. Rev.*, **A46**, 7942 (1992); Ellison, *Phys. Rev. Lett.*, **70**, 591 (1993); Syphers, M., *Phys. Rev. Lett.*, **71**, 720 (1993); *D. li Phys. Rev.*, **E48**, 3 (1993); *D. li, Phys. Rev.*, **E48**, R1638 (1993); Huang, H., *Phys. Rev.*, **E48**, 4678 (1993); Wang, Y., *Phys. Rev.*, **E49**, 1610 (1994); Wang, Y., *Phys. Rev.*, **E49**, 5697 (1994); Lee, S.Y., *Phys. Rev.*, **E49**, 5717 (1994); Ellison, M., *Phys. Rev.*, **E50**, 4051 (1994); Liu, L.Y., *Phys. Rev.*, **E50**, R3344 (1994).

3. Keil, E., and Schnell, W., CERN Report SI/BR/72-5, 1972; Boussard, D., CERN Report Lab II/RF/Int./75-2, 1975.

4. Gluckstern, R.L., Cheng, W.-H., and Ye, H., *Phys. Rev. Lett.*, **75**, 2835 (1995); Gluckstern, R.L., Cheng, W.-H., Kurennoy, S.S., and Ye, H., *Phys. Rev.*, **E54**, 6788 (1996).

5. Lee, S.Y., and Riabko, A., *Phys. Rev.*, **E51**, 1609 (1995).

6. Riabko, A., Ellison, M., Kang, X., Lee, S.Y., Li, D., Liu, J.Y., Pei, X., and Wang, L., *Phys. Rev.*, **E51**, 3529 (1995).

7. Kurennoy, S.S., private communication.

8. Forest, E., and Berz, M., LBNL Report, LBNL-25609, ESG-46, 1989.

Numerical Simulations of the UCLA Experiments on a High Gain SASE FEL

E.L. Saldin*, E.A. Schneidmiller* and M.V. Yurkov†

*Automatic Systems Corporation, 443050 Samara, Russia
†Joint Institute for Nuclear Research, Dubna, 141980 Moscow Region, Russia

Abstract. In this paper we present theoretical analysis of a recent SASE FEL experiment performed by UCLA/LANL/RRCKI/SLAC team reporting on a high power gain of about 10^5 at the wavelength of 12 μm The region of physical parameters of this experiment (as well as of future X-ray FELs) does not allow to apply available analytical techniques for quantitative description of the obtained results. The analysis presented in this paper is based on the results produced by a three-dimensional, time-dependent FEL simulation code FAST. It is shown that within the limit of accuracy of the experiment obtained data fully agree with the results of numerical simulations.

I INTRODUCTION

Recently it has been reported on an operation of a high gain FEL amplifier starting from noise [1] (see Table 1). The power gain of larger than 10^5 has been obtained. Parameters of the output radiation have been measured with a high accuracy. In particular, it has been demonstrated that the fluctuations of the energy in the radiation pulse follows the gamma distribution predicted in ref. [2,3]. Thorough analysis shows that despite this experiment has been performed at a relatively long wavelength, the physical processes in this SASE FEL are similar to those expected to occur in a short wavelength SASE FEL. This is connected with the fact that the aperture of the vacuum chamber is significantly larger than the transverse size of the beam radiation mode. As a result, the waveguide effects do not influence the amplification process. Also, the effects of coherent synchrotron radiation are not important at the chosen parameters of the experiment. Helpful factor simplifying analysis of this experiment consists in careful choice of experimental conditions. Namely, undulator field has been tuned precisely and distortions of the electron beam trajectory are significantly less than typical transverse size of the beam radiation mode. There is also significant safety margin with respect to the energy spread and the emiitance of the electron beam.

CP468, *Nonlinear and Collective Phenomena in Beam Physics–1998 Workshop*,
edited by S. Chattopadhyay, M. Cornacchia, and C. Pellegrini

TABLE 1. Parameters of the UCLA/LANL SASE FEL [1]

Electron beam	
Energy [MeV]	18
Charge per micropulse [nC]	0.3 – 2.2
Transverse spot size (σ) [μm]	115 – 145
Energy spread (rms) [%]	~ 0.3
Pulse length (FWHM) [ps]	7 – 13
Peak current [A]	40 – 170
Undulator	
Period [cm]	2.05
Number of periods	98
Undulator parameter K	1
Betatron wavelength [m]	1.2
FEL	
Radiation wavelength [μm]	12
Power gain length (at 2.2 nC) [cm]	~ 14

In this paper we perform theoretical analysis of the high gain SASE FEL experiment [1]. The analysis is based on the results produced by a three-dimensional, time-dependent FEL simulation code FAST [4]. Parameters of the electron beam and of the undulator [1] have been used as the input parameters for the numerical simulation code without making any corrections as it is described in section 3. It is shown that the obtained experimental results are in good agreement with the results of numerical calculations. Statistical simulations of the energy fluctuations in the radiation pulse performed over several thousands shots give the result identical to the experimental one. Namely, fluctuations of the energy in the radiation pulse follow gamma distribution with the value of parameter $M \simeq 8$.

II REGION OF PHYSICAL PARAMETERS

We begin our analysis within the framework of the steady-state approximation in order to estimate the power of different physical effects influencing the operation of the FEL amplifier. Numerical solution of the corresponding eigenvalue equation [5–7] shows that the main physical effects defining the operation of the UCLA/LANL SASE FEL are the diffraction effects and the space charge effects. It is seen from the plot in Fig. 1 that the field gain length should be about 30 cm. Also, analysis of the imaginary part of the eigenvalue shows that the slippage effect will be suppressed by a factor of four with respect to kinematic slippage due to the fact that the group velocity of the amplified wave, $\partial\omega/\partial k$, is less than the velocity of light c.

322

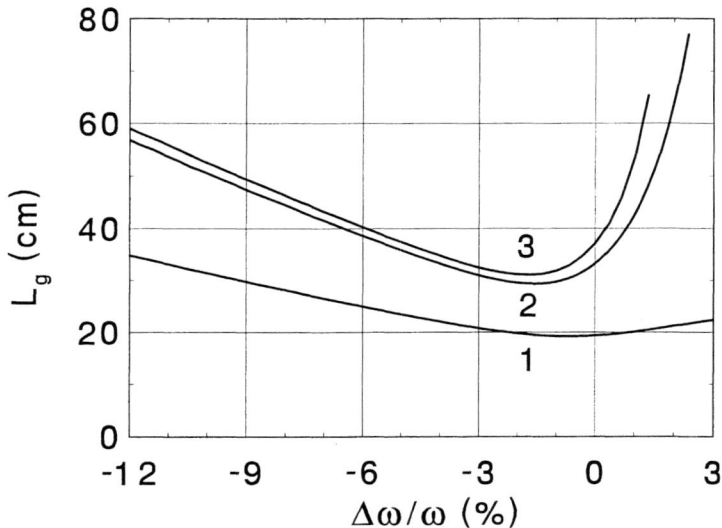

FIGURE 1. The field gain length versus the frequency deviation from the resonance value. Calculations have been performed in the steady-state approximation. Curve 1 is calculated with the only diffraction effects taken into account, curve 2 includes also the space charge effects, and curve 3 is calculated taking into account all the effects (diffraction, space charge and energy spread).

III   NUMERICAL ANALYSIS OF THE EXPERIMENT

Numerical simulations have been performed with three-dimensional, time-dependent FEL simulation code FAST [4]. The equations of motion (kinetic equation or equations for macroparticle motion) and Maxwell's equations are solved simultaneously taking into account the slippage effect. Radiation fields are calculated using integral solution of Maxwell's equations. The code allows one to simulate the radiation from the electron bunch of any transverse and longitudinal bunch shape; to simulate simultaneously external seed with superimposed shot noise in the electron beam; to take into account energy spread in the electron beam and the space charge fields; and to simulate high-gain, high-efficiency FEL amplifier with tapered undulator. The code is extremely fast thus allowing to perform precise statistical calculations.

To calculate averaged characteristics of the FEL amplifier we performed several thousands statistically independent runs. Input data for the numerical simulation code are the value of the undulator period and the peak field, the value of the bunch charge, the value of the bunch length and the bunch radius. When performing simulations we used two models of the axial profile of the bunch current, a gaussian one

$$I(z) = \frac{Qc}{\sqrt{2\pi}\sigma_z} \exp[-\frac{z^2}{2\sigma_z^2}] \, ,$$

with $\sigma_z = \sigma_z^{\mathrm{HWHM}}/\sqrt{2\ln 2}$, and a parabolic one:

$$I(z) = \frac{3Qc}{4\sigma_z}[1 - \frac{z^2}{\sigma_z^2}] \, , \qquad |z| < \sigma_z \, ,$$

with $\sigma_z = \sqrt{2}\sigma_z^{\mathrm{HWHM}}$. Transverse distribution of the beam current density assumed to be gaussian:

$$j(z,r) = \frac{I(z)}{2\pi\sigma_r^2} \exp[-\frac{r^2}{2\sigma_r^2}] \, ,$$

with $\sigma_r = \sigma_r^{\mathrm{HWHM}}/\sqrt{2\ln 2}$. HWHM values are defined by fitting formulae presented in ref. [1]: $\sigma_{z,r}^{\mathrm{HWHM}} = \sqrt{a^2 + (bQ)^2}$. Parameters for the spot size are $a = 120$ μm, $b = 38$ μm/nC, and for the pulse length $a = 3$ ps, $b = 2.2$ ps/nC.

In Fig. 2 we present typical time structure of the radiation pulse at the undulator exit at the value of the bunch charge of 2.2 nC. Averaging over one thousand of

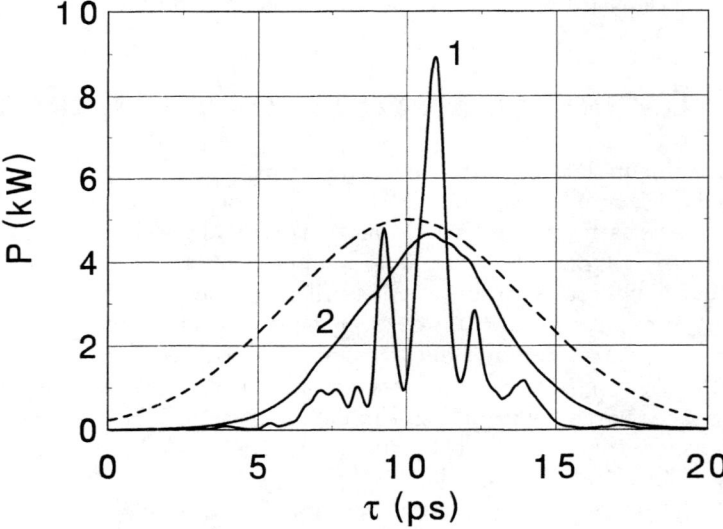

FIGURE 2. Typical time structure of the radiation pulse at the undulator exit (curve 1) and time structure of the radiation pulse averaged over 1000 statistically independent runs (curve 2). Dashed line presents axial profile of the beam current. Charge in the electron bunch is 2.2 nC.

independent shots gives the value of the averaged radiation pulse shape (curve 2 in Fig. 2). Analysis of the latter plot allows one to obtain the value of the "effective" shot noise power at the undulator entrance to be used in the steady-state codes for obtaining approximate value of the average output power [8,9]. The value of the "effective" power of shot noise is of about 20 mW in the case under study. It is seen also from Fig. 2 that the slippage of the radiation with respect to the electron bunch is significantly smaller than kinematic one. This is connected with the fact that the group velocity of the spikes, $\partial\omega/\partial k$, is less than the velocity of light c as it has been mentioned in section 2.

In Fig. 3 we present comparison of experimental and simulation results for different values of the charge. Fig. 4 presents the dependency of the energy in the radiation pulse on the transverse and longitudinal beam sizes at a fixed value of the bunch charge of 1 nC. It is seen that within uncertainties in measuring these values [1,10,11], numerical and experimental results agree rather well. One can also obtain unusual behaviour of the energy in the radiation pulse on the value of the transverse beam size. Namely, the radiation energy is increased at the increase of the transverse beam size. This effect is connected with the fact which we have mentioned in section 2 that UCLA/LANL FEL amplifier operates in the regime of a strong influence of the space charge effects. In the case under study the diffrac-

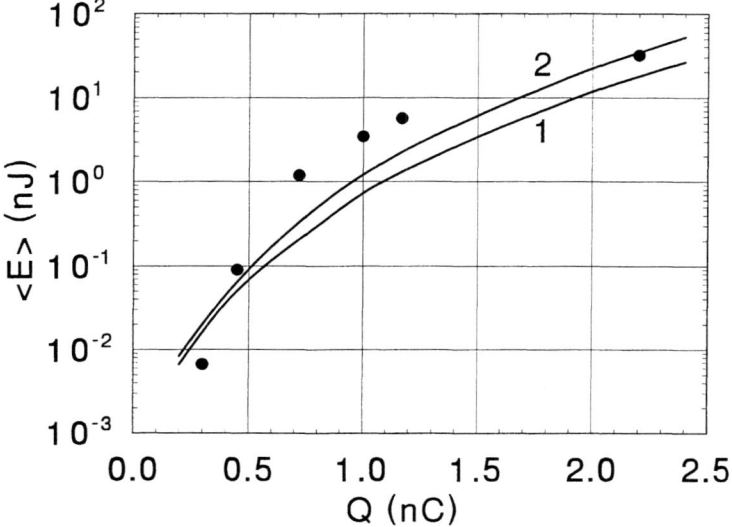

FIGURE 3. Dependence of the averaged energy in the radiation pulse versus the bunch charge. Curves 1 and 2 correspond to the gaussian and the parabolic axial beam profiles. The circles are experimental results [1].

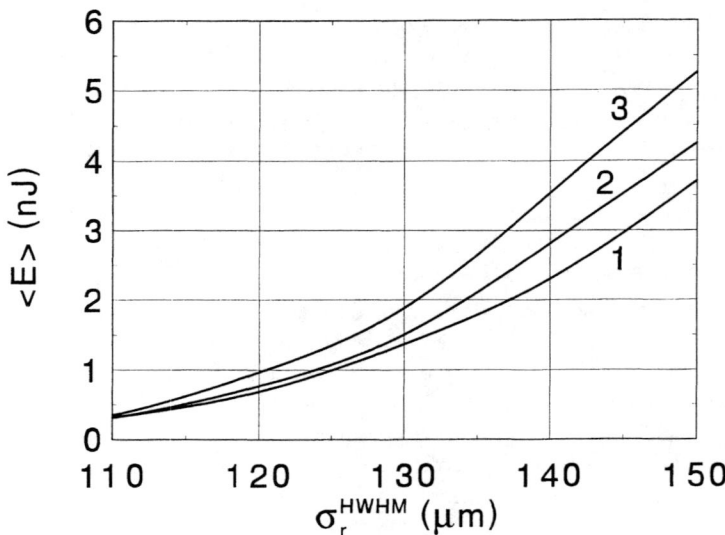

FIGURE 4. Dependence of the averaged energy in the radiation pulse on the transverse bunch size. Curves 1, 2 and 3 correspond to the values of the longitudinal HWHM bunch size σ_z^{HWHM} of 3.9 ps, 3.7 ps and 3.5 ps, respectively. The bunch charge is 1 nC.

tion parameter B is much less than unity, and increasing of the beam size results only in logarithmic decreasing of the field gain due to diffraction effects [5,6]. On the other hand, there is strong influence of the space charge fields. Increasing of the transverse beam size leads to quadratic decrease of the space charge parameter which results in the increase of the field gain. In the region of parameters traced in Fig. 4 this effect dominates above the diffraction effects. As a result, the field gain and the energy in the radiation pulse grow with the increase of the transverse size of the electron beam. Calculations show that such a tendency will take place up to the value of diffraction parameter $B \sim 0.3$. Above this pont the space charge effect becomes to be a small perturbation to the FEL process and increase of the transverse beam size will lead to the decrease of the field gain (see, e.g. ref. [12]).

So, we see that there is good agreement between measurements of the energy in the radiation pulse and the simulation results. Indeed, in all the region of the charge of the electron beam (see Fig. 3), the relative difference recalculated in the units of the gain length is less than one power gain length. Taking into account good agreement of the energy in the radiation pulse, we can expect much better agreement with the probability distribution of the energy in the radiation pulse. We performed the corresponding simulations at the value of the bunch charge of 2.2 nC. Simulations show that the probability distribution of the energy in the radiation pulse is quite close to the gamma distribution

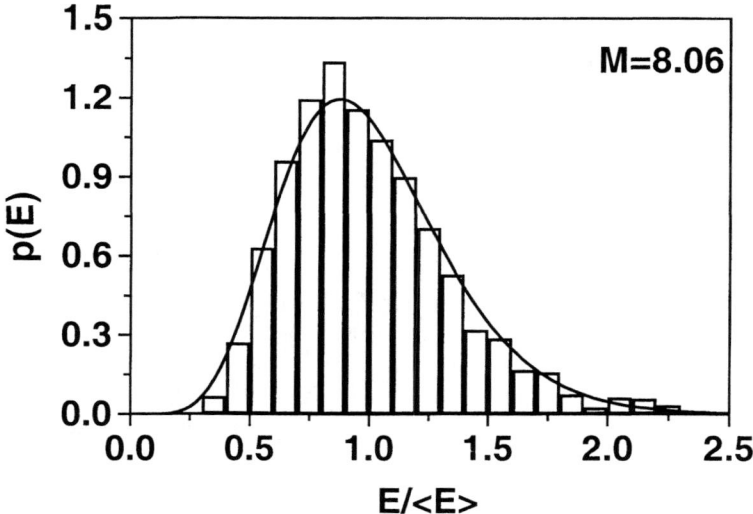

FIGURE 5. Probability distribution of the energy in the radiation pulse at the bunch charge of 2.2 nC calculated over 2400 statistically independent runs. Solid curve presents gamma distribution with $M = 8.06$.

$$p(E) = \frac{M^M}{\Gamma(M)} \left(\frac{E}{<E>} \right)^{M-1} \frac{1}{<E>} \exp\left(-M \frac{E}{<E>} \right) \,,$$

where $\Gamma(M)$ is the gamma function of argument M. Parameter of the distribution is equal to $M = 1/\sigma^2$, where $\sigma^2 = \langle E^2 - \langle E \rangle^2 \rangle / \langle E \rangle^2$ is the normalized dispersion of the energy distribution. The first set of runs has been performed at fixed values of the bunch charge, of the transverse beam size and of the longitudinal beam size. Simulations with the gaussian and the parabolic bunch profiles give very close results, namely that $M \simeq 11.5$ and $\sigma \simeq 30$ %. On the other hand, experimental result gives the value of the fluctuations of about $\sigma \simeq 37$ %. Such a visible difference indicates that shot-to-shot fluctuations of the beam parameters contribute to the fluctuations of the radiation energy. To take these fluctuations into account we performed 2400 statistically independent runs. During each run the fluctuations have been introduced in the following limits: ± 0.75 % for the bunch charge, ± 5.5 % for the transverse beam size and ± 6 % for the bunch length [1]. The results of the simulations are shown in Fig. 5. The probability distribution of the radiation energy follows the gamma distribution with $\sigma \simeq 35$ % and $M \simeq 8$. This result is in a good agreement with the experiment.

IV THE FIRST UCLA EXPERIMENT ON SASE FEL

Analysis of the high gain SASE experiment [1] shows that the FEL theory correctly predicts output characteristics of the radiation. In this section we present the comparison of the experimental and theoretical results for the SASE FEL operating at a relatively low gain. At present there are available relible experimental results obtained by the UCLA/RRCKI group [13]. Parameters of the experimet are presented in Table 2. Analysis of this experiment shows that the region of physical parameters for this experiment is similar to the high gain experiment described in the previous sections, i.e. the main physical effects defining the operation of the SASE FEL are the diffraction effects and the space charge effects.

In the same way as it has been described above, we performed large number of statistically independent runs for specific value of the bunch charge of 0.56 nC. Parameters of the simulations have been optimized in order to make the final result be insensitive to the details of simulations. The size of axial division of the electron beam is equal to the radiation wavelength, 16 μm. Total number of axial slices of the bunch is equal to 250. The number of the radial nodes is equal to 16, and the number of azimuthal nodes is equal to 100. Calculations of the radiation field have been performed taking into account 17 azimuthal harmonics of the beam modulation. The output of the program is three-dimensional mesh of the electromagnetic field (250 longitudinal slices ×16 radial nodes × 100 azimuthal nodes) at the exit of the undulator. Typical result of the program output for one slice is shown in Fig. 6. Figure 7 presents angular distribution of the instantaneous radiation intensity in far zone corresponding to the distribution presented in Fig. 6. The observation angle in Fig 7 is defined as $\theta = |\vec{r}|/R$, where R is the distance from the undulator exit to the observation plane and $\vec{r}$ is the transverse position of the observation point on the observation plane. The origin of coordinates is located at the undulator axis.

It is seen from Figs. 6 and 7 that there is no full transverse coherence of the

TABLE 2. Parameters of the first UCLA SASE FEL experiment (Phys. Rev. Lett. **80**(1998)289)

Electron beam	
Energy [MeV]	13
Charge per micropulse [nC]	0.2 – 0.6
Emittance (normalized rms) [mm mrad]	8 – 10
Energy spread (rms) [%]	0.08 – 0.14
Pulse length (rms) [ps]	2 – 3
Peak current [A]	38 – 83
Undulator	
Period [cm]	1.5
Number of periods	40
Peak magnetic field [T]	0.75
FEL	
Radiation wavelength [μm]	16

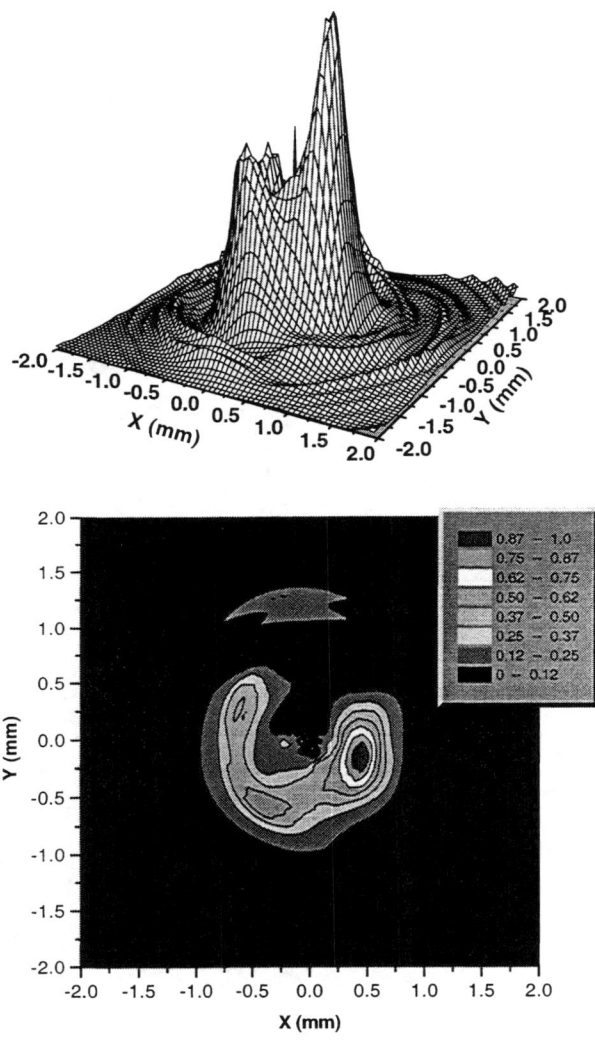

FIGURE 6. Distribution of the instanteneous radiation intensity in near zone.

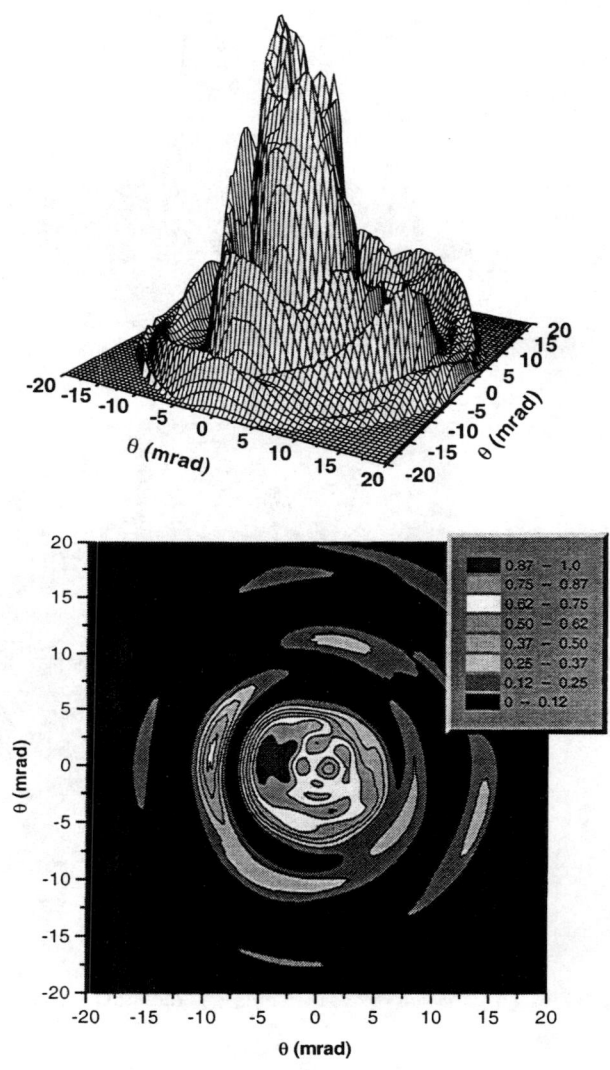

FIGURE 7. Angular distribution of the instanteneous radiation intensity in far zone.

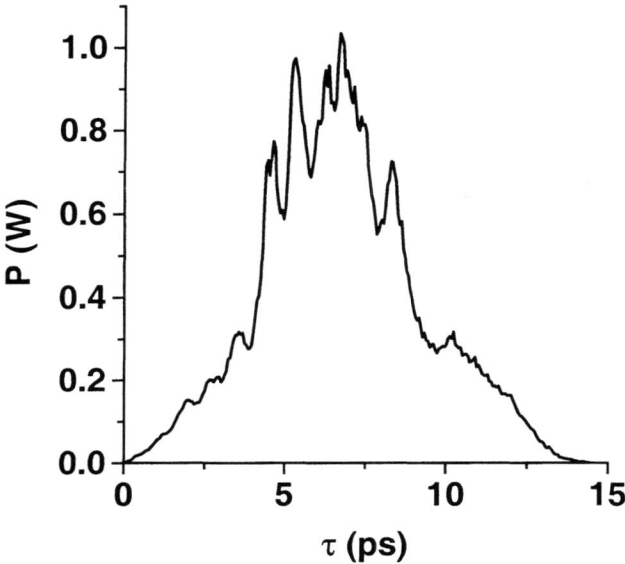

FIGURE 8. Typical structure of the radiation pulse from the SASE FEL in the first UCLA experiment.

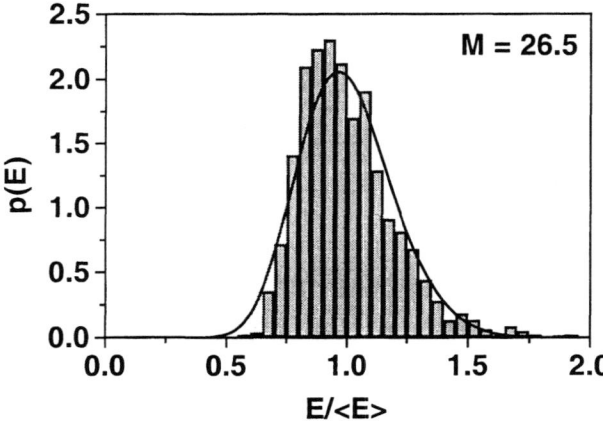

FIGURE 9. Probability distribution of the energy in the radiation pulse at the bunch charge of 0.56 nC calculated over 2000 statistically independent runs. Solid curve presents gamma distribution with $M = 26.5$.

output radiation which is connected with relatively low gain in the first UCLA SASE FEL. Since the process of amplification starts from shot noise, there is only partial transverse coherence of the radiation and several transverse radiation modes contribute to the radiation power at the initial stage of amplification. Statistical analysis shows that the average number of the transverse radiation modes is about of three in the case under study.

Experimental technique used in the first UCLA experiment did not allow to resolve instantaneous structure of the radiation field, and only total energy in the radiation pulse within certain solid angle has been measured. This experimental situation has been simulated in the following way. During each simulation run we calculated the field distributions in far zone (see Fig. 7). Then we integrated the intensity distribution of the radiation over transverse coordinate within the solid angle covered by the detector [13]. Such a procedure allows one to obtain time structure of the radiation pulse (see Fig. 8). Integration of the radiation power over the pulse length gives the value of the radiation energy in the pulse. Averaging of the latter value over many shots gives the value of about 6 pJ which is in a good agreement with the experimental measurements. It should be noticed that the time structure of the radiation pulse presented in Fig. 8 is typical for the SASE FEL operating at a relatively low gain. One can obtain that it differs visibly from that of the high gain SASE FEL (see Fig. 2). As we mentioned above, this difference is connected with relatively low gain when large number of transverse modes contribute to the radiation power.

The energy in the radiation pulse fluctuates from one shot to another. To find the probability distribution of these fluctuations, we performed two thousands of statistically independent runs. Simulations shows that the probability distribution of the energy fluctuations follows the gamma distribution (see Fig. 9) with the value of parameter M equal to 26.5. This is also in perfect agreement with the experimental data giving the value of $M \simeq 27$.

V CONCLUSION

In conclusion we should like to notice that there is no doubt that UCLA/LANL experiment [1] is a proof-of-principle of a high-gain SASE FEL. Despite it has been performed at a relatively long wavelength, the physics of its operation is described with the same equations as future VUV and X-ray SASE FELs. All the simulations presented in this paper have been performed with the simulation code developed for simulation of short-wavelength SASE FELs. It is seen that there is good agreement between theoretical predictions and experimental results, which forms reliable base for future design of short-wavelength SASE FELs.

ACKNOWLEDGMENTS

We are extremely grateful to C. Pellegrini, M. Hogan and A. Varfolomeev for providing us with experimental results and fruitful discussions. We wish to thank B. Faatz, J. Feldhaus, J. Krzywinski, G. Materlik, T. Möller, C. Pagani, J. Pflüger, S. Reiche, J. Roßbach and J.R. Schneider for many useful discussions.

REFERENCES

1. Hogan M., Pellegrini C., Rosenzweig J., Anderson A., Frigola P., Tremaine A., Fortgang C., Nguyen D., Sheffield R., Kinross-Wright J., Varfolomeev A., Varfolomeev A.A., Tolmachev S., and Carr R., *Phys. Rev. Lett.* **81**, 4867 (1998).
2. Saldin E.L., Schneidmiller E.A., and Yurkov M.V., *DESY Print TESLA-FEL* **97-02** (1997).
3. Saldin E.L., Schneidmiller E.A., and Yurkov M.V., *Opt. Commun.* **148**, 383 (1998).
4. Saldin E.L., Schneidmiller E.A., and Yurkov M.V., *"FAST: Three dimensional, time-dependent FEL simulation code"*, presented at the 20 th FEL Conference, Williamsburg (1998).
5. Saldin E.L., Schneidmiller E.A., and Yurkov M.V., *Phys. Rep.* **260**, 187 (1995).
6. Saldin E.L., Schneidmiller E.A., and Yurkov M.V., *Opt. Commun.* **97**, 272 (1993).
7. Saldin E.L., Schneidmiller E.A., and Yurkov M.V., *DESY Print TESLA-FEL* **95-02** (1995).
8. Kim K.J., *Nucl. Instrum. and Methods* **A250**, 396 (1986).
9. Wang J.M., and Yu H.L., *Nucl. Instrum. and Methods* **A250**, 484 (1986).
10. Sheffield R.L. et al., *Nucl. Instrum. and Methods* **A341**, 371 (1994).
11. Kong S.H. et al., *Nucl. Instrum. and Methods* **A358**, 284 (1995).
12. Saldin E.L., Schneidmiller E.A., and Yurkov M.V., *Proceedings of the Fifth European Particle Accelerator Conference*, Bristol: Institute of Physics Publishing, 1996, Vol.1, pp. 471-473.
13. Hogan M., Pellegrini C., Rosenzweig J., Travish G., Varfolomeev A., Anderson S., Bishofberger K., Frigola P., Murokh A., Osmanov N., Reche S., and Tremaine A., *Phys. Rev. Lett.* **80**, 289 (1998).

Effect of Centrifugal Transverse Wakefield for Microbunch in Bend

G. V. Stupakov

Stanford Linear Accelerator Center Stanford University, Stanford, CA 94309

Abstract. We calculate centrifugal force for a short bunch in vacuum moving in a circular orbit and estimate the emittance growth of the beam in a bend due to this force.

INTRODUCTION

Many of the basic features of the coherent synchrotron radiation (CSR) of short bunches and its effect on beam dynamics in accelerators are now well established [1–4]. The effect is usually described in terms of the longitudinal force, or wakefield, that causes the energy loss in the beam, and also redistributes the energy between the particles by accelerating the head and decelerating the tail of the bunch. Coherent radiation becomes most important for short bunches and high currents. More subtle features of CSR such as transition effect due to the entrance to and exit from the bend [5], CSR force in the undulator [6], and shielding due to the close metallic boundaries [7] have been also studied.

Much less is known about the transverse force in a short bunch moving on a circular orbit. The problem has been treated in several papers beginning from R. Talman's work [8], who pointed out that the centrifugal force of a rotating bunch can result in a noticeable tune shift of betatron oscillations. Later, an important correction to the Talman paper has been added in Ref. [9], where it was shown that due to the energy variation in the bunch, the effect of the transverse force proportional to R^{-1} is cancelled, and the residual effect is of the order of R^{-2}, that is much smaller than originally predicted. Recently, however, Derbenev and Shiltsev [10] found the centrifugal force of the order of R^{-1} that differs from Talman's result by a logarithmic factor only.

Taking into account the existing controversy in the literature, in this paper, we consider the transverse force in a bunch based on simple physical arguments, starting from a dc beam. We will derive the centrifugal force for a relativistic coasting beam in vacuum, and then generalize the result for a short bunch, and estimate its effect on the emittance growth in a bend.

CP468, *Nonlinear and Collective Phenomena in Beam Physics–1998 Workshop,*
edited by S. Chattopadhyay, M. Cornacchia, and C. Pellegrini
1999 The American Institute of Physics 1-56396-862-2

Throughout this paper we assume ultrarelativistic beam, $v = c$, moving on a circular orbit of radius R.

LIENARD-WIECHERT POTENTIALS AND FIELDS

The electromagnetic field of a point charge moving in vacuum, as is well known, can be found using Lienard-Wiechert potentials, and the fields can be explicitly expressed in terms of particle's velocity and acceleration at the retarded time [11]. We will use the coordinate system shown in Fig. 1. For $\gamma = \infty$, the fields of the

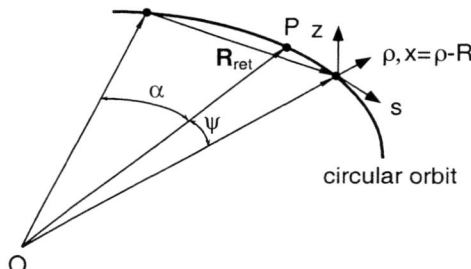

FIGURE 1. Coordinate system. Shown are locations of the particle at time t, the point P, the observation point in front of the particle at an angle ψ, and the point where the radiation occurred at the retarded time (at an angle α behind the particle). Vector $\boldsymbol{R}_{\mathrm{ret}}$ connects the radiation point with the observation point.

charge are given by the following equations,

$$
\boldsymbol{E} = \frac{q}{c} \frac{\boldsymbol{R}_{\mathrm{ret}} \times \left[(\boldsymbol{R}_{\mathrm{ret}} - R_{\mathrm{ret}}\boldsymbol{\beta}_{\mathrm{ret}}) \times \dot{\boldsymbol{\beta}}_{\mathrm{ret}} \right]}{(\boldsymbol{R}_{\mathrm{ret}} - R_{\mathrm{ret}}\boldsymbol{\beta}_{\mathrm{ret}})^3},
\tag{1}
$$

$$
\boldsymbol{H} = \frac{1}{R_{\mathrm{ret}}} [\boldsymbol{E} \times \boldsymbol{R}_{\mathrm{ret}}],
\tag{2}
$$

where the distance $\boldsymbol{R}_{\mathrm{ret}}$ shown in Fig. 1 connects the position of the particle at the radiation time t_{ret} and the observation point, $R_{\mathrm{ret}} = c(t - t_{\mathrm{ret}})$, $\boldsymbol{\beta}_{\mathrm{ret}}$ is the unit vector ($v = c$) directed along the particle velocity at the radiation time, and $\dot{\boldsymbol{\beta}}_{\mathrm{ret}}$ is the derivative of the velocity at that time.

If the transverse size of the bunch σ_r is much smaller than its length σ_z, only the field on the orbit interacts with the beam. In this case, the observation point can be chosen on the circle, as shown in Fig. 1, and Eqs. (1) and (2) can be written as (see Fig. 1 for notation)

$$E_s(\psi) = \frac{q}{R^2} \frac{2 \sin \frac{1}{2}(\alpha + \psi)}{[\alpha - \sin(\alpha + \psi)]^3}$$
$$\times \left\{ 2(\sin \frac{1}{2}(\alpha + \psi))^3 \left[\cos \frac{1}{2}(\alpha + \psi) - \cos(\alpha + \psi) \right] \right.$$
$$\left. - \sin(\alpha + \psi)[\alpha - \sin(\alpha + \psi)] \right\}, \tag{3}$$

$$E_\rho(\psi) = \frac{q}{R^2} \frac{2 \sin \frac{1}{2}(\alpha + \psi)}{[\alpha - \sin(\alpha + \psi)]^3}$$
$$\times \left\{ 2(\sin \frac{1}{2}(\alpha + \psi))^3 \left[\sin \frac{1}{2}(\alpha + \psi) - \sin(\alpha + \psi) \right] \right.$$
$$\left. + \cos(\alpha + \psi)[\alpha - \sin(\alpha + \psi)] \right\}, \tag{4}$$

$$H_z(\psi) = E_s \sin \frac{1}{2}(\alpha + \psi) - E_\rho \cos \frac{1}{2}(\alpha + \psi), \tag{5}$$

where $E_s(\psi)$ is the longitudinal and $E_\rho(\psi)$ – radial components of the electric field, $H_z(\psi)$ is the vertical magnetic field, $\psi = s/R$, and the angle α is related to the position of the observation point by equation

$$\alpha = 2|\sin \frac{1}{2}(\psi + \alpha)|. \tag{6}$$

The plot of the longitudinal field as a function of the position on the circle is shown in Fig. 2.

The electric field E_s per unit charge is equal to the longitudinal wake w. For small distances, $s \ll R$, one finds from Eqs. (3) and (6),

$$w(s) = \frac{1}{q} E_s(s) \approx \begin{cases} 2(3s)^{-4/3} R^{-2/3}, & s > 0, \\ 0, & s < 0. \end{cases} \tag{7}$$

For a short bunch ($\sigma_z \ll R$) with a given charge distribution $\lambda(s)$ ($\int \lambda(s)ds = 1$), the longitudinal wake for the bunch is defined as a convolution with the distribution function,

$$w_{\text{bunch}}(s) = \int_{-\infty}^{\infty} w(s - s')\lambda(s')ds' \tag{8}$$

$$= \frac{2}{3^{3/4} R^{2/3}} \int_s^{\infty} \frac{\lambda(s')ds'}{(s' - s)^{4/3}}. \tag{9}$$

The problem here is that the above integral diverges when $s' \to s$. To overcome this difficulty, we can use a trick and integrate the last equation by parts (neglecting the nonintegral term!):

$$w_{\text{bunch}}(s) = \frac{2}{3^{1/4} R^{2/3}} \int_0^{\infty} \frac{ds'}{(s' - s)^{1/3}} \frac{d\lambda(s')}{ds'}. \tag{10}$$

336

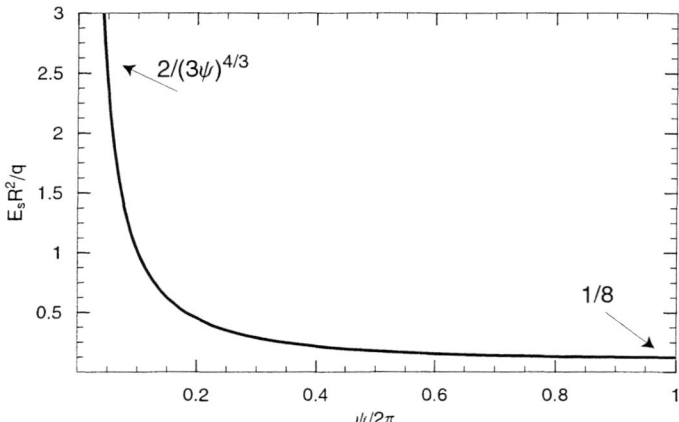

FIGURE 2. Longitudinal electric field on a circular orbit as a function of angle ψ counted from the location of the particle. The arrows show asymptotic expressions for the field in the vicinity of the particle.

Now the integral converges, and gives the right result for the wake [2,3]. The justification for this trick can be found in a more accurate consideration of the fields in a small vicinity of the particle [4,5].

We also mention here that although we obtained the above result assuming an infinitely thin beam, the applicability condition for the longitudinal wake is actually very mild , $\sigma_r/\sigma_z \ll (R/\sigma_z)^{1/3}$ [3]. At this point we are tempted to apply the same approach for the calculation of the transverse force in a thin bunch. The transverse force (per unit charge) F_ρ for an ultrarelativistic bunch is $F_\rho = E_\rho + H_z$. The plots of E_ρ and H_z are shown in Fig. 3, and the transverse force as a function of angle ψ on the circular orbit is shown in Fig. 4. Asymptotically, for small positive ψ in front of the particle $F_\rho \approx q/3R^2\psi$; behind the particle, for negative small ψ, $F_\rho \approx q/R^2|\psi|$. Again, if we want to convolve this force with the bunch distribution and to find the transverse wakefield for the bunch, as we did above for the longitudinal wake, the integral would diverge, and there is no trick that could make it convergent. As we will see in the next section, there is a profound reason for such divergence: the transverse force depends on the beam radius σ_r that we neglected in the above consideration.

TRANSVERSE FORCE – COASTING BEAM

To make our consideration of the transverse force as simple as possible we begin here from a problem of a coasting relativistic dc beam of radius a, shown in Fig. 5. To find the transverse force f_ρ acting on a unit length of the beam in this case,

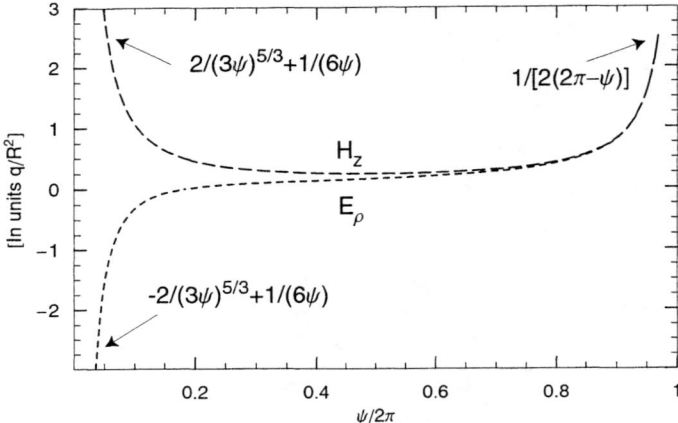

FIGURE 3. Radial electric field E_ρ and vertical magnetic field H_z on the orbit. Arrows show the asymptotic expressions for the fields near the particle.

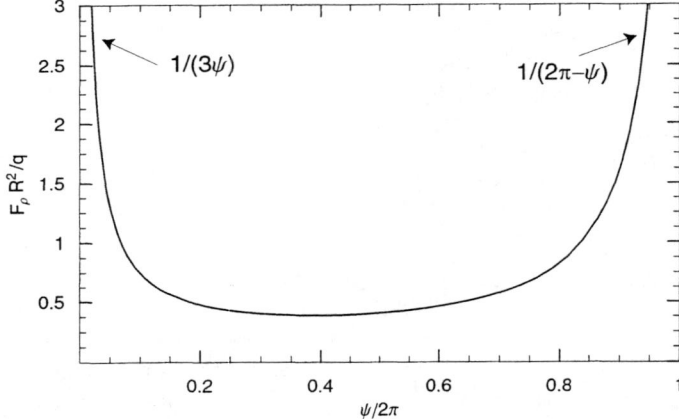

FIGURE 4. Transverse force per unit charge F_ρ as a function of angle ψ. Arrows show asymptotic expressions for the force near the particle.

we will use the energy principle that relates the force to the variation of the energy of the system under infinitesimally small displacement [12]. Since the electric and magnetic fields of a coasting beam are time independent, the electromagnetic energy of the beam is the sum of the electrostatic and magnetic energies. To find them, we need to know the capacitance and inductance of a charged rotating ring, which can be found in textbooks on electrodynamics (see, e.g., [12]).

338

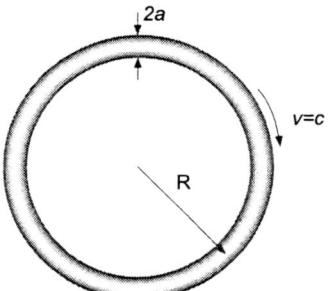

FIGURE 5. Coasting beam of radius a moving along a circular orbit in vacuum.

The inductance of a circular current in vacuum is given by the following formula,

$$L = 4\pi R \left(\ln \frac{8R}{a} - \frac{7}{4} \right). \tag{11}$$

Differentiating the magnetic energy $LI^2/2c^2$ with respect to the circumference of the beam, we obtain the radial magnetic force per unit length of the ring,

$$f_m = \frac{\partial (LI^2/2c^2)}{\partial (2\pi R)} = \frac{I^2}{Rc^2} \left(\ln \frac{8R}{a} - \frac{3}{4} \right). \tag{12}$$

Analogously, the capacitance C of the uniformly charged ring is

$$C^{-1} = \frac{1}{\pi R} \left(\ln \frac{8R}{a} + \frac{1}{4} \right), \tag{13}$$

and the electric force per unit length is equal to the derivative of the electrostatic energy with respect to the circumference (with the minus sign),

$$f_e = -\frac{\partial (Q^2/2C)}{\partial (2\pi R)} = \frac{Q^2}{4\pi^2 R^3} \left(\ln \frac{8R}{a} - \frac{3}{4} \right). \tag{14}$$

Taking into account that for an ultrarelativistic beam $Q = 2\pi RI/c$, we find that the electric and magnetic forces are equal, which is expected in the limit $v = c$.

Adding Eq. (12) and (14) gives the total force,

$$f_\rho = f_m + f_e = \frac{I^2}{Rc^2} \left(2\ln \frac{8R}{a} - \frac{3}{2} \right). \tag{15}$$

We can also easily find the force, if the beam propagates in a circular pipe of radius b. In this case the capacitance of the ring is,

$$C^{-1} = \frac{1}{\pi R} \left(\ln \frac{b}{a} + \frac{1}{4} \right), \tag{16}$$

339

with the electric force

$$f_e = -\frac{\partial(Q^2/2C)}{\partial(2\pi R)} = \frac{Q^2}{4\pi^2 R^3}\left(\ln\frac{b}{a} - \frac{3}{4}\right). \tag{17}$$

Calculation of the magnetic force in this case shows that, as above, it is equal to the electric one, and the total force is twice the electric force,

$$f_\rho = f_m + f_e = 2f_e = \frac{I^2}{Rc^2}\left(2\ln\frac{b}{a} - \frac{3}{2}\right). \tag{18}$$

The above derivation is very simple, but it does not tell how the centrifugal force varies in the cross section of the beam. To answer this question, we need to find the electric and magnetic fields inside the beam. For the beam in vacuum, we will find these fields using a perturbation theory in small parameter a/R.

As a first step in calculations, we need the electrostatic potential ϕ and the vector potential A_θ at distances far from the center of the beam in comparison with the beam radius, $r = \sqrt{x^2 + z^2} \gg a$, but close, relative to the orbit size, $R \gg r$. This approximation is equivalent to the limit of infinitely thin beam, $a \to 0$, and the result can be found in textbooks (see, e.g. [12]),

$$\phi(\rho, z) = 2\mu\sqrt{\frac{R}{\rho}}\kappa K(\kappa^2), \tag{19}$$

$$A_\theta(\rho, z) = \left(\frac{2}{\kappa^2} - 1\right)\phi - 4\mu\sqrt{\frac{R}{\rho}}\kappa^{-1}E(\kappa^2), \tag{20}$$

where $\kappa^2 = 4\rho R/[(\rho+R)^2 + z^2]$, ρ is the radius counted from the center of the orbit, and μ is the charge per unit length. Expanding this expression in the vicinity of the beam, $x = \rho - R$, $r = \sqrt{x^2 + z^2} \ll R$, one finds

$$\phi = -2\mu\left(1 - \frac{x}{2R}\right)\ln\frac{\sqrt{x^2 + z^2}}{8R} + \mu\frac{x}{R} + \ldots, \tag{21}$$

$$A_\theta = \phi - 4\mu\left(1 - \frac{x}{2R}\right) + \ldots. \tag{22}$$

To find the potential ϕ inside the beam we need to solve the Poisson equation,

$$\Delta\phi = \frac{1}{\rho}\frac{\partial}{\partial\rho}\rho\frac{\partial\phi}{\partial\rho} + \frac{\partial^2\phi}{\partial z^2} = -4\pi en(\rho, z), \tag{23}$$

where n is the particle density in the beam. Using the coordinate x, $\rho = R+x$, $x \ll R$, we can expand the first term in the equation keeping only linear terms in R^{-1},

$$\frac{\partial^2 \phi}{\partial x^2} + \frac{\partial^2 \phi}{\partial z^2} = -4\pi en - \frac{1}{R}\frac{\partial \phi}{\partial x} + O(R^{-2}). \tag{24}$$

Let us assume that $\phi = \phi_0 + \phi_1$, where ϕ_0 is the potential in the limit $R \to \infty$, and ϕ_1 is linear in R^{-1}, $\phi_1 \ll \phi_0$. In the zeroth approximation, we have

$$\frac{\partial^2 \phi_0}{\partial x^2} + \frac{\partial^2 \phi_0}{\partial z^2} = -4\pi en. \tag{25}$$

For a constant density beam, $n = $ const for $r < a$, the solution of Eq. (25) is

$$\phi_0 = -\mu\frac{r^2}{a^2}, \qquad r < a,$$
$$\phi_0 = -2\mu\left(\ln\frac{r}{a} + \frac{1}{2}\right), \qquad , r > a, \tag{26}$$

where $\mu = \pi a^2 en$ is the beam charge per unit length. In the first order, the potential ϕ_1 satisfies the equation

$$\frac{\partial^2 \phi_1}{\partial x^2} + \frac{\partial^2 \phi_1}{\partial z^2} = -\frac{1}{R}\frac{\partial \phi_0}{\partial x}. \tag{27}$$

The solution can be found by solving Eq. (27) and using as a boundary condition the asymptotic behaviour for large r, given by Eq. (21). The potential ϕ_1 inside the beam is,

$$\phi_1 = \frac{\mu}{R}x\left(1 + \frac{r^2}{4a^2} + \ln\frac{a}{8R}\right), \qquad r < a. \tag{28}$$

In a similar fashion, we can find the vector potential A_θ that satisfies the equation

$$\Delta A_\theta = -4\pi en, \tag{29}$$

but has a different asymptotic condition at large r, Eq. (22). The result is

$$A_{\theta 0} = -\mu\left(\frac{r^2}{a^2} + 2\ln\frac{a}{8R} - 3\right), \qquad r < a,$$
$$A_{\theta 1} = \frac{\mu}{R}x\left(3 + \frac{r^2}{4a^2} + \ln\frac{a}{8R}\right), \qquad r < a. \tag{30}$$

It is interesting to note, that although ϕ and A_θ satisfy the same equation (see Eq. (23), and Eq. (29)), $A_\theta \neq \phi$ due to different asymptotic conditions at large r.

Using Eqs. (28) and Eq. (30) for the potentials, one can find the fields inside the beam and calculate the distribution of the transverse force over the cross section of the bunch. This force, as a function of radius r is given by the following equation,

$$F_\rho = \frac{\mu}{R}\left(-1 - \frac{r^2}{a^2} + 2\ln\frac{8R}{a}\right). \tag{31}$$

We see that the force has a parabolic profile with the maximum value on the axis of the beam. Averaging this force over the cross section yields

$$\bar{F}_\rho = \frac{\mu}{R}\left(2\ln\frac{8R}{a} - \frac{3}{2}\right). \tag{32}$$

To compare this result with Eq. (15), we need to take into account that $\mu = I/c$ and the force per units length of the beam f_ρ equals the force per unit charge F_ρ multiplied by I/c. With those factors, we conclude that both results agree with each other.

Using Eq. (31) and (32) we can also find the relative difference between the force and its average value

$$\frac{F_\rho - \bar{F}_\rho}{\bar{F}_\rho} = \frac{1 - 2r^2/a^2}{4\ln(8R/a) - 3}. \tag{33}$$

For a thin bunch, when $\ln(8R/a) \gg 1$, the variation of the force in the cross section is relatively small.

At this point, it is instructive to consider the transverse particle motion in a coasting beam under the influence of the centrifugal force. Such motion in the horizontal plane, $z = 0$, is governed by the following equation

$$x'' + Kx = \frac{eF_\rho(x)}{E} + \frac{1}{R(s)}\frac{\Delta E}{E}, \tag{34}$$

where K is the external focusing, and ΔE is the particle energy variation arising due to the potential inside the beam, $\Delta E = -e\phi(x)$,

$$x'' + Kx = \frac{e}{E}\left(F_\rho - \frac{\phi}{R}\right). \tag{35}$$

For $z = 0$, from Eqs. (26) and (31) we have

$$\phi = -\frac{\mu x^2}{a^2}, \tag{36}$$

$$F_\rho = \frac{\mu}{R}\left(2\ln\frac{8R}{a} - 1 - \frac{x^2}{a^2}\right).$$

We see that $F_\rho - \phi/R$ does not depend on x, which means that the centrifugal force does not contribute to the betatron tune in first order in R^{-1}, in agreement with Ref. [9].

SHORT BUNCH

To calculate the transverse force for a short bunch, we will use both Lienard-Wiechert fields and the result found in the previous section for a coasting beam. We will assume that the bunch density is constant in the cross section within the radius $\sigma_r = a$, and the longitudinal charge distribution per unit length is given by $\mu(s)$ with the rms bunch length σ_z.

To find the force F_ρ acting on unit charge in the bunch at point $s = s_0$, we select a small slice of the bunch of length Δs, such that $\sigma_z \gg \Delta s \gg a$, with a local density $\mu(s_0)$, and calculate the contribution to the force separately from the slice and from the rest of the beam. First, let us find the contribution to the force from the bunch excluding the slice. This can be done by integrating the force shown in Fig. 4. Since the force from a particle of charge q located at point s ahead of the point s_0 is equal to $q/R(s - s_0)$, the contribution from the part of the bunch in front of the slice is

$$\frac{1}{R}\int_{s_0 + \Delta s/2}^{\infty} \frac{\mu(s')ds'}{s - s'} = -\frac{\mu(s_0)}{R}\ln\frac{\Delta s}{2R} - \frac{1}{R}\int_{s_0}^{\infty}\ln\left(\frac{s' - s}{R}\right)\frac{d\mu(s')}{ds'}ds'. \tag{37}$$

In the second integral, we extended the integration region from s_0, because Δs is small, and the integral converges at the lower limit. Similarly, the force from a particle behind the point s_0 is equal to $q/3R(s_0 - s)$, and the contribution from the part of the bunch behind the slice is

$$\frac{1}{3R}\int_{-\infty}^{s_0 - \Delta s/2} \frac{\mu(s')ds'}{s - s'} = -\frac{\mu(s_0)}{3R}\ln\frac{\Delta s}{2R} + \frac{1}{3R}\int_{-\infty}^{s_0}\ln\left(\frac{s - s'}{R}\right)\frac{d\mu(s')}{ds'}ds'. \tag{38}$$

To find the contribution to the force from the slice itself, we will use the following trick. Consider a circular dc beam of density $\mu = \mu(s_0)$ and select a slice Δs with its center located at $\psi = 0$. From the previous section, we know that the force in this case does not depend on position and is given by Eq. (32). If we subtract from this force the contribution $\tilde{F}$ of the part of the circle external to the slice, that is the part occupying the region $\Delta\psi/2 < \psi < 2\pi - \Delta\psi/2$, where $\Delta\psi = \Delta s/R$, we will find the force of the slice itself. The quantity $\tilde{F}$ is equal

$$\tilde{F} = \int_{\Delta\psi/2}^{2\pi - \Delta\psi/2} F_\rho(\psi)d\psi = \int_{\Delta\psi/2}^{2\pi - \Delta\psi/2} [E_\rho(\psi) + H_z(\psi)]d\psi, \tag{39}$$

where $E_\rho(\psi)$ and $H_z(\psi)$ are given by Eqs. (4) and (5) with q substituted by the charge per unit angle μR. The result of the integration in the limit of small Δs is

$$\tilde{F} = \int_{\Delta\psi/2}^{2\pi - \Delta\psi/2} F_\rho(\psi)d\psi = \frac{\mu}{R}\left(A + \frac{4}{3}\ln\frac{R}{\Delta s}\right), \tag{40}$$

where the constant $A = 3.33$ was found from numerical integration. If we now subtract this result from the force of the dc current, Eq. (32), the difference

343

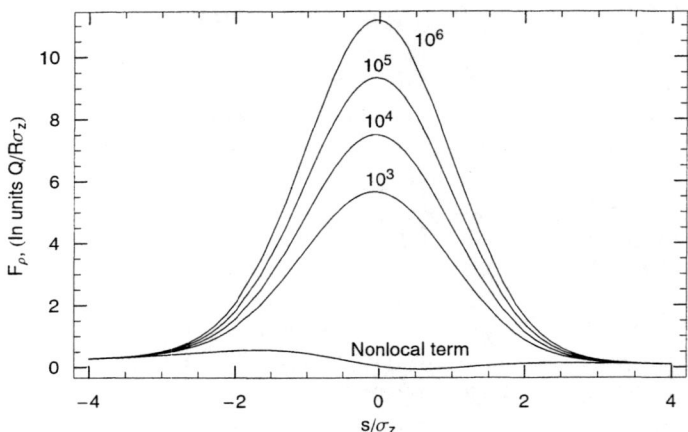

FIGURE 6. Transverse force measured in units $Q/R\sigma_z$ for a short Gaussian bunch for several different ratios R/a indicated by a number near the curve. The bottom curve shows the contribution of the second term in Eq. (42) which does not depend on the ratio R/a.

$$\frac{\mu(s_0)}{R}\left(2\ln\frac{8R}{a}-\frac{3}{2}\right)-\frac{\mu(s_0)}{R}\left(A+\frac{4}{3}\ln\frac{R}{\Delta s}\right) \tag{41}$$

is equal to the force induced by the slice at its center.

Summing Eqs. (37), (38) and (41) gives the total transverse force in a short bunch:

$$\bar{F}_\rho(s)=\frac{\mu(s)}{R}\left(2\ln\frac{8R}{a}-3.91\right)$$
$$+\frac{1}{R}\int_0^\infty \ln\xi\left(\frac{1}{3}\mu'(s-\xi)-\mu'(s+\xi)\right)d\xi. \tag{42}$$

As we see from this equation, the first term in this force is local – it depends on the charge density at the observation point. This term also logarithmically depends on the bunch radius a. The second term involves the distribution of the charge in the bunch. The plot of the force for a Gaussian bunch for several different ratios R/a is shown in Fig. 6. As we see, for large values of R/a, typical in accelerators with short bunches, the dominant contribution comes from the term that is proportional to the local current in the bunch.

NUMERICAL ESTIMATES

In this section, we estimate the effect of the transverse force on the emittance growth of a bunch passing through a bend and compare it with the emittance growth due to the longitudinal CSR force. We start from the longitudinal CSR

344

wake. When the beam passes through the magnet, the energy within the bunch changes, and due to the variation of the energy, the deflection angle $\Delta x'$ for different slices of the bunch also varies. This variation is given by the following formula,

$$\Delta x'(s) = \frac{\theta w_{\text{bunch}}(s) N e^2 L_b}{2E}, \tag{43}$$

where θ is the deflection angle for the nominal energy, $\theta \approx L_b/R$, L_b is the length of the bend, and E is the beam energy. For a short bend, the variance of the deflection angle is proportional to the increase of the projected emittance $\Delta \epsilon_N$ (we assume that $\Delta \epsilon_N \ll \epsilon_N$),

$$\Delta \epsilon_N = \frac{1}{2} \gamma \beta \langle ((\Delta x' - \langle \Delta x' \rangle))^2 \rangle, \tag{44}$$

where β is the beta function at the location of the bend. For a Gaussian bunch, using Eq. (10), we find

$$\Delta \epsilon_N = 7.5 \times 10^{-3} \frac{\beta}{\gamma} \left(\frac{N r_e L_b^2}{R^{5/3} \sigma_z^{4/3}} \right)^2. \tag{45}$$

To estimate the effect numerically, we will use the parameters of one of the magnets of the LCLS bunch compressor [13]: $\theta = 3.6°$, $L_b = 1.5$ m, $E = 6$ GeV, $\sigma_z = 60$ μm, $N = 6 \times 10^9$, $R = 24$ m, $\beta = 10$ m. Putting these numbers into Eq. (45) gives

$$\Delta \epsilon_N = 4.2 \times 10^{-8} \text{ m}, \tag{46}$$

which is about 4% of the nominal emittance in the LCLS [14].

To estimate the emittance growth due to the *transverse* wake, we note that the transverse force in the bend deflects the slice by

$$\Delta x'(s) = \frac{1}{E} \bar{F}_\rho(s), e L_b \tag{47}$$

where $\bar{F}_\rho$ is the centrifugal force averaged over the cross section. Again, using Eq. (44) we find

$$\Delta \epsilon_N \approx 2.5 \times 10^{-2} \frac{\beta}{\gamma} \left(\frac{\Lambda N r_e L_b}{R \sigma_z} \right)^2, \tag{48}$$

where Λ is a logarithmic factor equal to $\ln(8R/a)$ if the beam travels in vacuum. To approximately take into account the effect of the walls of the vacuum chamber, we will use for Λ the value $\ln(b/a)$ following from Eq. (18). Assuming $b/a = 150$, that gives $\Lambda = 5$, we find

$$\Delta \epsilon_N = 1.6 \times 10^{-7} \text{ m}, \tag{49}$$

that is about four times larger than the longitudinal effect.

CONCLUSION

We found that the centrifugal force for a short bunch is approximately given by

$$\bar{F}_\rho(s) \approx \Lambda \frac{\lambda(s)}{R},\tag{50}$$

where Λ is a logarithmic factor, typically of the order of several units. The presence of the conducting walls does not eliminate the transverse force and only modifies the factor Λ. Although the centrifugal force does not contribute to the tune shift in a circular accelerator, it does effect the transverse motion of the beam. One of the examples, considered in this paper, is the emittance growth of a short bunch in a magnetic compressor. In this case, the ratio of the emittance growth due to the centrifugal force and that due to unshielded CSR wake is of the order

$$\Lambda^2 \left(\frac{R}{L_b}\right)^2 \left(\frac{\sigma_z}{R}\right)^{2/3}.\tag{51}$$

The relative role of the transverse force becomes essential for short bends (although the gross emittance increase goes down with L_b).

ACKNOWLEDGEMENTS

This work was supported by Department of Energy contract DE-AC03-76SF00515.

REFERENCES

1. Iogansen, L. V., and Rabinovich, M. S., *Sov. Phys. JETP* **37**, 83 (1960).
2. Murphy, J. B., Krinsky, S., and Gluckstern, R. L., "Longitudinal Wakefield for Synchrotron Radiation" in *Proc. IEEE Particle Accelerator Conference and International Conference on High-Energy Accelerators, Dallas, 1995* (IEEE, Piscataway, NJ, 1996), pp. 2980–2982.
3. Derbenev, Y. S., Rossbach, J., Saldin, E. L., and Shiltsev, V. D., DESY FEL Report TESLA-FEL 95-05, DESY, Hamburg, Germany (unpublished).
4. Murphy, J. B., Krinsky, S., and Gluckstern, R. L., *Part. Accel.* **57**, 9 (1997).
5. Saldin, E. L., Schneidmiller, E. A., and Yurkov, M. V., DESY FEL Report TESLA-FEL 96-14, DESY, Hamburg, Germany (unpublished).
6. Saldin, E. L., Schneidmiller, E. A., and Yurkov, M. V., Technical Report No. DESY-TESLA-FEL-97-08, DESY, Hamburg, Germany (unpublished).
7. Kheifets, S. A., and Zotter, B., "Shielding effects on coherent synchrotron radiation", in *Proc. Micro Bunches Workshop, Upton, 1995*, No. 367 in *AIP Conference Proceedings*, edited by E. B. Blum, M. Dienes, and J. B. Murphy (American Institute of Physics, New York, 1995), pp. 424–434.

8. Talman, R., *Phys. Rev. Lett.* **56**, 1429 (1986).

9. Lee, E. P., *Part. Accel.* **25**, 241 (1990).

10. Derbenev, Y. S., and Shiltsev, V. D., Technical Report No. SLAC-PUB-7181, SLAC, Stanford, CA, USA (unpublished).

11. Jackson, J. D., *Classical Electrodynamics*, 2nd ed. (Wiley, New York, 1975).

12. Landau, L. D., and Lifschitz, E. M., *Electrodynamics of Continuous Media* (Pergamon, NY, 1960).

13. LCLS Design Study Group, Technical Report No. SLAC-R-521, SLAC (unpublished).

14. Emma, P., and Brinkmann, R., Technical Report No. SLAC-PUB-7554, SLAC, Stanford, CA, USA (unpublished).

AUTHOR INDEX

A

Abe, H., 101
Anderson, D., 197
Asaka, T., 101

B

Bazzani, A., 3, 15
Berntson, A., 197
Bongini, L., 3, 173

C

Communian, M., 15
Corbett, J., 3, 25
Cornacchia, M., 129

D

Dattoli, G., 293
Decking, W., 119
Dome, G., 3

E

Emma, P., 143

F

Fedele, R., 197
Fedorova, A., 3, 48, 69
Franchetti, G., 173, 185
Freguglia, P., 3, 94

H

Hanaki, H., 101
Hasse, R. W., 219
Helm, R., 143

J

Johansson, S., 197

K

Koutin, S. V., 285

L

Lebedev, A. N., 285
Lebedev, V. A., 224
Lee, S. Y., 300
Lindau, I., 11
Lisak, M., 197

M

Mezi, L., 293
Migliorati, M., 293
Mizuno, A., 101
Monteiro, S., 237

N

Ng, K. Y., 3, 300
Nosochkov, Y., 25, 143

O

Ohmi, K., 3
Owen, H., 3

P

Palumbo, L., 293
Papaphilippou, Y., 3, 107
Parsa, Z., 48, 69
Pellegrini, C., 129, 237
Pisent, A., 15
Pitthan, R., 143